Programming Finite Elements in Java™

Gennadiy Nikishkov

Programming Finite Elements in Java™

 Springer

Gennadiy Nikishkov, DSc, PhD
The University of Aizu
Aizu-Wakamatsu, 965-8580
Japan

ISBN 978-1-4471-5759-5   ISBN 978-1-84882-972-5 (eBook)
DOI 10.1007/978-1-84882-972-5
Springer London Dordrecht Heidelberg New York

British Library Cataloguing in Publication Data
A catalogue record for this book is available from the British Library

Cover design: eStudioCalamar, Figueres/Berlin

Springer is part of Springer Science+Business Media (www.springer.com)

Preface

The finite element method can be applied to problems in various fields of science and engineering. It is well established and its algorithms are presented in numerous publications. Many books are devoted to different aspects of the finite element method. Still, algorithms of the finite element method are difficult to understand, and programming of the finite element techniques is complicated.

Objective

This book focuses on algorithms of the finite element method and their programming. First, general equations of the finite element method for solving solid mechanics and thermal conductivity problems are introduced. Then, algorithms of the finite element method and their programming in Java^{TM} are considered. In addition to solution methods, the book presents algorithms and programming approaches for mesh generation and visualization.

Why Java?

The Java language is selected for its numerous advantages: an object-oriented paradigm, multiplatform support, ease of development, reliability and stability, the ability to use legacy C or C++ code, good documentation, development-tool availability, etc. The Java runtime environment always checks subscript legitimacy to ensure that each subscript is equal to or greater than zero and less than the number of elements in the array. Even this simple feature means a lot to developers. As a result, Java programs are less susceptible to bugs and security flaws. Java also provides application programming interfaces (APIs) for development of GUI, and three-dimensional graphics applications.

I started programming finite elements in Fortran. Later I used Pascal, C, and C++, before settling on Java. Comparing these languages I found that programming finite elements in Java is not just efficient because the productivity is higher and the code contains fewer errors, but is also more pleasant.

An opinion exists that Java is not suitable for computational modeling and finite element programming because of its slow execution speed. It is true that Java is slower than C in performing "multiply-add" arithmetic inside double and triple loops. However, tuning of important Java code fragments provides computational speed comparable to that of C.

The attractive features of Java prevail over some of its drawbacks. In my opinion, Java is good for both learning finite element programming and for finite element code development with easy debugging, modification, and support. Further, Java is easy to understand even for those who do not program in Java. In most cases, methods performing computations can be easily used with minimum modification to procedures written in other languages such as C or C++.

For Whom Is This Book Written?

This book is an introductory text about finite element algorithms and especially finite element programming using an object-oriented approach. All important aspects of finite element techniques are considered – finite element solution, generation of finite element meshes, and visualization of finite element models and results.

The book is useful for graduate and undergraduate students for self-study of finite element algorithms and programming techniques. It can be used as a textbook for introductory graduate courses or in advanced undergraduate courses. I hope that the book will be interesting to researchers and engineers who are already familiar with finite element algorithms and codes, since the programming approaches of this book differ from other publications.

Organization

The book is organized into four parts. Part I covers general formulation of the finite element method. Chapter 1 introduces the finite element formulation in the one-dimensional case. Both the Galerkin method and variational formulations are considered. Chapter 2 presents finite element equations for heat transfer problems derived with the use of the Galerkin approach. Chapter 3 contains variational formulation of general finite element equations for solid mechanics problems. An object-oriented approach to development of the finite element code is discussed in Chapter 4.

Part II is devoted to algorithms and programming of the finite element solution of solid mechanics problems. Chapter 5 considers the class structure of the finite element processor code. Data structures of the finite element model and corresponding Java class are presented in Chapter 6. Relations for elastic material and the corresponding class are given in Chapter 7. Chapters 8 and 9 describe an abstract class for a finite element and a class for numerical integration. Chapters 10–13 present algorithms and programming implementation for two- and three-dimensional isoparametric quadratic elements. Assembly and solution of the finite element equation system are discussed in Chapters 14–16. Chapter 17 is devoted to assembly of the global load vector. Computing stress increments is presented in Chapter 18. Solution and programming implementation of elastic–plastic problems is discussed in Chapter 19.

Part III focuses on mesh generation for solution of two- and three-dimensional finite element problems. The block decomposition method used for mesh generation and general organization of the mesh generator is given in Chapter 20. Chapter 21 presents two-dimensional mesh generators. Chapter 22 describes three-dimensional mesh generation by sweeping a two-dimensional mesh. Chapters 23–25 contain algorithms and classes for pasting mesh blocks and various operations on mesh blocks, including their pasting for creation of a complex mesh of simple blocks.

Part IV describes algorithms for visualization of finite element methods and results. Chapter 26 introduces the Java 3D^{TM} API, which is used for rendering three-dimensional objects. Visualization algorithms for higher-order finite elements and visualization code structure are presented in Chapter 27. A scene graph for visualization of meshes and results is discussed in Chapter 28. Chapter 29 describes algorithms for creation of the model surface. Chapters 30 and 31 are devoted to subdivision of the model surface into polygons. Chapter 32 presents Java classes for results field, color-gradation strip, mouse interaction and lights.

Appendices A, B, and C contain brief instructions for preparing data for finite element programs that perform problem solution, mesh generation, and visualization of models and results. Appendix D provides examples of finite element analysis: mesh generation, problem solution, and visualization of results.

How This Book Differs from Others

There are many books about the finite element method. Some of them contain finite element program segments. Two qualities distinguish this book from other books on the finite element method.

First, programming in this book is based upon the Java programming language. In my opinion, Java is well suited for explaining programming of the finite element method. It allows compact and simple code to be written. This helps greatly because the reader has a chance to actually read finite element programs and to understand them.

Secondly, algorithms presented in this book are tightly connected to programming. The book is written around one finite element Java program, which includes solution of solid mechanics boundary value problems, mesh generation, and visualization of finite element models and results. Presentation of computational algorithms is followed by a Java class or group of Java methods and then accompanied by code explanation.

Web Resources

The Java programs and examples presented in this book are available on the Web: `http://www.springer.com/978-1-84882-971-8`. Comments, suggestions, and corrections are welcome by e-mail: `fem.java@gmail.com`.

About the Author

Gennadiy Nikishkov got his Ph.D. and D.Sc. degrees from the Moscow Engineering Physics Institute (Technical University) in computational mechanics. He held a Professor position at the Moscow Engineering Physics Institute. He also had visiting positions at Georgia Institute of Technology (USA), Karlsruhe Research Center (Germany), RIKEN Institute of Physical and Chemical Research (Japan), GKSS Research Center (Germany), and the University of California at Los Angeles (USA). Currently, he is a Professor at the University of Aizu (Japan). His research interests include computational mechanics, computational fracture mechanics, computational nanomechanics, development of finite element and boundary element codes, scientific visualization, and computer graphics.

Acknowledgments

The author is grateful to Prof. Michael Cohen, University of Aizu, Japan, for his attentive reading of the book manuscript and his valuable suggestions, and to Dr. Yuriy Nikishkov, Georgia Institute of Technology, USA, for his advice related to development of the Java program. I also acknowledge my colleagues who contributed to this book through many discussions on computational methods over the years.

I thank the anonymous reviewers for their precious comments. Special thanks go to my editor Oliver Jackson and to Ms. Aislinn Bunning and Ms. Sorina Moosdorf for their help, which greatly improved the quality of the book. Finally, I express my loving appreciation to my wife Valentina for her support and encouragement.

Aizu-Wakamatsu, Japan   *Gennadiy Nikishkov*
July 2009

Contents

Part I
Finite Element Formulation

Chapter 1
Introduction

Abstract This introductory chapter presents the main steps of the finite element analysis. Derivation of the finite element equations is demonstrated on a second-order differential equation using the Galerkin approach in a one-dimensional case. The same one-dimensional example is considered using variational formulation. Determination of quadratic shape functions for a one-dimensional finite element with three nodes is explained.

1.1 Basic Ideas of FEM

The finite element method (FEM) is a computational technique for solving problems that are described by partial differential equations or can be formulated as functional minimization. A domain of interest is represented as an assembly of *finite elements*. Interpolation functions in finite elements are determined in terms of nodal values of a physical field that is sought. A continuous physical problem is transformed into a discretized finite element problem with unknown nodal values. For a linear problem a system of linear algebraic equations should be assembled and solved. Values within finite elements can be recovered by interpolating nodal values.

Two features of the FEM are worth mentioning:

1. *Piece-wise approximation* of physical fields on finite elements provides good precision even with simple approximating (interpolation) functions. By increasing the number of elements and nodes we can achieve an arbitrary precision of results.

2. *Locality of approximation* leads to sparse equation systems for a discretized problem. This helps to solve problems with a very large number of nodal unknowns using reasonable memory and computing time.

Theory, practice, and programming of the finite element method are described in many texts, among them are the comprehensive books of Bathe [3], Hughes [16], and Zienkiewicz [32, 33], the textbook of Fish and Belytschko [10], and the collection of Fortran programs of Smith and Griffiths [29].

The main steps of the finite element solution procedure are listed below.

1. *Discretize the domain.* The first step is to divide a solution region (domain) into finite elements connected at nodes. Because of the large amount of data, the finite element mesh is typically generated by a preprocessor program. The description of a mesh consists of several arrays, the main of which are nodal coordinates and element connectivities.

2. *Determine interpolation functions.* Interpolation functions are used to interpolate the field variables over the element. Usually, polynomials are selected as interpolation functions. The degree of the polynomial depends on the number of nodes belonging to the element. Interpolation functions are commonly called shape functions since they are also used for definition of the element shape.

3. *Compute the element matrices and vectors.* The matrix equation for the finite element is established that relates the nodal values of the unknown function to known parameters. For this task different approaches can be used; the most convenient are: the variational approach and the Galerkin method.

4. *Assemble the element equations.* To find the global equation system for the whole solution region we must assemble all the element equations. In other words we must properly combine local element equations for all elements used for discretization. Element connectivities are used for the assembly process. Before solution, boundary conditions (which are not accounted for in the element equations) should be imposed.

5. *Solve the global equation system.* The finite element global equation system is typically sparse, symmetric and positive-definite. Direct and iterative methods can be used for solution. The nodal values of the sought function are produced as a result of the solution.

6. *Compute additional results.* In many cases we need to calculate additional parameters. For example, in mechanical problems strains and stresses are of interest in addition to displacements, which are obtained after solution of the global equation system. While the primary result function (displacements) is continuous, its derivatives (strains and stresses) have discontinuities at element boundaries.

1.2 Formulation of Finite Element Equations

Several approaches can be used to transform the physical formulation of the problem to its finite element discrete analog. If the physical formulation of the problem is known as a differential equation then the most popular method of its finite element formulation is the *Galerkin method.* If the physical problem can be formulated as minimization of a functional then *variational formulation* of the finite element equations is usually used.

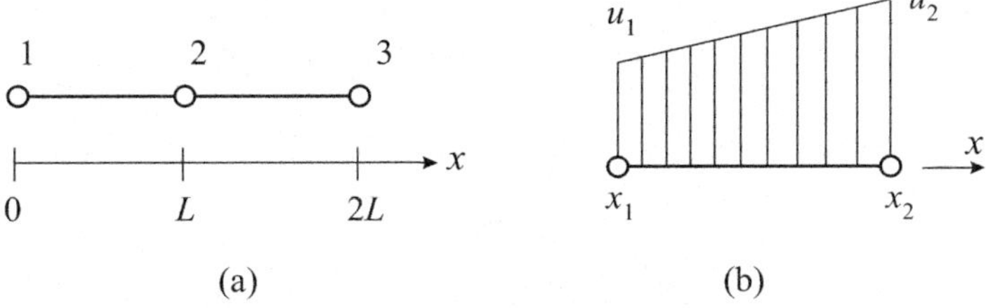

Fig. 1.1 Two one-dimensional linear elements (a) and function interpolation inside element (b)

1.2.1 Galerkin Method

Let us use a simple one-dimensional example for the explanation of finite element formulation using the Galerkin method. Suppose that we need to solve numerically the following differential equation:

$$a\frac{d^2u}{dx^2} + b = 0, \quad 0 \le x \le 2L, \tag{1.1}$$

with boundary conditions

$$u|_{x=0} = 0,$$

$$a\frac{du}{dx}\Big|_{x=2L} = R, \tag{1.2}$$

where $u = u(x)$ is an unknown function. We shall solve the problem using two linear one-dimensional finite elements, as shown in Figure 1.1a.

First, consider a finite element presented in Figure 1.1b. The element has two nodes and approximation of the function $u(x)$ can be done as follows:

$$u = N_1 u_1 + N_2 u_2 = [N]\{u\},$$

$$[N] = [N_1 \ N_2], \tag{1.3}$$

$$\{u\} = \{u_1 \ u_2\},$$

where N_i are the so-called *shape functions*

$$N_1 = 1 - \frac{x - x_1}{x_2 - x_1},$$

$$\tag{1.4}$$

$$N_2 = \frac{x - x_1}{x_2 - x_1},$$

which are used for interpolation of $u(x)$ using its nodal values. Nodal values u_1 and u_2 are unknowns that should be determined from the discrete global equation system.

After substituting u expressed through its nodal values and shape functions, in the differential equation, it has the following approximate form:

$$a\frac{d^2}{dx^2}[N]\{u\} + b = \psi, \tag{1.5}$$

where ψ is a non-zero residual due to the approximate representation of a function inside a finite element. The Galerkin method provides residual minimization by multiplying the terms of the above equation by shape functions, integrating over the element, and equating to zero:

$$\int_{x_1}^{x_2} [N]^T a \frac{d^2}{dx^2}[N]\{u\}dx + \int_{x_1}^{x_2} [N]^T b \, dx = 0. \tag{1.6}$$

Integration by parts leads to the following discrete form of the differential equation for the finite element:

$$\int_{x_1}^{x_2} \left[\frac{dN}{dx}\right]^T a \left[\frac{dN}{dx}\right] dx \{u\} - \int_{x_1}^{x_2} [N]^T b \, dx$$

$$- \begin{Bmatrix} 0 \\ 1 \end{Bmatrix} a\frac{du}{dx}|_{x=x_2} + \begin{Bmatrix} 1 \\ 0 \end{Bmatrix} a\frac{du}{dx}|_{x=x_1} = 0. \tag{1.7}$$

Usually, such relations for a finite element are presented as:

$$[k]\{u\} = \{f\},$$

$$[k] = \int_{x_1}^{x_2} \left[\frac{dN}{dx}\right]^T a \left[\frac{dN}{dx}\right] dx, \tag{1.8}$$

$$\{f\} = \int_{x_1}^{x_2} [N]^T b \, dx + \begin{Bmatrix} 0 \\ 1 \end{Bmatrix} a\frac{du}{dx}|_{x=x_2} - \begin{Bmatrix} 1 \\ 0 \end{Bmatrix} a\frac{du}{dx}|_{x=x_1}.$$

In solid mechanics $[k]$ is called the *stiffness matrix* and $\{f\}$ is called the *load vector*. In the considered simple case for two finite elements of length L, the stiffness matrices and the load vectors can be easily calculated:

$$[k_1] = [k_2] = \frac{a}{L}\begin{bmatrix} 1 & -1 \\ -1 & 1 \end{bmatrix},$$

$$\{f_1\} = \frac{bL}{2}\begin{Bmatrix} 1 \\ 1 \end{Bmatrix}, \quad \{f_2\} = \frac{bL}{2}\begin{Bmatrix} 1 \\ 1 \end{Bmatrix} + \begin{Bmatrix} 0 \\ R \end{Bmatrix}. \tag{1.9}$$

The above relations provide finite element equations for the two separate finite elements. A global equation system for the domain with two elements and three nodes can be obtained by an assembly of element equations. In our simple case it is clear that elements interact with each other at the node with global number 2. The assembled global equation system is:

$$\frac{a}{L}\begin{bmatrix} 1 & -1 & 0 \\ -1 & 2 & -1 \\ 0 & -1 & 1 \end{bmatrix}\begin{Bmatrix} u_1 \\ u_2 \\ u_3 \end{Bmatrix} = \frac{bL}{2}\begin{Bmatrix} 1 \\ 2 \\ 1 \end{Bmatrix} + \begin{Bmatrix} 0 \\ 0 \\ R \end{Bmatrix}. \tag{1.10}$$

After application of the boundary condition $u(x = 0) = 0$, the final appearance of the global equation system is:

$$\frac{a}{L}\begin{bmatrix} 1 & 0 & 0 \\ 0 & 2 & -1 \\ 0 & -1 & 1 \end{bmatrix}\begin{Bmatrix} u_1 \\ u_2 \\ u_3 \end{Bmatrix} = \frac{bL}{2}\begin{Bmatrix} 0 \\ 2 \\ 1 \end{Bmatrix} + \begin{Bmatrix} 0 \\ 0 \\ R \end{Bmatrix}. \tag{1.11}$$

When applying the boundary condition $u_1 = 0$ we put zeros in the first row of the equation system matrix and in the right-hand side; put zeros in the first column of the matrix and, finally, place unit value on the main diagonal.

Nodal values u_i are obtained as results of solution of the linear algebraic equation system. The value of u at any point inside a finite element can be calculated using the shape functions. The finite element solution of the differential equation is shown in Figure 1.2 for $a = 1, b = 1, L = 1$, and $R = 1$.

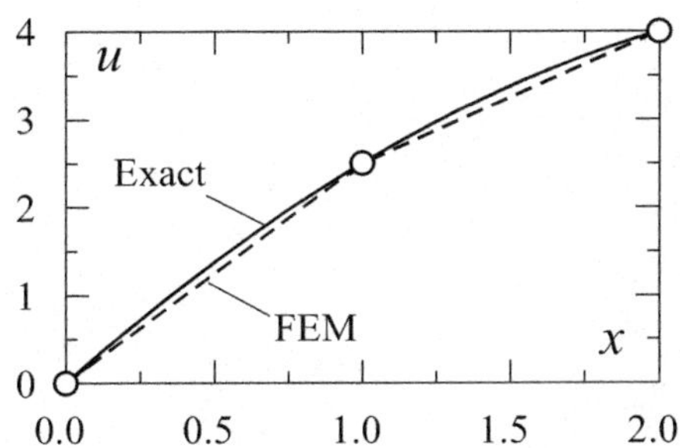

Fig. 1.2 Comparison of finite element solution and exact solution

The exact solution is a quadratic function. The finite element solution with the use of the simplest element is piece-wise linear. A more precise finite element solution can be obtained by increasing the number of simple elements or with the use of elements with more complicated shape functions. It is worth noting that at nodes the finite element method provides exact values of u (only for this particular problem). Finite elements with linear shape functions produce exact nodal values if the sought solution is quadratic. Quadratic elements give exact nodal values for the cubic solution.

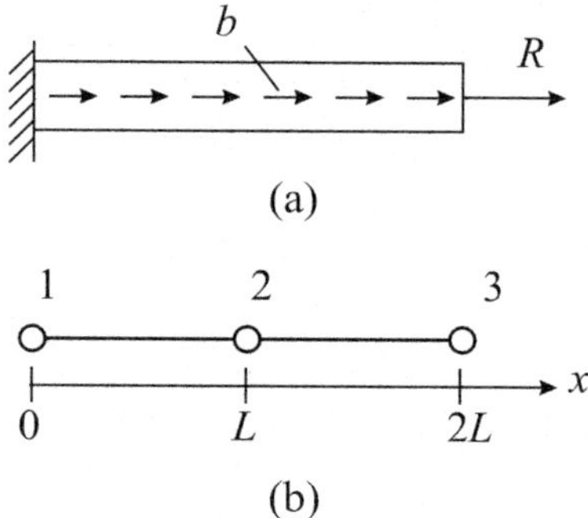

(a)

(b)

Fig. 1.3 One-dimensional bar subjected to a distributed load and a concentrated load (a) discretized with two linear elements (b)

1.2.2 Variational Formulation

The differential equation (1.1) with boundary conditions (1.2) and parameter $a = EA$ has the following physical meaning in solid mechanics. It describes the tension of the one-dimensional bar with cross-sectional area A made of material with the elasticity modulus E and subjected to a distributed load b and a concentrated load R at one end, as shown in Figure 1.3a.

Such problems can be solved using variational formulation by minimizing the potential energy functional Π over two elements of Figure 1.3b:

$$\Pi = \int_L \frac{1}{2} a \left(\frac{du}{dx}\right)^2 dx - \int_L bu\,dx - Ru|_{x=2L},$$

$$u|_{x=0} = 0. \tag{1.12}$$

Using representation of $\{u\}$ with shape functions (1.3) and (1.4) we can write the value of potential energy for the second finite element as:

$$\Pi_e = \int_{x_1}^{x_2} \frac{1}{2} a \{u\}^{\mathrm{T}} \left[\frac{dN}{dx}\right]^{\mathrm{T}} \left[\frac{dN}{dx}\right] \{u\} dx$$

$$- \int_{x_1}^{x_2} \{u\}^{\mathrm{T}} [N]^{\mathrm{T}} b\,dx - \{u\}^{\mathrm{T}} \begin{Bmatrix} 0 \\ R \end{Bmatrix}. \tag{1.13}$$

The condition for the minimum of Π is:

$$\delta\Pi = \frac{\partial \Pi}{\partial u_1} \delta u_1 + \ldots + \frac{\partial \Pi}{\partial u_n} \delta u_n = 0, \tag{1.14}$$

which is equivalent to

$$\frac{\partial \Pi}{\partial u_i} = 0 , \quad i = 1 \dots n. \tag{1.15}$$

It is easy to check that differentiation of Π with respect to u_i gives the following finite element equilibrium equation:

$$\int_{x_1}^{x_2} \left[\frac{dN}{dx}\right]^{\mathrm{T}} EA \left[\frac{dN}{dx}\right] dx\{u\} - \int_{x_1}^{x_2} [N]^{\mathrm{T}} b dx - \left\{\begin{matrix} 0 \\ R \end{matrix}\right\}. \tag{1.16}$$

This expression coincides with the equation obtained by the Galerkin method. Having two approaches to derivation of finite element equations, we can use the Galerkin method when a differential equation for the problem is known; when the problem is formulated as functional minimization, then it is possible to employ the variational approach.

1.3 Example of Shape-function Determination

Obtain shape functions for the one-dimensional quadratic element with three nodes depicted in Figure 1.4. Use local coordinate system $-1 \le \xi \le 1$.

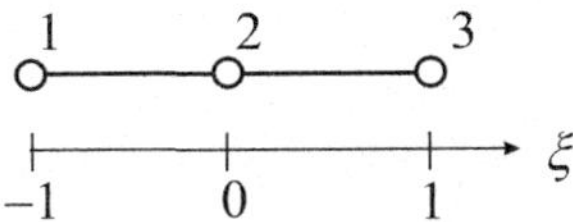

Fig. 1.4 One-dimensional quadratic element with three nodes

Solution

With shape functions, any field inside the element is expressible as:

$$u(\xi) = \sum N_i u_i , \quad i = 1, 2, 3.$$

At each node the approximated function should be equal to its nodal value:

$$u(-1) = u_1, \quad u(0) = u_2, \quad u(1) = u_3.$$

Since the element has three nodes the shape functions can be quadratic polynomials (with three coefficients). The shape function N_1 can be written as:

$$N_1 = \alpha_1 + \alpha_2 \xi + \alpha_3 \xi^2.$$

Unknown coefficients α_i are obtained from the following system of equations:

$$N_1(-1) = \alpha_1 - \alpha_2 + \alpha_3 = 1,$$
$$N_1(0) = \alpha_1 = 0,$$
$$N_1(1) = \alpha_1 + \alpha_2 + \alpha_3 = 0.$$

The solution is: $\alpha_1 = 0$, $\alpha_2 = -1/2$, $\alpha_3 = 1/2$. Thus, the shape function N_1 is:

$$N_1 = -\frac{1}{2}\xi(1-\xi).$$

Similarly determined shape functions N_2 and N_3 are equal to:

$$N_2 = 1 - \xi^2,$$
$$N_3 = \frac{1}{2}\xi(1+\xi).$$

It is possible to avoid solution of the equation system if we write down the sought formula for a shape function in the following form:

$$N_i = a_1(a_2 + \xi)(a_3 + \xi).$$

Coefficients a_1, a_2 and a_3 are determined from the condition that the shape function is equal to one at its own node and it is equal to zero at all other nodes. For example, it is easy to get a_2 and a_3 for shape function N_1 by equating to zero the expressions in braces:

$$\text{at } \xi = 0:\quad a_2 + 0 = 0,$$
$$\text{at } \xi = 1:\quad a_3 + 1 = 0.$$

We find that $a_2 = 0$, $a_3 = -1$ and function N_1 has the appearance

$$N_1 = a_1\xi(-1+\xi).$$

Coefficient a_1 is determined from the condition:

$$\text{at } \xi = -1:\quad a_1(-1)(-1-1) = 1.$$

Thus, we get $a_1 = 1/2$ and shape function N_1 has been found.

Problems

1.1. The formula for integration by parts for functions u and v is

$$\int_{x_1}^{x_2} u\,dv = (uv)\big|_{x_1}^{x_2} - \int_{x_1}^{x_2} v\,du.$$

Using this formula, show how to transform Equation 1.6 into Equation 1.7.

1.2. Check that the determinant of the matrix in Equation 1.10 is equal to zero and that the determinant of the matrix in Equation 1.11 is nonzero. Try to explain the physical reasons for this.

1.3. Derive finite element equations for the following heat-conduction problem.

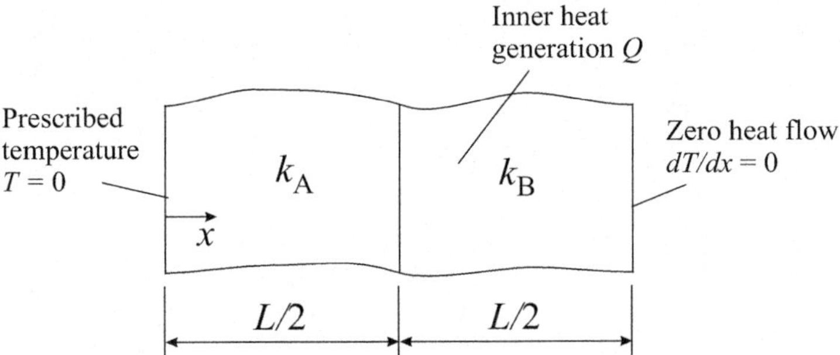

The body consists of two materials A and B with thermal conductivity coefficients k_A and k_B. Material B generates heat with volume rate Q. Constant zero temperature $T = 0$ is supported on the left boundary. Zero heat flow $dT/dx = 0$ is specified on the right boundary.

Use two linear one-dimensional elements to obtain equations for unknown nodal temperatures T_1, T_2 and T_3. Base your derivation on minimization of the functional:

$$\Pi = \int_L \frac{1}{2} k \left(\frac{dT}{dx} \right)^2 dx - \int_L T Q dx,$$

where k is the thermal-conductivity coefficient.

1.4. Find shape functions N_1 and N_2 for the one-dimensional linear element shown below. Use the local coordinate system $-1 \le \xi \le 1$.

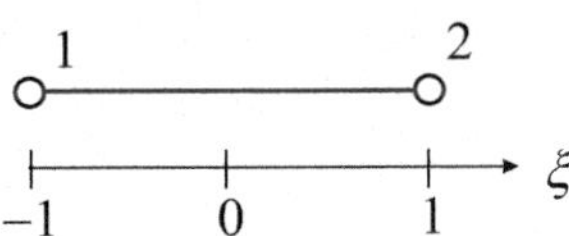

1.5. Determine shape functions N_1, N_2 and N_3 for the one-dimensional quadratic element shown below with intermediate node 2 placed at $\xi = -0.5$.

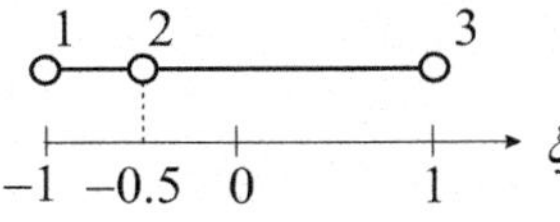

Analyze the behavior of the shape functions when the location of node 2 approaches that of node 1.

Chapter 2
Finite Element Equations for Heat Transfer

Abstract Solution of heat transfer problems is considered. Finite element equations are obtained using the Galerkin method. The conductivity matrix for a triangular finite element is calculated.

2.1 Problem Statement

Let us consider an isotropic body with temperature-dependent heat transfer. A basic equation of heat transfer has the following form [15]:

$$-\left(\frac{\partial q_x}{\partial x} + \frac{\partial q_y}{\partial y} + \frac{\partial q_z}{\partial z}\right) + Q = \rho c \frac{\partial T}{\partial t}. \tag{2.1}$$

Here, q_x, q_y and q_z are components of heat flow through the unit area; $Q = Q(x,y,z,t)$ is the inner heat-generation rate per unit volume; ρ is material density; c is heat capacity; T is temperature and t is time. According to Fourier's law the components of heat flow can be expressed as follows:

$$q_x = -k\frac{\partial T}{\partial x},$$

$$q_y = -k\frac{\partial T}{\partial y}, \tag{2.2}$$

$$q_z = -k\frac{\partial T}{\partial z},$$

where k is the thermal-conductivity coefficient of the media. Substitution of Fourier's relations gives the following basic heat transfer equation:

$$\frac{\partial}{\partial x}\left(k\frac{\partial T}{\partial x}\right) + \frac{\partial}{\partial y}\left(k\frac{\partial T}{\partial y}\right) + \frac{\partial}{\partial z}\left(k\frac{\partial T}{\partial z}\right) + Q = \rho c \frac{\partial T}{\partial t}. \tag{2.3}$$

It is assumed that the *boundary conditions* can be of the following types:

1. Specified temperature

$$T_s = T_1(x,y,z,t) \text{ on } S_1 \ .$$

2. Specified heat flow

$$q_x n_x + q_y n_y + q_z n_z = -q_s \text{ on } S_2 \ .$$

3. Convection boundary conditions

$$q_x n_x + q_y n_y + q_z n_z = h(T_s - T_e) \text{ on } S_3 \ ,$$

4. Radiation

$$q_x n_x + q_y n_y + q_z n_z = \sigma \varepsilon T_s^4 - \alpha q_r \text{ on } S_4,$$

where h is the convection coefficient; T_s is an unknown surface temperature; T_e is a convective exchange temperature; σ is the Stefan–Boltzmann constant; ε is the surface emission coefficient; α is the surface absorption coefficient, and q_r is the incident radiant heat flow per unit surface area. For transient problems it is necessary to specify an initial temperature field for a body at the time $t = 0$:

$$T(x,y,z,0) = T_0(x,y,z). \tag{2.4}$$

2.2 Finite Element Discretization of Heat Transfer Equations

A domain V is divided into finite elements connected at nodes. We shall write all the relations for a finite element. Global equations for the domain can be assembled from finite element equations using connectivity information.

Shape functions N_i are used for interpolation of temperature inside a finite element:

$$T = [N]\{T\},$$

$$[N] = \begin{bmatrix} N_1 & N_2 & \ldots \end{bmatrix}, \tag{2.5}$$

$$\{T\} = \{T_1 \ T_2 \ \ldots\} \ .$$

Differentiation of the temperature-interpolation equation gives the following interpolation relation for temperature gradients:

$$\begin{Bmatrix} \dfrac{\partial T}{\partial x} \\ \dfrac{\partial T}{\partial y} \\ \dfrac{\partial T}{\partial z} \end{Bmatrix} = \begin{bmatrix} \dfrac{\partial N_1}{\partial x} & \dfrac{\partial N_2}{\partial x} & \cdots \\ \dfrac{\partial N_1}{\partial y} & \dfrac{\partial N_2}{\partial y} & \cdots \\ \dfrac{\partial N_1}{\partial z} & \dfrac{\partial N_2}{\partial z} & \cdots \end{bmatrix} \{T\} = [B]\{T\}. \tag{2.6}$$

Here, $\{T\}$ is a vector of temperatures at nodes, $[N]$ is a matrix of shape functions, and $[B]$ is a matrix for temperature-gradient interpolation.

Using the Galerkin method, we can rewrite the basic heat transfer equation in the following form:

$$\int_V \left(\frac{\partial q_x}{\partial x} + \frac{\partial q_y}{\partial y} + \frac{\partial q_z}{\partial z} - Q + \rho c \frac{\partial T}{\partial t} \right) N_i dV = 0. \tag{2.7}$$

Applying the divergence theorem to the first three terms, we arrive at the relations:

$$\int_V \rho c \frac{\partial T}{\partial t} N_i dV - \int_V \left[\frac{\partial N_i}{\partial x} \; \frac{\partial N_i}{\partial y} \; \frac{\partial N_i}{\partial z} \right] \{q\} dV$$

$$= \int_V Q N_i dV - \int_S \{q\}^{\mathrm{T}} \{n\} N_i dS, \tag{2.8}$$

$$\{q\}^{\mathrm{T}} = \left[q_x \; q_y \; q_z \right],$$

$$\{n\}^{\mathrm{T}} = \left[n_x \; n_y \; n_z \right],$$

where $\{n\}$ is an outer normal to the surface of the body. After insertion of boundary conditions into the above equation, the discretized equations are as follows:

$$\int_V \rho c \frac{\partial T}{\partial t} N_i dV - \int_V \left[\frac{\partial N_i}{\partial x} \; \frac{\partial N_i}{\partial y} \; \frac{\partial N_i}{\partial z} \right] \{q\} dV$$

$$= \int_V Q N_i dV - \int_{S_1} \{q\}^{\mathrm{T}} \{n\} N_i dS \tag{2.9}$$

$$+ \int_{S_2} q_{\mathrm{s}} N_i dS - \int_{S_3} h(T - T_{\mathrm{e}}) N_i dS - \int_{S_4} (\sigma \varepsilon T^4 - \alpha q_{\mathrm{r}}) N_i dS.$$

It is worth noting that

$$\{q\} = -k[B]\{T\}. \tag{2.10}$$

The discretized finite element equations for heat transfer problems have the following form:

$$[C]\{\dot{T}\} + ([K_c] + [K_h] + [K_r])\{T\}$$

$$= \{R_T\} + \{R_Q\} + \{R_q\} + \{R_h\} + \{R_r\}, \tag{2.11}$$

$$[C] = \int_V \rho c [N]^{\mathrm{T}} [N] dV,$$

$$[K_c] = \int_V k [B]^{\mathrm{T}} [B] dV,$$

$$[K_h] = \int_{S_3} h [N]^{\mathrm{T}} [N] dS,$$

$$[K_r]\{T\} = \int_{S_4} \sigma \varepsilon T^4 [N]^{\mathrm{T}} dS,$$

$$\{R_T\} = -\int_{S_1} \{q\}^{\mathrm{T}} \{n\} [N]^{\mathrm{T}} dS,$$

$$\{R_Q\} = \int_V Q [N]^{\mathrm{T}} dV,$$

$$\{R_q\} = \int_{S_2} q_{\mathrm{s}} [N]^{\mathrm{T}} dS,$$

$$\{R_h\} = \int_{S_3} h T_{\mathrm{e}} [N]^{\mathrm{T}} dS,$$

$$\{R_r\} = \int_{S_4} \alpha q_{\mathrm{r}} [N]^{\mathrm{T}} dS. \tag{2.12}$$

Here, $\{\dot{T}\}$ is a nodal vector of temperature derivatives with respect to time.

2.3 Different Type Problems

Equations for different types of problems can be deducted from the above general equation:

Stationary linear problem

$$([K_c] + [K_h])\{T\} = \{R_Q\} + \{R_q\} + \{R_h\}. \tag{2.13}$$

Stationary nonlinear problem

$$([K_c] + [K_h] + [K_r])\{T\}$$
$$= \{R_Q(T)\} + \{R_q(T)\} + \{R_h(T)\} + \{R_r(T)\}. \tag{2.14}$$

Transient linear problem

$$[C]\{\dot{T}(t)\} + ([K_c] + [K_h(t)])\{T(t)\}$$
$$= \{R_Q(t)\} + \{R_q(t)\} + \{R_h(t)\}.$$

(2.15)

Transient nonlinear problem

$$[C(T)]\{\dot{T}\} + ([K_c(T)] + [K_h(T,t)] + [K_r(T)])\{T\}$$
$$= \{R_Q(T,t)\} + \{R_q(T,t)\} + \{R_h(T,t)\} + \{R_r(T,t)\}.$$

(2.16)

2.4 Triangular Element

Calculation of element conductivity matrix $[k_c]$ and heat flow vector $\{r_q\}$ is illustrated for a two-dimensional triangular element with three nodes. A simple triangular finite element is shown in Figure 2.1. The temperature distribution $T(x,y)$ inside the triangular element is described by linear interpolation of its nodal values:

$$T(x,y) = N_1(x,y)T_1 + N_2(x,y)T_2 + N_3(x,y)T_3,$$
$$N_i(x,y) = \alpha_i + \beta_i x + \gamma_i y.$$

(2.17)

Interpolation functions (usually called shape functions) $N_i(x,y)$ should satisfy the following conditions:

$$T(x_i,y_i) = T_i , \quad i = 1,\ 2,\ 3.$$

(2.18)

Solution of the above equation system provides expressions for the shape functions:

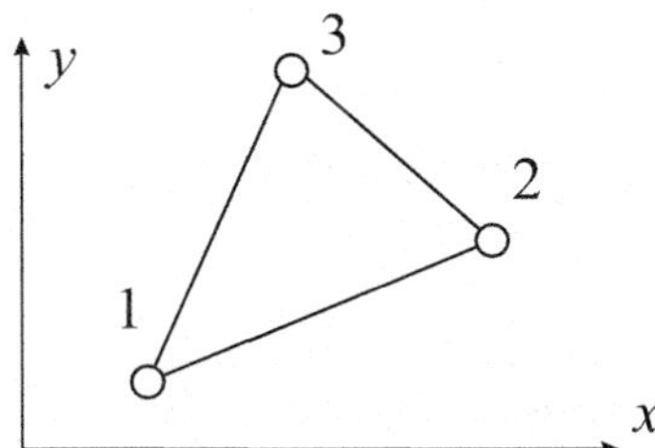

Fig. 2.1 Triangular finite element

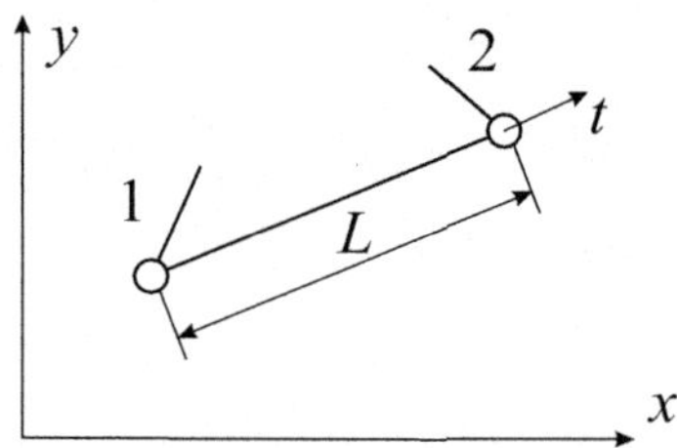

Fig. 2.2 Integration along an element side

$$N_i = \frac{1}{2\Delta}(a_i + b_i x + c_i y),$$

$$a_i = x_{i+1} y_{i+2} - x_{i+2} y_{i+1},$$

$$b_i = y_{i+1} - y_{i+2}, \tag{2.19}$$

$$c_i = x_{i+2} - x_{i+1},$$

$$\Delta = \frac{1}{2}(x_2 y_3 + x_3 y_1 + x_1 y_2 - x_2 y_1 - x_3 y_2 - x_1 y_3),$$

where Δ is the element area.

The conductivity matrix of the triangular element is determined by integration over element area A (assuming that the element has unit thickness),

$$[k_c] = \int_A k[B]^{\mathrm{T}}[B]\,dxdy. \tag{2.20}$$

The temperature differentiation matrix $[B]$ has expression

$$[B] = \begin{bmatrix} \dfrac{\partial N_1}{\partial x} & \dfrac{\partial N_2}{\partial x} & \dfrac{\partial N_3}{\partial x} \\[2mm] \dfrac{\partial N_1}{\partial y} & \dfrac{\partial N_2}{\partial y} & \dfrac{\partial N_3}{\partial y} \end{bmatrix} = \frac{1}{2\Delta} \begin{bmatrix} b_1 & b_2 & b_3 \\ c_1 & c_2 & c_3 \end{bmatrix}. \tag{2.21}$$

Since the temperature differentiation matrix does not depend on coordinates, integration of the conductivity matrix is simple;

$$[k_c] = \frac{k}{4\Delta} \begin{bmatrix} b_1^2 + c_1^2 & b_1 b_2 + c_1 c_2 & b_1 b_3 + c_1 c_3 \\ b_1 b_2 + c_1 c_2 & b_2^2 + c_2^2 & b_2 b_3 + c_2 c_3 \\ b_1 b_3 + c_1 c_3 & b_2 b_3 + c_2 c_3 & b_3^2 + c_3^2 \end{bmatrix}. \tag{2.22}$$

The heat-flow vector $\{r_q\}$ is evaluated by integration over the element side, as shown in Figure 2.2

$${r_q} = - \int_L q_s [N]^T dL = - \int_0^1 q_s [N_1 \ N_2]^T L dt. \qquad (2.23)$$

Here, integration over an element side L is replaced by integration using variable t ranging from 0 to 1. Shape functions N_1 and N_2 on element side 1–2 can be expressed through t:

$$N_1 = 1 - t, \quad N_2 = t. \qquad (2.24)$$

After integration with substituting integration limits, the heat-flow vector equals

$$\{r_q\} = -q_s \frac{L}{2} \begin{bmatrix} 1 \\ 1 \end{bmatrix}. \qquad (2.25)$$

Element matrices and vectors are calculated for all elements in a mesh and assembled into the global equation system. After application of prescribed temperatures, solution of the global equation system produces temperatures at nodes.

Problems

2.1. Calculate matrix $[k_h]$ describing convection boundary conditions

$$[k_h] = \int_L h [N]^T [N] dL$$

for a side of a triangular element (see Figure 2.2).

2.2. Obtain shape functions N_1, N_2, N_3 and N_4 for the square element shown below.

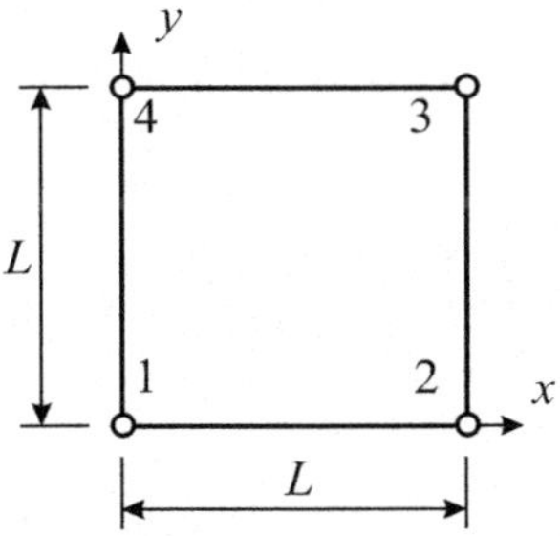

Assume that its size is $L = 1$ and that shape functions can be represented as $N_i = a_1 (a_2 + x)(a_3 + y)$.

2.3. For the square element of the previous problem, estimate the heat-generation vector

$$\{r_Q\} = \int_V Q [N]^T dV.$$

Use the shape functions obtained in the previous problem.

Chapter 3
FEM for Solid Mechanics Problems

Abstract Finite element equations for elasticity problems are derived from the variational principle based on minimum potential energy. A stiffness matrix for a simple triangular element is obtained. It is shown that assembly of a global finite element matrix and vectors can be performed through matrix multiplications using element matrices and vectors.

3.1 Problem Statement

Let us start consideration of solid mechanics problems with a three-dimensional elastic body subjected to surface forces p^S, body forces p^V, and temperature field T, as shown in Figure 3.1. In addition, displacements are specified on some surface area. For a given geometry of the body, applied loads, displacement boundary conditions, temperature field, and material stress–strain law, it is necessary to determine the displacement field for the body. The corresponding strains and stresses are also of interest.

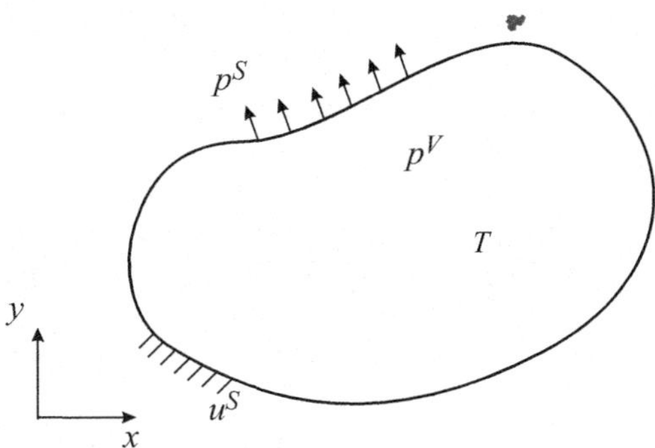

Fig. 3.1 Elastic body subjected to surface forces p^S, body forces p^V and temperature field T with displacements specified as u^S

Displacements along coordinate axes x, y, and z are defined by the displacement vector $\{u\}$:

$$\{u\} = \{u \ v \ w\}. \tag{3.1}$$

Six different strain components can be placed in the strain vector $\{\varepsilon\}$:

$$\{\varepsilon\} = \{\varepsilon_x \ \varepsilon_y \ \varepsilon_z \ \gamma_{xy} \ \gamma_{yz} \ \gamma_{zx}\}. \tag{3.2}$$

For small strains the relationship between strains and displacements is:

$$\{\varepsilon\} = [D]\{u\}, \tag{3.3}$$

where $[D]$ is the matrix differentiation operator:

$$[D] = \begin{bmatrix} \dfrac{\partial}{\partial x} & 0 & 0 \\ 0 & \dfrac{\partial}{\partial y} & 0 \\ 0 & 0 & \dfrac{\partial}{\partial z} \\ \dfrac{\partial}{\partial y} & \dfrac{\partial}{\partial x} & 0 \\ 0 & \dfrac{\partial}{\partial z} & \dfrac{\partial}{\partial y} \\ \dfrac{\partial}{\partial z} & 0 & \dfrac{\partial}{\partial x} \end{bmatrix}. \tag{3.4}$$

Six different stress components form the stress vector $\{\sigma\}$:

$$\{\sigma\} = \{\sigma_x \ \sigma_y \ \sigma_z \ \tau_{xy} \ \tau_{yz} \ \tau_{zx}\}, \tag{3.5}$$

which are related to strains for an elastic body by Hooke's law:

$$\{\sigma\} = [E]\{\varepsilon^e\} = [E](\{\varepsilon\} - \{\varepsilon^t\}),$$
$$\{\varepsilon^t\} = \{\alpha T \ \alpha T \ \alpha T \ 0 \ 0 \ 0\}. \tag{3.6}$$

Here, $[E]$ is the elasticity matrix depending on elastic material properties, $\{\varepsilon^e\}$ is the elastic part of strains, $\{\varepsilon^t\}$ is the thermal part of strains, α is the thermal expansion coefficient, and T is temperature.

The purpose of finite element solution of an elastic problem is to find such a displacement field that provides a minimum to the functional of total potential energy Π:

$$\Pi = \int_V \frac{1}{2}\{\varepsilon^e\}^{\mathrm{T}}\{\sigma\}dV - \int_V \{u\}^{\mathrm{T}}\{p^V\}dV - \int_S \{u\}^{\mathrm{T}}\{p^S\}dS. \tag{3.7}$$

Here, $\{p^V\} = \{p_x^V \ p_y^V \ p_z^V\}$ is the vector of body force and $\{p^S\} = \{p_x^S \ p_y^S \ p_z^S\}$ is the vector of surface forces. Prescribed displacements are specified on the part of the body surface where surface forces are absent.

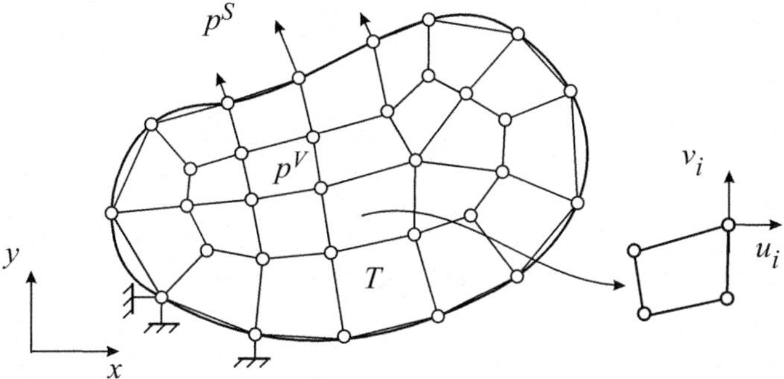

Fig. 3.2 Discretized representation of the problem is achieved through subdivision of the solution domain into finite elements. All quantities in the discretized problem should be related to nodal points

Displacement boundary conditions are not present in the functional of Π. Therefore, displacement boundary conditions should be implemented after assembly of finite element equations.

3.2 Finite Element Equations

In the finite element method the solution domain is divided into a set of subdomains of a simple shape that are called finite elements. Subdivision of a two-dimensional domain into simple quadrilateral elements is shown in Figure 3.2. Subdivision leads to a discretized representation of the problem. Instead of an infinite number of points in a continuum problem we now have the discretized problem with a finite number of nodal points. All quantities in the discretized problem should be related to nodal points.

In order to establish finite element equations for the considered problem, we first derive element equations and then show how to assemble them into global equations.

Let us consider some general three-dimensional finite element having the vector of nodal displacements $\{q\}$:

$$\{q\} = \{u_1 \ v_1 \ w_1 \ u_2 \ v_2 \ w_2 \ ...\}. \tag{3.8}$$

Displacements at some point inside a finite element $\{u\}$ can be determined with the use of nodal displacements $\{q\}$ and shape functions N_i:

$$u = \sum N_i u_i,$$

$$v = \sum N_i v_i, \tag{3.9}$$

$$w = \sum N_i w_i.$$

These relations can be rewritten in a matrix form as follows:

$$\{u\} = [N]\{q\},$$

$$[N] = \begin{bmatrix} N_1 & 0 & 0 & N_2 & \dots \\ 0 & N_1 & 0 & 0 & \dots \\ 0 & 0 & N_1 & 0 & \dots \end{bmatrix}. \tag{3.10}$$

Strains can also be determined through displacements at nodal points:

$$\{\varepsilon\} = [B]\{q\},$$

$$[B] = [D][N] = [B_1 \ \ B_2 \ \ B_3 \ \ \dots]. \tag{3.11}$$

Matrix $[B]$ is called the displacement differentiation matrix. It can be obtained by differentiation of displacements expressed through shape functions and nodal displacements:

$$[B_i] = \begin{bmatrix} \dfrac{\partial N_i}{\partial x} & 0 & 0 \\[2mm] 0 & \dfrac{\partial N_i}{\partial y} & 0 \\[2mm] 0 & 0 & \dfrac{\partial N_i}{\partial z} \\[2mm] \dfrac{\partial N_i}{\partial y} & \dfrac{\partial N_i}{\partial x} & 0 \\[2mm] 0 & \dfrac{\partial N_i}{\partial z} & \dfrac{\partial N_i}{\partial y} \\[2mm] \dfrac{\partial N_i}{\partial z} & 0 & \dfrac{\partial N_i}{\partial x} \end{bmatrix}. \tag{3.12}$$

Now, using the relations for stresses and strains we are able to express the total potential energy through nodal displacements:

$$\Pi = \int_V \frac{1}{2}([B]\{q\} - \{\varepsilon^t\})^{\mathrm{T}}[E]([B]\{q\} - \{\varepsilon^t\})dV$$

$$- \int_V ([N]\{q\})^{\mathrm{T}}\{p^V\}dV - \int_S ([N]\{q\})^{\mathrm{T}}\{p^S\}dS. \tag{3.13}$$

After performing multiplications the expression for the potential energy becomes

$$\Pi = \int_V \frac{1}{2}\{q\}^{\mathrm{T}}[B]^{\mathrm{T}}[E][B]\{q\}dV - \int_V \{q\}^{\mathrm{T}}[B]^{\mathrm{T}}[E]\{\varepsilon^t\}dV$$

$$+ \int_V \frac{1}{2}\{\varepsilon^t\}^{\mathrm{T}}[E]\{\varepsilon^t\}dV \tag{3.14}$$

$$- \int_V \{q\}^{\mathrm{T}}[N]^{\mathrm{T}}\{p^V\}dV - \int_S \{q\}^{\mathrm{T}}[N]^{\mathrm{T}}\{p^S\}dS.$$

Nodal displacements $\{q\}$ that correspond to the minimum of the functional Π are determined by the conditions:

$$\left\{\frac{\partial\Pi}{\partial q}\right\} = 0. \tag{3.15}$$

Differentiation of Π with respect to nodal displacements $\{q\}$ produces the following equilibrium equations for a finite element:

$$\int_V [B]^{\mathrm{T}}[E][B]dV\{q\} - \int_V [B]^{\mathrm{T}}[E]\{\varepsilon^{\mathrm{t}}\}dV$$
$$- \int_V [N]^{\mathrm{T}}\{p^V\}dV - \int_S [N]^{\mathrm{T}}\{p^S\}dS = 0, \tag{3.16}$$

which is usually presented in the following form:

$$[k]\{q\} = \{f\}, \tag{3.17}$$

$$\{f\} = \{p\} + \{h\}, \tag{3.18}$$

$$[k] = \int_V [B]^{\mathrm{T}}[E][B]dV, \tag{3.19}$$

$$\{p\} = \int_V [N]^{\mathrm{T}}\{p^V\}dV + \int_S [N]^{\mathrm{T}}\{p^S\}dS, \tag{3.20}$$

$$\{h\} = \int_V [B]^{\mathrm{T}}[E]\{\varepsilon^{\mathrm{t}}\}dV. \tag{3.21}$$

Here, $[k]$ is the element stiffness matrix, $\{f\}$ is the load vector, $\{p\}$ is the vector of actual forces, and $\{h\}$ is the thermal vector, which represents fictitious forces for modeling thermal expansion.

Equation 3.17 represents element equilibrium expressed through nodal displacements. Applying Hooke's law (3.6) in Equation (3.16), the element equilibrium equation can be expressed through stresses:

$$\int_V [B]^{\mathrm{T}}\{\sigma\}dV - \{p\} = 0. \tag{3.22}$$

The above stress equilibrium equation is valid for both linear elastic and nonlinear nonelastic problems. In nonlinear problems the residual vector of the stress equilibrium equation is used for organization of the iteration procedure for finding the problem solution.

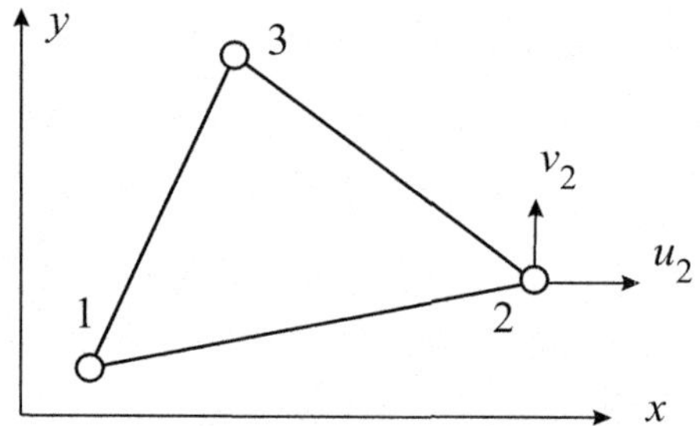

Fig. 3.3 A triangular finite element is the simplest two-dimensional element

3.3 Stiffness Matrix of a Triangular Element

To illustrate implementation of the finite element equations for particular elements, let us consider an algorithm of stiffness matrix calculation for a simple triangular element.

The triangular finite element was the first finite element proposed for continuous problems. Because of simplicity it can be used as an introduction to other elements. A triangular finite element in the coordinate system xy is shown in Figure 3.3. Since the element has three nodes, linear approximation of displacements u and v is selected:

$$u(x,y) = N_1 u_1 + N_2 u_2 + N_3 u_3,$$

$$v(x,y) = N_1 v_1 + N_2 v_2 + N_3 v_3, \tag{3.23}$$

$$N_i = \alpha_i + \beta_i x + \gamma_i y.$$

Shape functions $N_i(x,y)$ can be determined from the following equation system:

$$u(x_i, y_i) = u_i, \quad i = 1, 2, 3. \tag{3.24}$$

Shape functions for the triangular element can be presented as:

$$N_i = \frac{1}{2\Delta}(a_i + b_i x + c_i y),$$

$$a_i = x_{i+1} y_{i+2} - x_{i+2} y_{i+1},$$

$$b_i = y_{i+1} - y_{i+2}, \tag{3.25}$$

$$c_i = x_{i+2} - x_{i+1},$$

$$\Delta = \frac{1}{2}(x_2 y_3 + x_3 y_1 + x_1 y_2 - x_2 y_1 - x_3 y_2 - x_1 y_3).$$

Here, Δ is the element area. The matrix $[B]$ for interpolating strains using nodal displacements is

$$[B] = \frac{1}{2\Delta} \begin{bmatrix} b_1 & 0 & b_2 & 0 & b_3 & 0 \\ 0 & c_1 & 0 & c_2 & 0 & c_3 \\ c_1 & b_1 & c_2 & b_2 & c_3 & b_3 \end{bmatrix}. \tag{3.26}$$

The elasticity matrix $[E]$ has the following appearance for the plane stress problem:

$$[E] = \frac{E}{1-v^2} \begin{bmatrix} 1 & v & 0 \\ v & 1 & 0 \\ 0 & 0 & \dfrac{1-v}{2} \end{bmatrix}, \tag{3.27}$$

where E is the elasticity modulus and v is Poisson's ratio. For the plane strain problem the elasticity matrix is

$$[E] = \frac{E}{(1+v)(1-2v)} \begin{bmatrix} 1-v & v & 0 \\ v & 1-v & 0 \\ 0 & 0 & \dfrac{1-v}{2} \end{bmatrix}. \tag{3.28}$$

The stiffness matrix for the three-node triangular element can be calculated as

$$[k] = \int_V [B]^{\mathrm{T}}[E][B]dV = [B]^{\mathrm{T}}[E][B]\Delta. \tag{3.29}$$

Here, it was taken into account that both matrices $[B]$ and $[E]$ do not depend on coordinates. It was assumed that the element has unit thickness. Since the matrix $[B]$ is constant inside the element, the strains and stresses are also constant inside the triangular element.

3.4 Assembly of the Global Equation System

The aim of assembly is to form the global equation system

$$[K]\{Q\} = \{F\} \tag{3.30}$$

using element equations

$$[k_i]\{q_i\} = \{f_i\}. \tag{3.31}$$

Here, $[k_i]$, q_i and f_i are the stiffness matrix, the displacement vector and the load vector of the ith finite element; $[K]$, $\{Q\}$ and $\{F\}$ are the global stiffness matrix, displacement vector, and load vector, respectively.

In order to derive an assembly algorithm let us present the total potential energy for the body as a sum of element potential energies π_i:

$$\Pi = \sum \pi_i = \sum \frac{1}{2}\{q_i\}^{\mathrm{T}}[k_i]\{q_i\} - \sum \{q_i\}^{\mathrm{T}}\{f_i\} + \sum E_i^0, \qquad (3.32)$$

where E_i^0 is the fraction of potential energy related to free thermal expansion:

$$E_i^0 = \int_{V_i} \frac{1}{2}\{\varepsilon^t\}^{\mathrm{T}}[E]\{\varepsilon^t\}dV. \qquad (3.33)$$

Let us introduce the following vectors and a matrix where element vectors and matrices are simply placed:

$$\{Q_d\} = \{\{q_1\}\ \{q_2\}\},$$

$$\{F_d\} = \{\{f_1\}\ \{f_2\}\ ...\}, \qquad (3.34)$$

$$[K_d] = \begin{bmatrix} [k_1] & 0 & 0 \\ 0 & [k_2] & 0 \\ 0 & 0 & ... \end{bmatrix}. \qquad (3.35)$$

It is evident that it is easy to find matrix $[A]$ such that

$$\{Q_d\} = [A]\{Q\},$$

$$\{F_d\} = [A]\{F\}. \qquad (3.36)$$

The total potential energy for the body can be rewritten in the following form:

$$\Pi = \frac{1}{2}\{Q_d\}^{\mathrm{T}}[K_d]\{Q_d\} - \{Q_d\}^{\mathrm{T}}\{F_d\} + \sum E_i^0$$

$$= \frac{1}{2}\{Q\}^{\mathrm{T}}[A]^{\mathrm{T}}[K_d][A]\{Q\} - \{Q\}^{\mathrm{T}}[A]^{\mathrm{T}}\{F_d\} + \sum E_i^0. \qquad (3.37)$$

Using the condition of minimum total potential energy

$$\left\{\frac{\partial \Pi}{\partial Q}\right\} = 0 \qquad (3.38)$$

we arrive at the following global equation system:

$$[A]^{\mathrm{T}}[K_d][A]\{Q\} - [A]^{\mathrm{T}}\{F_d\} = 0. \qquad (3.39)$$

The last equation shows that algorithms of the assembly of the global stiffness matrix and global load vector are:

$$[K] = [A]^{\mathrm{T}}[K_d][A],$$

$$\{F\} = [A]^{\mathrm{T}}\{F_d\}. \qquad (3.40)$$

Here, $[A]$ is the sparse matrix providing transformation from global to local enumeration. The fraction of nonzero (unit) entries in the matrix $[A]$ is very small. Because of this, the matrix $[A]$ is never used explicitly in actual computer programs.

3.5 Example of the Global Matrix Assembly

Express a matrix $[A]$ that relates local (element) and global (domain) node numbers for the finite element mesh shown in Figure 3.4b.

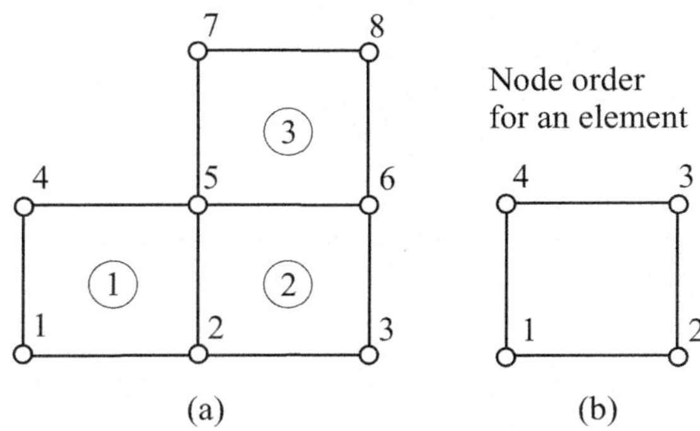

Fig. 3.4 Finite element mesh with global node numbers (a) and typical element with local node numbers (b)

Solution

To make the matrix representation compact let us assume that each node has one degree of freedom (note that in two-dimensional solid mechanics problems there are two degrees of freedom at each node). The displacement vector $\{q\}_1$ for element 1 using local numbering is simply

$$\{q\}_1 = \{q_1 \quad q_2 \quad q_3 \quad q_4\}.$$

The same displacement vector for element 1 using global node numbers is

$$\{q\}_1 = \{Q_1 \quad Q_2 \quad Q_5 \quad Q_4\}.$$

Matrix $[A]$ relates element and global nodal values in the following way:

$$\{Q_d\} = [A]\{Q\},$$

where $\{Q\}$ is a global vector of nodal values and $\{Q_d\}$ is a vector containing all the element vectors. The explicit rewriting of the above relation yields:

$$\begin{Bmatrix} Q_1 \\ Q_2 \\ Q_5 \\ Q_4 \\ Q_2 \\ Q_3 \\ Q_6 \\ Q_5 \\ Q_5 \\ Q_6 \\ Q_8 \\ Q_7 \end{Bmatrix} = \begin{bmatrix} 1 & 0 & 0 & 0 & 0 & 0 & 0 & 0 \\ 0 & 1 & 0 & 0 & 0 & 0 & 0 & 0 \\ 0 & 0 & 0 & 0 & 1 & 0 & 0 & 0 \\ 0 & 0 & 0 & 1 & 0 & 0 & 0 & 0 \\ 0 & 1 & 0 & 0 & 0 & 0 & 0 & 0 \\ 0 & 0 & 1 & 0 & 0 & 0 & 0 & 0 \\ 0 & 0 & 0 & 0 & 0 & 1 & 0 & 0 \\ 0 & 0 & 0 & 0 & 1 & 0 & 0 & 0 \\ 0 & 0 & 0 & 0 & 1 & 0 & 0 & 0 \\ 0 & 0 & 0 & 0 & 0 & 1 & 0 & 0 \\ 0 & 0 & 0 & 0 & 0 & 0 & 0 & 1 \\ 0 & 0 & 0 & 0 & 0 & 0 & 1 & 0 \end{bmatrix} \begin{Bmatrix} Q_1 \\ Q_2 \\ Q_3 \\ Q_4 \\ Q_5 \\ Q_6 \\ Q_7 \\ Q_8 \end{Bmatrix}.$$

Problems

3.1. In transforming Equation 3.13 into Equation 3.14 we used the fact that the transpose of the product of two matrices is equivalent to the product of their transposes in reversed order $([A][B])^{\mathrm{T}} = [B]^{\mathrm{T}}[A]^{\mathrm{T}}$. Show that this matrix identity is true.

3.2. In equilibrium equation (3.16) the nodal displacement vector $\{q\}$ is moved outside the integral. Explain why this is possible.

3.3. Show that the sum of element shape functions is unity at any point of the element:

$$\sum_i N_i = 1.$$

3.4. Show that the element stiffness matrix $[k]$ obtained from the principle of minimum potential energy is a positive-definite matrix, satisfying the inequality

$$\{v\}^{\mathrm{T}}[k]\{v\} > 0$$

for any nonzero vector v.
Hint: express the elastic energy of the finite element through its stiffness matrix and displacement vector.

3.5. Prove that the element stiffness matrix $[k]$ is symmetric:

$$k_{ij} = k_{ji}.$$

Hint: use the reciprocity theorem.

3.6. Explain why the row sums of the element stiffness matrix coefficients are equal to zero:

$$\sum_j k_{ij} = 0 \text{ for any row } j.$$

Hint: consider translation of the element as a rigid body.

3.7. In the previous three problems you proved that the element stiffness matrix has the following properties:

- it is positive-definite;
- it is symmetric; and
- the sum of coefficients in any row is zero.

A global stiffness matrix is assembled from element stiffness matrices. Does the global stiffness matrix possess the same properties if displacement boundary conditions are not applied?

3.8. Why is the element stiffness matrix singular in our finite element formulation? Singularity of the element stiffness matrix means that its determinant is equal to zero:

$$\det[k] = |k| = 0.$$

Hint: consider element rigid displacement (translation or rotation) without application of displacement boundary conditions.

Chapter 4
Finite Element Program

Abstract The object-oriented approach to finite element programming is briefly reviewed. Requirements for the finite element program under development are presented. The general structure of the finite element code is discussed. The packages and classes of the Java^{TM} finite element system `Jfea` considered in this book are listed.

4.1 Object-oriented Approach to Finite Element Programming

We now apply the equations and principles of the previous chapter to development of a finite element program for solid mechanics problems. We start with the general plan of the finite element code.

Finite element programs were traditionally developed in the Fortran and C languages, which support procedural programming. During the last fifteen years, finite element development has gradually shifted towards an object-oriented approach. Forde *et al.* [11], in one of the first publications on the object-oriented approach to the finite element development, presented the essential finite element classes such as elements, nodes, displacement, and force boundary conditions, vectors and matrices. Several authors described a detailed finite element architecture using an object-oriented approach. Zimmermann *et al.* [34] and Commend *et al.* [6] proposed the basics of object-oriented class structures for elastic and elastic–plastic structural problems. A flexible object-oriented approach that isolates numerical modules from a structural model is presented by Archer *et al.* [2]. Macki devoted numerous papers and a book [18] to various aspects of finite element object-oriented programming including creation of interactive codes with a graphical user interface (GUI). Extensive bibliographical information on the object-oriented approach in FEM and BEM is collected by Mackerle [19].

Mostly, object-oriented finite element algorithms have been implemented in C++ programming language. It was shown that an object-oriented approach with the C++ programming language could be used without sacrificing computational efficiency

[8, 34] compared to Fortran. A paper by Akin *et al.* [1] advocates employing Fortran 90 for object-oriented development of finite element codes since the authors consider Fortran execution faster than C++.

The Java language, introduced by Sun Microsystems, possesses features that make it attractive for use in computational modeling. Java is a simple language (simpler than C++). It has a rich collection of libraries implementing various APIs . With Java it is easy to create GUIs and to communicate with other computers over a network. Java has built-in garbage collection, preventing memory leaks. Another advantage of Java is its portability. Java virtual machines (JVM) are developed for all major computer systems. JVM is embedded in most popular Web browsers. Java applets can be downloaded through the Internet and executed within a Web browser. Useful for object-oriented design Java features are packages for organizing classes and prohibition of class multiple inheritance. This allows cleaner object-oriented design in comparison to C++. Despite its attractive features, Java is rarely used in finite element analysis. Few publications can be found on object-oriented Java finite element programs [9]. Previously, Java had a reputation as a relatively slow language because Java bytecode is interpreted by the JVM during execution, but modern just-in-time compilers used in the JVM make the efficiency of Java code comparable to that of C or C++ [20].

4.2 Requirements for the Finite Element Application

A software development process starts with creation of a list of requirements. While we are not going to demonstrate here the entire documentation resource that should accompany software development, brief requirements can help us to understand the computer program we are going to develop.

4.2.1 Overall Description

The program system `Jfea` is designed as educational software helping to understand algorithms and programming techniques of the finite element method. The program should solve two- and three-dimensional solid mechanics problems, including elastic problems and elastic–plastic problems. It should provide all solution stages: finite element mesh generation, boundary value problem solution, and visualization of finite element models and results. The program should be compact and understandable by the reader. At the same time it should implement real finite element analysis and allow its further extension. The Java programming language is selected for development of the `Jfea` finite element system.

4.2.2 User Description

Typical users of this program are students and researchers learning the finite element method programming or deepening their knowledge of the subject. The users know a programming language, which is used for code development, and the basics of the finite element method. They should be able to solve solid mechanics problems using the `Jfea` program. They should also be able to read, understand, and modify the code.

4.2.3 User Interface

Most computer applications have GUIs. However, programming GUIs leads to a large amount of code. Since our purpose is explanation of finite element programming and we want our code to be compact, its user interface will be based on text data files. A simple GUI will be used during visualization of finite element models and results.

4.2.4 Functions

The main functions of our finite element system `Jfea` [21] are to perform three tasks of the finite element analysis:

- preprocessing (finite element model generation);
- processing (problem solution); and
- postprocessing (results calculation and visualization).

The mesh-generation program `Jmgen` creates two- and three-dimensional meshes. A solution domain is divided into blocks of simple shape. A finite element mesh is generated inside blocks. Pasting of blocks allows creation of finite element models of complicated shape.

Finite element solution of elastic and elastic–plastic boundary value problems is performed by the program `Jfem`. Two main element types are used for problem discretization – the two-dimensional quadrilateral elements with eight nodes and three-dimensional hexahedral elements with twenty nodes. It should be possible to add new finite element types by adding new Java classes. The program includes two equation solvers implementing direct and iterative methods. It should be possible to add other equation solvers.

The postprocessing program `Jvis` creates continuous fields of results and performs visualization of finite element models and result fields as contours on a model surface. Visualization algorithms should take into account the fact that element edges and surfaces can be curved. The Java 3D API is used for real-time rendering of three-dimensional graphics.

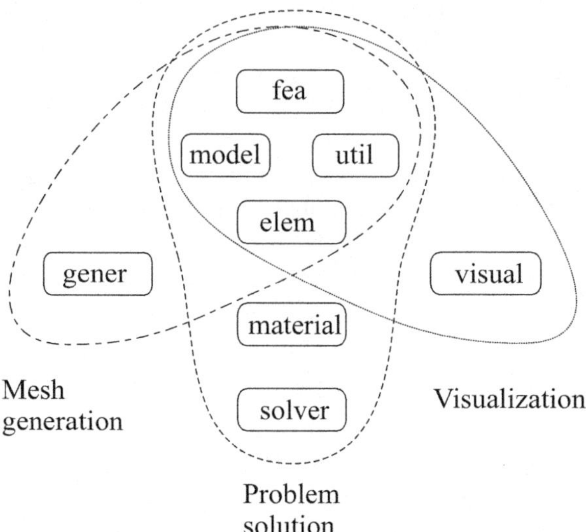

Fig. 4.1 Packages of the finite element Java code. Eight class packages are used for three tasks of finite element analysis – mesh generation, problem solution and visualization

4.2.5 Other Requirements

Since our finite element program is presented in this book we relax some normal requirements related to documentation and error diagnostics. To keep the source code brief we do not include special Java comments that can be used for automatic generation of program documentation. We shall check data for possible errors, but our error control is limited and the usual reaction to error discovery is program termination with display of an error message.

4.3 General Structure of the Finite Element Code

During program development, three tasks of the finite element analysis – preprocessing, processing, and postprocessing – are often implemented as three separate computer programs. Since the tasks have many common data structures and methods, the three modules contain duplicated or similar code fragments complicating support and modification.

In the Java language it is possible to have several main methods. The code (classes) can be organized into packages. A package is a named collection of classes providing encapsulation and modularity, which can eliminate code duplication and provide a means for easy code reuse.

Our Jfea finite element system is organized into eight class packages, as shown in Figure 4.1. The packages include the following classes.

Package `fea` – main classes:

Class `FE` – symbolic constants;
Class `Jfem` – main class for solution of elastic and elastic–plastic problems (finite element processor);
Class `Jmgen` – main class for mesh generation (preprocessor);
Class `Jvis` – main class for visualization of models and results (postprocessor).

Package `model` – finite element model and loading:

Class `Dof` – degree of freedom;
Class `ElemFaceLoad` – element face loading;
Class `FeLoad` – load increment for the finite element model;
Class `FeLoadData` – load data for the finite element model;
Class `FeModel` – description of the finite element model;
Class `FeModelData` – data for the finite element model;
Class `FeStress` – computing stress increment.

Package `util` – utility classes:

Class `FePrintWriter` – helper class for organizing printing to a file;
Class `FeScanner` – scanning finite element data;
Class `GaussRule` – several Gauss integration rules;
Class `UTIL` – printing error messages, dates, etc.

Package `elem` – finite elements:

Abstract class `Element` – finite element;
Class `ElementQuad2D` – two-dimensional quadratic isoparametric element;
Class `ElementQuad3D` – three-dimensional quadratic isoparametric element;
Class `ShapeQuad2D` – two-dimensional quadratic shape functions and their derivatives;
Class `ShapeQuad3D` – three-dimensional quadratic shape functions and their derivatives;
Class `StressContainer` – stresses and equivalent strains at integration point.

Package `material` – constitutive relations for materials:

Class `Material` – material constitutive relations;
Class `ElasticMaterial` – constitutive relations for an elastic material;
Class `ElasticPlasticMaterial` – constitutive relations for an elastic–plastic material.

Package `solver` – assembly and solution of global finite element equation systems:

Abstract class `Solver` – solution of the global equation system;

Class `SolverLDU` – profile LDU (lower, diagonal and upper matrix decomposition) symmetric solver;
Class `SolverPCG` – preconditioned conjugate gradient solver with sparse-row format storage.

Package `gener` – mesh generators:

Class `connect` – paste two mesh blocks;
Class `copy` – copy mesh block;
Class `genquad8` – generate a mesh inside a macroelement with eight nodes;
Class `readmesh` – read mesh data from a text file;
Class `rectangle` – generate mesh inside a rectangle;
Class `sweep` – generate a three-dimensional mesh by sweeping a two-dimensional mesh;
Class `transform` – mesh transformations (translate, scale, rotate);
Class `writemesh` – write a mesh to a text file.

Package `visual` – visualization of models and results:

Class `ColorScale` – two-dimensional texture for drawing contours;
Class `FaceSubdivision` – subdivision of an element face into triangles;
Class `J3dScene` – Java3D scene graph for visualization;
Class `Lights` – lights and background;
Class `MouseInteraction` – mouse behaviors for interaction;
Class `ResultAtNodes` – computing result values at nodes of the finite element model;
Class `SurfaceGeometry` – surface geometry for visualization: faces, edges and nodes;
Class `SurfaceSubGeometry` – element subfaces, subedges and nodes;
Class `VisData` – visualization parameters.

Classes from four packages `fea`, `model`, `util` and `elem` are employed for all three tasks of finite element analysis – mesh generation, problem solution and visualization. Other packages contain specific classes for tasks. Package `gener` is used for mesh generation. Packages `material` and `solver` are specific for problem solution. Package `visual` is designed for the visualization stage of the finite element analysis.

We begin with the development of the classes related to the solution of solid mechanics problems since this is the main part of the finite element analysis. Later, preprocessing and postprocessing parts are considered.

Problems

4.1. Explain the notion of "class" in object-oriented programming languages. Give examples of classes related to computational methods.

4.2. Explain what a Java package is, and how to place several classes in one package.

4.3. Name primitive data types in Java. Which floating-point data type is most useful in numerical computations? Provide reasons for your choice.

4.4. Analyze packages of the finite element system given in this chapter. Propose another package structure for a finite element program.

Part II
Finite Element Solution

Chapter 5
Finite Element Processor

Abstract Creation of the finite element processor, a Java^{TM} program that solves two- and three-dimensional solid mechanics problems, is discussed. A general finite element solution procedure and class structure of Java program Jfem are introduced. Use of free data format for data input is adopted. Input data for finite element analysis, which includes model data and load data, is described. A data scanner class is presented.

5.1 Class Structure

The finite element processor obtains the solution of the boundary value problem. It is the main part of the finite element analysis.

Typical finite element solution flow is composed of data input, assembly of the global equation system, solution of the equation system, computing stresses and results output. A pseudocode expression of the finite element processor is given below.

```
Read data and create finite element model
Assemble global stiffness matrix
do while Load data is available
    Read load data
    Assemble load vector
    do
        Solve global equation system
        Compute stress increment
    while not in equilibrium
    Accumulate and write results
end do
```

The finite element processor uses classes from the following packages: fea, model, util, element, material and solver. Major classes of the finite

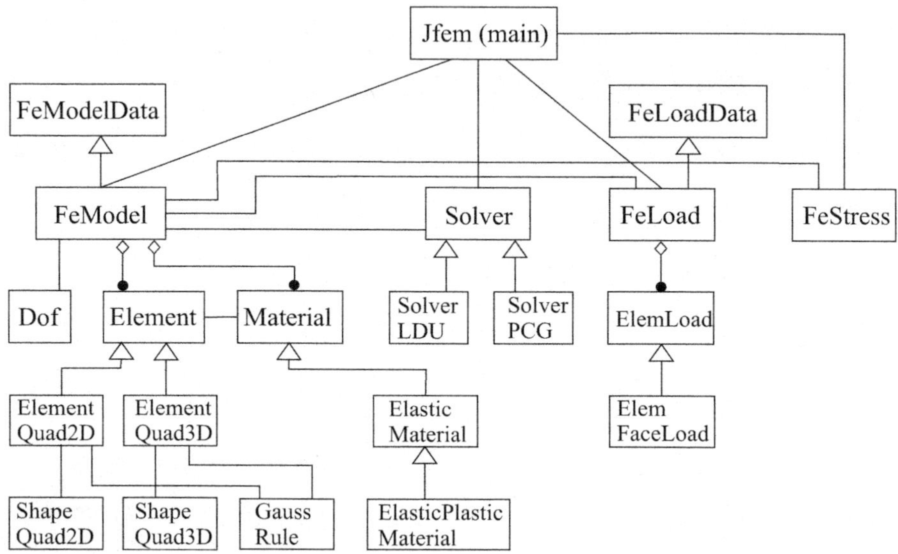

Fig. 5.1 Class diagram of the finite element processor

element processor are shown in Figure 5.1. The diagram shows that the main class
Jfem refers directly to four classes – FeModel (finite element model), Solver
(assembly and solution of the finite element equations), FeLoad (load case), and
FeStress (computing stresses in finite elements). The FeModel object contains
many elements (Element objects) and one or several materials (Material objects). Class Element is abstract and is implemented by classes for different types
of finite elements. Class Material has subclasses for elastic and elastic–plastic
materials. Abstract class Solver allows implementing different methods for solution of finite element equation system. Object FeLoad comprises element loads
ElemLoad. Data from the finite element model is used at all solution stages. So,
classes Solver, FeLoad, and FeStress are linked to class FeModel.

The main class Jfem contains the main method of the finite element processor.
The main class can be coded in Java as follows.

```java
1 package fea;
2
3 import elem.*;
4 import model.*;
5 import solver.*;
6 import util.*;
7 import java.io.*;
8
9 // Main class of the finite element processor
10 public class Jfem {
11
12     private static FeScanner RD;
13     private static PrintWriter PR;
```

```java
14      public static String fileOut;
15
16      public static void main(String[] args) {
17
18          if (args.length == 0) {
19              System.out.println(
20                      "Usage: java fea.JFEM FileIn [FileOut]\n");
21              return;
22          }
23          FE.main = FE.JFEM;
24
25          RD = new FeScanner(args[0]);
26
27          fileOut = (args.length==1) ? args[0]+".lst" : args[1];
28          PR = new FePrintWriter().getPrinter(fileOut);
29
30          PR.println("fea.JFEM: FE code. Data file: " + args[0]);
31          System.out.println("fea.JFEM: FE code. Data file: "
32                      + args[0]);
33
34          new Jfem();
35          PR.close();
36      }
37
38      public Jfem () {
39
40          UTIL.printDate(PR);
41
42          FeModel fem = new FeModel(RD, PR);
43          Element.fem = fem;
44
45          fem.readData();
46
47          PR.printf("\nNumber of elements    nEl = %d\n"+
48                      "Number of nodes       nNod = %d\n"+
49                      "Number of dimensions nDim = %d\n",
50                      fem.nEl, fem.nNod, fem.nDim);
51
52          long t0 = System.currentTimeMillis();
53
54          Solver solver = Solver.newSolver(fem);
55          solver.assembleGSM();
56
57          PR.printf("Memory for global matrix: %7.2f MB\n",
58                      Solver.lengthOfGSM*8.0e-6);
59
60          FeLoad load = new FeLoad(fem);
61          Element.load = load;
62
63          FeStress stress = new FeStress(fem);
64
65          // Load step loop
66          while (load.readData( )) {
67              load.assembleRHS();
```

```
68              int iter = 0;
69              // Equilibrium iterations
70              do {
71                  iter++;
72                  int its = solver.solve(FeLoad.RHS);
73                  if (its > 0) PR.printf(
74                      "Solver: %d iterations\n", its);
75                  stress.computeIncrement();
76              } while (!stress.equilibrium(iter));
77
78              stress.accumulate();
79              stress.writeResults();
80              PR.printf("Loadstep %s", FeLoad.loadStepName);
81              if (iter>1) PR.printf(" %5d iterations, " +
82                  "Relative residual norm = %10.5f",
83                  iter, FeStress.relResidNorm);
84              PR.printf("\n");
85          }
86
87          PR.printf("\nSolution time = %10.2f s\n",
88                  (System.currentTimeMillis()-t0)*0.001);
89      }
90
91 }
```

In the main method, we first check the case when no parameters are specified by the
user. If so, a message is printed that the code Jfem should be run with one or two
parameters (lines 18–22). Line 23 sets parameter main in class FE to JFEM, thus
making available the name of the main class of a running application to its methods.
Then, a scanner RD for reading input data from a specified ASCII file is constructed
in line 25 and a printer PR for saving information into an ASCII file is created in
lines 27 and 28. Finally, the main object Jfem is created to solve the finite element
problem. Statement 35 closes the printer file.

In the constructor Jfem, static method printDate records the current date and
time using print writer PR. Line 42 creates object FeModel that contains data and
methods related to a finite element model of the problem excluding the load model.
Method readData in line 45 reads data for the finite element model.

Depending on the specified data the finite element solver Solver is constructed
in line 54. Two types of solvers are available. A direct equation solver SolverLDU
performs solution of the finite element equation system using symmetric LDU decomposition of the matrix. An iterative solver SolverPCG is based on the preconditioned conjugate gradient method.

In line 55, the method assembleGSM assembles the global stiffness matrix for
the finite element model. Different storage formats of the global stiffness matrix are
used depending on the solver. Lines 60 and 63 construct objects FeLoad for a load
case and FeStress for computing stress increment.

The load step loop (lines 66–85) contains input of load data, performed by
method readData of class FeLoad. The method assembleRHS (line 67) assembles all load contributions into the right-hand side (RHS) of the global equation

system. The equilibrium iteration loop (lines 70–76) includes solution of the global equation system (method `solve` in line 62) and computation of a stress increment (method `computeIncrement` in line 75). The loop is finished when stresses are in equilibrium with the applied load (method `equilibrium` in line 76). Just one iteration is done for linear (elastic) problems, which we consider now. In nonlinear problems, such as elastic–plastic problems that are considered later, some number of iterations is necessary to achieve stress equilibrium. The method `accumulate` (line 78) adds increments of loads and stresses to their total values. The method `writeResults` (lines 79) records results into an output file.

Some symbolic constants are placed in class `FE` given below.

```
1  package fea;
2
3  // Symbolic constants
4  public class FE {
5
6      // Main method of application: JFEM/JMGEN/JVIS
7      public static int main;
8      public static final int JFEM = 0, JMGEN = 1, JVIS = 2;
9
10     public static final int maxNodesPerElem = 20;
11
12     // Big value for displacements boundary conditions
13     public static double bigValue = 1.0e64;
14     // Solution tuning
15     public static boolean tunedSolver = true;
16
17     // Error tolerance for PCG solution of FE equation system
18     public static final double epsPCG = 1.e-10;
19     // Constants for PCG solution method
20     public static int maxRow2D = 21,
21                       maxRow3D = 117,
22                       maxIterPcg = 10000;
23
24     // Integration scheme for elastic-plastic problems
25     public static boolean epIntegrationTANGENT = false;
26 }
```

All class fields (variables and constants) are declared as static and can be used without object creation. We will refer to this class fields in other parts of this book.

Class `FePrintWriter` is a helper class for organizing printing to a text file with a specified name. Object `PR` returned by method `getPrinter` can be used for printing.

```
1  package util;
2
3  import java.io.*;
4
5  // Finite element printer to file
6  public class FePrintWriter {
7      PrintWriter PR;
8      public PrintWriter getPrinter(String fileOut) {
9          try {
```

```
10                PR = new PrintWriter(
11                      new BufferedWriter(
12                            new FileWriter(fileOut)));
13          } catch (Exception e) {
14          UTIL.errorMsg("Cannot open output file: " + fileOut);
15          }
16          return PR;
17      }
18
19 }
```

Another class with methods used at many places of the Jfea system is class UTIL.

```
1 package util;
2
3 import java.util.Calendar;
4 import java.util.GregorianCalendar;
5 import java.io.PrintWriter;
6
7 // Miscellaneous static methods
8 public class UTIL {
9
10     // Print date and time.
11     // PR - PrintWriter for listing file
12     public static void printDate(PrintWriter PR) {
13
14         Calendar c = new GregorianCalendar();
15
16         PR.printf("Date: %d-%02d-%02d  Time: %02d:%02d:%02d\n",
17             c.get(Calendar.YEAR),   c.get(Calendar.MONTH)+1,
18             c.get(Calendar.DATE),   c.get(Calendar.HOUR_OF_DAY),
19             c.get(Calendar.MINUTE),c.get(Calendar.SECOND));
20     }
21
22     // Print error message and exit.
23     // message - error message that is printed.
24     public static void errorMsg(String message) {
25             System.out.println("=== ERROR: " + message);
26             System.exit(1);
27     }
28
29     // Transform text direction into integer.
30     // s - direction x/y/z/n.
31     // returns  integer direction 1/2/3/0, error: -1.
32     public static int direction(String s) {
33         if       (s.equals("x")) return 1;
34         else if (s.equals("y")) return 2;
35         else if (s.equals("z")) return 3;
36         else if (s.equals("n")) return 0;
37         else return -1;
38     }
39
40 }
```

The class contains three static methods. The method `printDate` outputs current date and time into a listing file. The method `errorMsg` displays an error message and stops program execution. The method `direction` transforms coordinate axes x, y, and z into their numerical values 1, 2, and 3; text n (normal direction) is interpreted as zero.

5.2 Problem Data

From the data point of view the finite element solution is transformation of input data into output data. We assume that input data is specified as an ASCII file. Such an input data file can be prepared manually. However, since data describing a finite element mesh is too large, it is usually generated by a preprocessor.

Let us restrict ourselves to elastic and elastic–plastic problems with displacement and force boundary conditions and a specified temperature field. The data can be divided into that related to the finite element model and that describing loading conditions. The finite element model data is not changed during problem solution since we suppose that the model shape, material properties, and displacement boundary conditions are constant. The loading conditions change with time. This means that it is possible to specify several loadings and treat them as load increments.

Description of the finite element model contains:

- scalar parameters (number of nodes, number of elements, etc.);
- material properties;
- nodal data (coordinates of nodal points);
- element data (element types, element materials, and connectivities);
- description of displacement boundary conditions.

Load data includes descriptions of:

- surface and concentrated loads;
- temperature field.

5.2.1 Data Statements

The data specification statement consists of the data item name, an equals sign, and the data value itself. Data items are specified in free format and in free order. Of course, if some data item is logically required before other data items, then data records should follow this logical order.

Data statement

A data statement has the following form:

```
<name> = <data>
```

Here, `<name>` is a data name and `<data>` is the data content. Data names are not case sensitive. Capital and small characters can be used for marking parts of the name. Blank and the equals sign are interpreted by the data parser as white space, so these elements are not allowed inside data names. Data on the right of an equals sign can contain one or more tokens. Tokens can be numbers or text literals. The number of tokens is predetermined by the data name.

A comment statement has the form:

```
# comment text
```

Several statements can be placed on one line. However, all text is considered as a comment after a comment sign # followed by a blank. A comment statement should be alone or it should be last on a line.

Including file

An input data file typically has a large size due to the large amount of information required for finite element mesh description. It is convenient to place large uniform data in separate files. This can be imported with the use of statement:

```
includeFile <fileName>
```

which includes all data contained in a file with name `<fileName>`.

End statement

Statement

```
end
```

is used as the last statement in the model data and in the load data. When we include a data part using directive `includeFile` it is possible to mark the end of this file by the directive `end`. If the word `end` is found then a read data method returns. Return also occurs when the end of a file is reached. However, placement of the `end` directive is necessary at the end of the finite element model data and at the end of load data.

5.2.2 Model Data

Parameters

The main problem parameters include:

```
nNod = <number>
```
– number of nodes;

```
nEl = <number>
```
– number of elements in the finite element model;

```
stressState = THREED/PLSTRAIN/PLSTRESS/AXISYM
```
– type of problem: THREED – three-dimensional problem, PLSTRAIN – plane strain two-dimensional problem, PLSTRESS – plane stress two-dimensional problem, AXISYM – axisymmetrical problem;

```
physLaw = ELASTIC/ELPLASTIC
```
– physical law for material behavior: elastic or elastic–plastic;

```
solver = LDU/PCG
```
– equation solver: LDU – direct solver based on LDU decomposition, PCG – preconditioned conjugate gradient iterative solver;

```
thermalLoading = N/Y
```
– existence of thermal loading: N – no, Y – yes.

Default parameter values are emboldened.

Material properties

For each elastic material, the following data should be specified:

```
material = matName E nu alpha
```

Here, matName is any name selected for referring to this material, E is the elasticity modulus, nu is Poisson's ratio, and alpha is a thermal-expansion coefficient.

For elastic–plastic material, the data statement contains three additional parameters related to elastic–plastic material behavior:

```
material = matName E nu alpha sY hardCoef hardPower
```

Additional parameters have the following interpretation: sY is a material yield stress, hardCoef is a hardening coefficient, and hardPower is a hardening power.

Finite element mesh

The finite element mesh is described by two arrays: nodal coordinates and element connectivities (including element type and element material). Nodal coordinates are specified by the statement:

```
nodCoord = <array>
```

For three-dimensional problems, nodal coordinates are specified as x_1 y_1 z_1 x_2 y_2 z_2, etc. A two-dimensional coordinate array has the appearance x_1 y_1 x_2 y_2 ...

Element data include element types, element materials, and element connectivities:

```
elCon = <array>
```

For each element the following data should be provided:

`ElType = QUAD8/HEX20` – element type (QUAD8 – two-dimensional element with eight nodes, HEX20 – twenty-node three-dimensional element),

`ElMat` – material name corresponding to that in data describing the material,

`ElemNodeNumbers` – node numbers belonging to the element.

Displacement boundary conditions

Displacement boundary conditions can be specified in two ways: direct specification of node numbers with constrained displacements or specification of a box for node definition.

The first style of displacement boundary condition specification has the following form:

```
constrDispl = Direct Value nNumbers NodNumbers[]
```

where `Direct = x/y/z` – direction of displacement constraint, `Value` – constrained displacemet value, `nNumbers` – number of items in the list of node numbers, `NodNumbers[]` – list of nodes where displacement boundary conditions are specified. The list can include positive integers as node numbers. If a range $n_1 - n_2$ is present in the list, then it is interpreted as a set of nodes from n_1 to n_2 (condition $n_1 < n_2$ should be fulfilled).

In many cases of flat or straight boundary segments, it is possible to define a box inside of which the nodes will be constrained. Such a form of displacement boundary conditions has the format:

```
boxConstrDispl = Direct Value BoxDiadonal[]
```

where `Direct = x/y/z` – direction of displacement constraint, `Value` – constrained displacemet value, and `BoxDiadonal[]` – coordinates of two points at the ends of a box diagonal. In the three-dimensional case the diagonal is specified as x_{min} y_{min} z_{min} x_{max} y_{max} z_{max}. In the two-dimensional case z-coordinates are absent.

5.2.3 Load Specification

Load can consist of one or several load steps. Load steps are considered as increments to the previous state. Each load step is defined by the following data.

Load step name

```
loadStep = LoadStepName
```

This statement should be the first statement of a load step. The specified load step name is used for identifying this load step. Result files have load label as their extensions.

Parameters

```
scaleLoad = Scale
```
 – load scaling parameter. The load vector of the current load step is obtained by multiplying the load vector from the previous step with the specified parameter `Scale`.

```
residTolerance = Tolerance
```
 – tolerance for ratio of a residual norm to the norm of force load. If the relative residual norm becomes less than the specified `Tolerance`, then equilibrium iterations are finished. The default value is `Tolerance = 0.01`.

```
maxiternumber = Number
```
 – maximum allowed number of equilibrium iterations (the default value is 100).

Parameters `scaleLoad`, `residTolerance`, and `maxiternumber` are useful for elastic–plastic problems. They are not used in elastic problems. It should be noted that parameters `residTolerance` and `maxiternumber` are valid for the next load step if they are not changed. All other loading parameters including `scaleLoad` do not exist in the beginning of each new load step and should be specified if necessary.

Nodal forces

```
nodForce = Direct Value nNumbers NodNumbers[]
```

This statement is used for specification of nodal forces. Here, `Direct = x/y/z` – direction of nodal forces, `Value` – force value, and `NodNumbers[]` – list of nodes where nodal forces are applied. The list can include positive integers as node numbers and range $n_1 - n_2$ for specifying nodes from–to.

Surface forces

```
surForce = Direct ElNumber nFaceNodes
          faceNodes[] forceAtFaceNodes[]
```

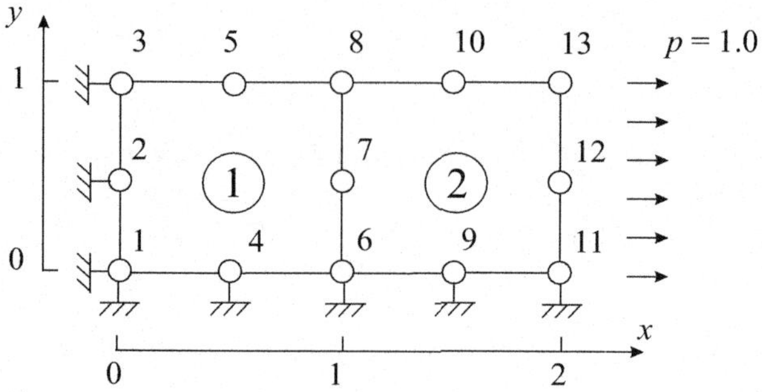

Fig. 5.2 Discrete model composed of two eight-node finite elements

Specification of distributed surface load consists of the following items:

$\texttt{Direct} = \texttt{x/y/z/n}$ – direction of surface load ($\texttt{n}$ means loading in the direction of the external normal to the surface),
$\texttt{ElNumber}$ – element number,
$\texttt{nFaceNodes}$ – number of nodes on the element face,
$\texttt{faceNodes[]}$ – node numbers defining the element face, and
$\texttt{forceAtFaceNodes[]}$ – intensities of distributed load at nodes.

Surface forces inside a box

```
boxSurForce = Direct Value BoxDiagonal[]
```

This statement allows one to apply a distributed surface load for all element faces, which are inside a box. Data include: $\texttt{Direct} = \texttt{x/y/z/n}$ – direction of surface load ($\texttt{n}$ – external normal direction), $\texttt{Value}$ – intensity of distributed load common to all faces, and $\texttt{BoxDiadonal[]}$ – coordinates of two points at ends of a box diagonal ($x_{\min}\, y_{\min}\, z_{\min}\, x_{\max}\, y_{\max}\, z_{\max}$).

```
nodTemp = NodeTemperatures[]
```

This statement is used for specifying temperatures at nodes.

5.2.4 Data Example

Let us consider numerical information for a simple tension problem. A finite element mesh consisting of two eight-node elements is depicted in Figure 5.2. Nodes 1, 2 and 3 are constrained in the x-direction. Nodes 1, 4, 6, 9 and 11 can not move in the y-direction. A distributed load of unit intensity is applied to side 11–12–13 of

element 2. Temperature $T = 20$ is applied to the specimen. The finite element model
can be described as follows:

```
# Number of nodes and number of elements
  nNod = 13    nEl = 2

# Plane stress state, 2D problem
 stressState = PLSTRESS

# Enable thermal loading option
  thermalLoading = Y

# Material properties:
# material name, elasticity modulus, Poisson's ratio and
#    coefficient of thermal expansion
  material = 1    1    0.3    0.1

# Nodal coordinates
  nodCoord = 0 0   0 0.5   0 1   0.5 0   0.5 1
             1 0   1 0.5   1 1   1.5 0   1.5 1
             2 0   2 0.5   2 1

# Element data: element type, material, connectivities
  elCon = QUAD8 1  1   4   6   7   8   5   3   2
          QUAD8 1  6   9  11  12  13  10   8   7

# Constraints: direction, value, number of constraints,
#    node numbers
  constrDispl = x 0.0 2   1 -3
  constrDispl = y 0.0 5   1 4 6 9 11

end

# Load
  loadStep = 1

# Surface load: direction, element number, number of face
#    nodes, face node numbers, intensities
  surForce = x   2 3 11 12 13   1 1 1

# Nodal temperatures
  nodTemp = 10 10 10 10 10 10 10 10 10 10 10 10 10

end
```

While the problem is simple the above data file illustrates the main principles of
data preparation for the finite element program Jfem.

Data for finite element analysis can be specified in many ways that will lead to
the same results. Let us illustrate this by several examples.

First, upper case and lower case characters in data name are not distinguished.
Because of this, names nNod, nnod, NNOD and other combinations of upper and
lower case characters are treated in the same way. Both blank and equal sign are

delimiters during data reading. So, the following three statements produce the same effect

```
nNod = 13
nNod=13
nNod 13
```

The order of data specification is to a certain extent arbitrary. The order is not rigid but the data sequence should correspond to the logical links between data items. For example, the number of nodes nNod should be specified anywhere before specification of the nodal coordinates array nodCoord since the length of this array is determined by nNod.

Specification of boundary conditions can be done in different ways. Displacement boundary conditions can be specified by explicit node numbers or by box specification. In addition, if node numbers are a sequence of numbers with step 1 then they can be defined using structure "from–to". In our simple problem of Figure 5.2 displacement constraints along x can be specified in the following three ways

```
constrDispl = x 0.0 2   1 -3
constrDispl = x 0.0 3   1 2 3
boxConstrDispl = x 0.0   -0.01 -0.01 0.01 1.01
```

When using a box for displacement constraint we define a rectangle that includes nodes 1, 2 and 3. The box boundary condition specification is especially efficient if a large number of nodes on a flat surface are to be constrained.

The data file shown above can be divided into several files. One main data file can have references to several other data files. Instead of the previous one data file we can prepare the following three data files.

File f.fem (main data file)

```
stressState = PLSTRESS      # Plane stress state
thermalLoading = Y          # Enable thermal loading

# Finite element mesh
 includeFile f.mesh

# Material properties:
 material = 1   1    0.3    0.1

# Constraints
 constrDispl = x 0.0 2   1 -3
 constrDispl = y 0.0 5   1 4 6 9 11

end

# Load case
 includeFile f.load

end
```

File `f.mesh` (finite element mesh)

```
    nNod = 13    nEl = 2   # Number of nodes and elements

  # Nodal coordinates
    nodCoord = 0 0   0 0.5   0 1   0.5 0   0.5 1
               1 0   1 0.5   1 1   1.5 0   1.5 1
               2 0   2 0.5   2 1

  # Element data: element type, material, connectivities
    elCon = QUAD8 1  1  4  6  7  8  5  3  2
            QUAD8 1  6  9 11 12 13 10  8  7
  end
```

File `f.load` (data for load step)

```
    loadStep = 1

  # Surface load: direction, element number, number of face
  #    nodes, face node numbers, intensities
    surForce = x  2 3 11 12 13  1 1 1

  # Nodal temperatures
    nodTemp = 10 10 10 10 10 10 10 10 10 10 10 10 10

  end
```

File `f.fem` includes two other files `f.mesh` and `f.load` containing data for a finite element mesh and a load step. The name of the main data file `f.fem` is passed to the Java program when we solve the boundary value problem.

5.3 Data Scanner

Java 5 introduced a simple text scanner that can parse primitive data types and strings using regular expressions. The scanner separates its input into tokens according to specified delimiters. The resulting tokens may be read into values of different types. Class `FeScanner`, which implements simple input operations for our finite element program, is shown below.

```
 1 package util;
 2
 3 import model.Dof;
 4 import java.util.Scanner;
 5 import java.util.ListIterator;
 6 import java.io.File;
 7
 8 // FEM data scanner. Delimiters: blank, =.
 9 public class FeScanner {
10
11     private Scanner es;
12
```

```java
13      // Constructs FE data scanner.
14      // fileIn - name of the file containing data.
15      public FeScanner(String fileIn) {
16
17          try {
18              es = new Scanner(new File(fileIn));
19          } catch (Exception e) {
20              UTIL.errorMsg("Input file not found: " + fileIn);
21          }
22          es.useDelimiter("\\s*=\\s*|\\s+");
23
24      }
25
26      // Returns  true if another token is available.
27      public boolean hasNext() { return es.hasNext(); }
28
29       // Returns  true if double is next in input.
30      public boolean hasNextDouble() {return es.hasNextDouble();}
31
32      // Gives the next token from this scanner.
33      public String next() { return es.next(); }
34
35      // Gives the next double from this scanner.
36      public double nextDouble() { return es.nextDouble(); }
37
38      // Reads the next integer.
39      // Generates an error if next token is not integer.
40      public int readInt() {
41          if (!es.hasNextInt()) UTIL.errorMsg(
42                  "Expected integer. Instead: "+es.next());
43          return es.nextInt();
44      }
45
46      // Reads the next double.
47      // Generates an error if next token is not double.
48      public double readDouble() {
49          if (!es.hasNextDouble()) UTIL.errorMsg(
50                  "Expected double. Instead: "+es.next());
51          return es.nextDouble();
52      }
53
54      // Advances the scanner past the current line.
55      public void nextLine() { es.nextLine(); }
56
57      // Moves to line which follows a line with the word.
58      public void moveAfterLineWithWord(String word) {
59
60          while (es.hasNext()) {
61              String varname = es.next().toLowerCase();
62              if (varname.equals("#")) { es.nextLine();
63                                                  continue; }
64              if (varname.equals(word)) {
65                  es.nextLine();
66                  return;
```

```
 67                    }
 68                }
 69            UTIL.errorMsg("moveAfterLineWithWord cannot find: "
 70                        + word);
 71        }
 72
 73        // Method reads < nNumbers numbers > and places resulting
 74        // degrees of freedom in a List data structure.
 75        // Here numbers is a sequence of the type n1 n2 -n3 ...
 76        // where n2 -n3 means from n2 to n3 inclusive.
 77        // it - list iterator.
 78        // dir - direction (1,2,3).
 79        // nDim - problem dimension (2/3).
 80        // sValue - specified value.
 81        // returns - modified list iterator it.
 82        public ListIterator readNumberList(ListIterator it,
 83                    int dir, int ndim, double sValue) {
 84            // number of items in the list
 85            int ndata = readInt();
 86            int i1, i2;
 87            i1 = i2 = readInt();
 88            for (int i=1; i<ndata; i++) {
 89                i2 = readInt();
 90                if (i2 > 0 && i1 >= 0) {
 91                    if (i1 > 0) {
 92                        it.add(new Dof(ndim*(i1-1)+dir, sValue));
 93                    }
 94                    i1 = i2;
 95                }
 96                else if (i2 < 0) {
 97                    for (int j=i1; j<=(-i2); j++) {
 98                        it.add(new Dof(ndim*(j-1)+dir, sValue));
 99                    }
100                    i1 = 0;
101                    i2 = 0;
102                }
103            }
104            if (i2 > 0) {
105                it.add(new Dof(ndim*(i2-1)+dir, sValue));
106            }
107            return it;
108        }
109
110        // Closes this scanner.
111        public void close() { es.close(); }
112
113 }
```

Here, constructor `FeScanner` creates a scanner for data in our finite element
program. ASCII data is read from a file with the name `fileIn`. The statement in
line 22 sets a blank and an equals sign as delimiters between data tokens.

Methods `hasNext` and `hasNextDouble` check if another token or another
double-precision number are available for input. They return true if another appro-

priate token is available. Method `next` returns the next token as a string if it is available.

Methods `next` and `nextDouble` return the next string and the next double values. To avoid exception it is necessary to check the availability of tokens of appropriate type.

Integer and double data items can be read with the help of methods `readInt` and `readDouble`. These methods first check the availability of correspondent data, then read data tokens and return numerical values. If an appropriate token is not available to the scanner, then an error message is written to the console and the program is stopped (method `errorMsg` in package `UTIL`).

Method `nextLine` advances the scanner past the current line and method `moveAfterLineWithWord` moves it to the line that follows a line with the word specified as a parameter.

In some cases of boundary conditions specifications it is convenient to use an expression "from–to". Method `readNumberList` (lines 82–108) is able to read and interpret a list

$$n_{\text{Numbers}}\ n_1\ n_2\ -n_3\ n_4\ \dots$$

Here, n_{Numbers} is the number of items in the list excluding n_{Numbers} itself, and n_i specify node numbers. Single positive numbers mean just node numbers, but a negative number $-n_i$ creates a pair with the previous positive number n_{i-1} and together they are interpreted as all integers from n_{i-1} to n_i. The method produces degree of freedom numbers using node numbers and direction `dir` (values 1, 2 and 3 correspond to axes x, y and z) and puts them in the `Dof` object. `Dof` is just a container for a degree of freedom. It contains a degree of freedom number and an associated double value, which can be displacement or force. The parameters of the method have the following meanings:

`it` – list iterator. Created degrees of freedom are added to a list with the list iterator `it`.

`dir` – direction. Direction values 1, 2, 3 correspond to coordinate directions x, y, and z.

`ndim` – problem dimension. Should be equal to 2 (two-dimensional problem) or 3 (three-dimensional problem).

`sValue` – the value that will be associated with all degrees of freedom read during this call of the method.

The method returns modified list iterator `it`.

Finally, method `close()` closes the scanner. The listing of class `Dof` is given below.

```
1 package model;
2
3 // Degree of freedom.
4 public class Dof {
5
6     public int dofNum;
7     public double value;
```

```
 8
 9        public Dof(int dofNum, double value) {
10            this.dofNum = dofNum;
11            this.value = value;
12        }
13
14  }
```

Problems

5.1. Prepare a data file for a three-dimensional mesh consisting of two twenty-node hexahedral elements.

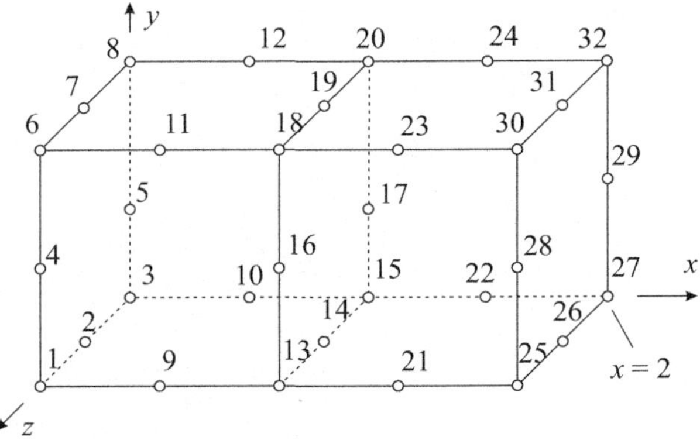

The length of all element edges is unity. A distributed load with intensity $p = 1$ normal to the surface is applied to element face $x = 2$. Specify the necessary displacement boundary conditions that prevent body movement and lead to stress $\sigma_x = 1$ (all other stresses are zero). An order of node numbering for the hexahedral element is given in Figure 12.1b.

5.2. For the mesh of the previous problem write down the data describing displacement constraint $u = 0$ for nodes located at plane $x = 0$. Use three ways for specification of this boundary condition: full list of constrained nodes, structure "from–to" and using a box.

5.3. Modify data of Problem 5.1 in such a way that it would describe a problem with two elements consisting of different materials.

Chapter 6
Finite Element Model

Abstract This chapter describes the finite element model. The finite element model contains nodal coordinates, element connectivities, material properties, and displacement boundary conditions. Class FeModelData is a container for the finite element model data. Class FeModel performs input of model data.

6.1 Data for the Finite Element Model

The finite element model contains all the data for a computational domain, which includes a finite element mesh, information regarding materials, and a description of displacement boundary conditions.

The data on the finite element model is contained in Java^{TM} class FeModelData belonging to package model.

```java
1 package model;
2
3 import elem.*;
4 import util.*;
5
6 import java.io.PrintWriter;
7 import java.util.HashMap;
8 import java.util.LinkedList;
9
10 // Finite element model data
11 public class FeModelData {
12
13     static FeScanner RD;
14     static PrintWriter PR;
15
16     // Problem dimension =2/3
17     public int nDim = 3;
18     // Number of degrees of freedom per node =2/3
19     public int nDf = 3;
20     // Number of nodes
```

```java
21      public int nNod;
22      // Number of elements
23      public int nEl;
24      // Number of degrees of freedom in the FE model
25      public int nEq;
26      // Elements
27      public Element elems[];
28      // Materials
29      public HashMap materials = new HashMap();
30      // Coordinates of nodes
31      private double xyz[];
32      // Constrained degrees of freedom
33      public  LinkedList defDs = new LinkedList();
34      public boolean thermalLoading;
35      static String varName;
36
37      public static enum StrStates {
38          plstrain, plstress, axisym, threed
39      }
40      public static StrStates stressState = StrStates.threed;
41
42      public static enum PhysLaws {
43          elastic, elplastic
44      }
45      public PhysLaws physLaw = PhysLaws.elastic;
46
47      // Input data names
48      enum vars {
49          nel, nnod, ndim, stressstate, physlaw, solver,
50          elcon, nodcoord, material,
51          constrdispl, boxconstrdispl, thermalloading,
52          includefile, user, end
53      }
54
55      // Allocation of nodal coordinate array
56      public void newCoordArray() {
57          xyz = new double[nNod*nDim];
58      }
59
60      // Set coordinates of node
61      public void setNodeCoords(int node, double[] xyzn) {
62          for (int i=0; i<nDim; i++) xyz[node*nDim+i] = xyzn[i];
63      }
64
65      // Set ith coordinates of node
66      public void setNodeCoord(int node, int i, double v) {
67          xyz[node*nDim+i] = v;
68      }
69
70      // Get coordinates of node
71      public double[] getNodeCoords(int node) {
72          double nodeCoord[] = new double[nDim];
73          for (int i=0; i<nDim; i++)
74              nodeCoord[i] = xyz[node*nDim+i];
```

```
75            return nodeCoord;
76        }
77
78        // Get ith coordinate of node
79        public double getNodeCoord(int node, int i) {
80            return xyz[node*nDim+i];
81        }
82
83 }
```

Class `FeModelData` contains scalars, arrays, and objects used for description of the finite element model. Data of the finite element model is declared in lines 16–34:

`nDim` – number of dimensions (2 or 3);

`nDf` – number of degrees of freedom per node (2 or 3);

`nNod` – number of nodes in the finite element model;

`nEl` – number of elements;

`nEq` – total number of degrees of freedom (`ndf`∗`nnod`);

`elems` – array of element objects;

`materials` – hash table of material objects;

`xyz` – array of nodal coordinates;

`defDs` – linked list containing constrained degrees of freedom;

`thermalLoading` = `true` when thermal loading exists.

Since the number of elements is known in advance (specified with the finite element mesh), a normal array is used for storing `Element` objects (line 27). Nodal coordinates are placed in a one-dimensional array `xyz`. We elected not to use special objects for nodes in order to conserve memory and to some extent computing time. The information describing node locations is very simple – just two or three double-precision numbers. We use `Element` objects for element description since element information is rather complicated. An element structure contains several arrays and objects.

Objects describing material properties are stored in a hash table because the number of materials is relatively small and material objects are accessed in an arbitrary order. Java class HashMap is used to store `Material` objects (line 29).

Java class `LinkedList` is employed for storing constrained degrees of freedom `defDs` (line 33). The linked list data structure is suitable for this purpose since the number of constrained displacements is not known in advance. It is determined by methods of constrained displacements specification in the data input file.

For data that can have several predetermined values, we use Java enum type. The enum is a flexible object-oriented enumerated-type facility, which allows one to create enumerated types with arbitrary methods and fields. The lookup among enum values is performed by the `valueOf` method. The `toString` method can be used to get `String` containing the text representation of a particular enum value.

The following data has enum type:

`strState` – problem stress state describing predetermined conditions for stresses and strains. It can have values:

`plstrain` – plane strain;

`plstress` – plane stress;

`axisym` – axisymmetrical problem;

`threed` – three-dimensional problem (default value).

`physLaw` – material physical law with permissible values: `elastic` for elastic material behavior (default) and `elplastic` for elastic–plastic material behavior.

`vars` – names of data items for processing input data file.

Nodal coordinates are stored in a one-dimensional array `xyz`, which is declared as private. Because of this, class *FeModelData* provides methods for working with nodal coordinates.

`newCoordArray` – this method allocates memory for nodal coordinate array `xyz`. The number of nodes `nNod` and problem dimension `nDim` should have values at the time of `xyz` allocation.

`setNodeCoords` – sets the nodal coordinates for node `node` as in array `xyzn`.

`setNodeCoord` – sets the `i`th coordinate for node `node`. Values `i = 0, 1, 2` correspond to axes `x, y, z`.

`getNodeCoords` – returns an array of coordinates for node `node`.

`getNodeCoord` – returns the `i`th coordinate for node `node`.

6.2 Class for the Finite Element Model

Class `FeModel` inherits data from class `FeModelData`. It implements methods for input and handling of data describing the finite element model. A listing of class `FeModel` is given below.

```
 1 package model;
 2
 3 import elem.*;
 4 import material.*;
 5 import util.*;
 6 import solver.*;
 7
 8 import java.io.PrintWriter;
 9 import java.util.ListIterator;
10
11 // Description of the finite element model
12 public class FeModel extends FeModelData {
13
14     static private int elCon[] = new int[20];
15     static private double box[][] = new double[2][3];
16     ListIterator it;
17
```

```
18      // Construct finite element model.
19      // RD - data scanner, PR - print writer.
20      public FeModel(FeScanner RD, PrintWriter PR) {
21          FeModelData.RD = RD;S
22          FeModelData.PR = PR;
23      }
24
25      // Read data for a finite element model
26      public void readData() {
27          readDataFile(RD);
28      }
29
30      private void readDataFile(FeScanner es) {
31
32          vars name = null;
33          String s;
34          Material mat;
35          it = defDs.listIterator(0);
36
37          while (es.hasNext()) {
38              varName = es.next();
39              String varname = varName.toLowerCase();
40              if (varName.equals("#")) {es.nextLine(); continue;}
41              try {
42                  name = vars.valueOf(varname);
43              } catch (Exception e) {
44                  UTIL.errorMsg(
45                      "Variable name is not found: "+varName);
46              }
47
48              switch (name) {
49
50              case nel:    nEl = es.readInt();
51                  break;
52
53              case nnod:   nNod = es.readInt();
54                  break;
55
56              case ndim:   nDim = es.readInt();
57                  nDf = nDim;
58                  break;
59
60              case stressstate:
61                  s = es.next().toLowerCase();
62                  try {
63                      stressState = StrStates.valueOf(s);
64                  } catch (Exception e) {
65                      UTIL.errorMsg(
66                          "stressState has forbidden value: "+s);
67                  }
68                  if (stressState != StrStates.threed)
69                      nDim = nDf = 2;
70                  else  nDim = nDf = 3;
71                  break;
```

```java
 72
 73         case physlaw:
 74             s = es.next().toLowerCase();
 75             try {
 76                 physLaw = PhysLaws.valueOf(s);
 77             } catch (Exception e) {
 78                 UTIL.errorMsg(
 79                         "physLaw has forbidden value: "+s);
 80             }
 81             break;
 82
 83         case solver:
 84             s = es.next().toLowerCase();
 85             try {
 86                 Solver.solver = Solver.Solvers.valueOf(s);
 87             } catch (Exception e) {
 88                 UTIL.errorMsg(
 89                         "solver has forbidden value: "+s);
 90             }
 91             break;
 92
 93         case elcon:
 94             readElemData(es);
 95             break;
 96
 97         case nodcoord:
 98             if (nNod == 0 || nDim == 0)
 99                 UTIL.errorMsg("nNod and nDim should be"
100                         +" specified before nodCoord");
101             nEq = nNod * nDim;
102             // Nodal coordinates
103             newCoordArray();
104             for (int i = 0; i < nNod; i++)
105                 for (int j = 0; j < nDim; j++)
106                     setNodeCoord(i, j, es.readDouble());
107             break;
108
109         case material:
110             String matname = es.next();
111             mat = Material.newMaterial(physLaw.toString(),
112                                 stressState.toString());
113             double e = es.readDouble();
114             double nu = es.readDouble();
115             double alpha = es.readDouble();
116             mat.setElasticProp(e, nu, alpha);
117             if (physLaw == PhysLaws.elplastic) {
118                 double sY = es.readDouble();
119                 double km = es.readDouble();
120                 double mm = es.readDouble();
121                 mat.setPlasticProp(sY, km, mm);
122             }
123             materials.put(matname, mat);
124             break;
125
```

```
126                 case constrdispl:
127                     readConstrDisplacements(es);
128                     break;
129
130                 case boxconstrdispl:
131                     createBoxConstrDisplacements(es);
132                     break;
133
134                 case thermalloading:
135                     s = es.next();
136                     if (s.toLowerCase().equals("y"))
137                         thermalLoading = true;
138                     else if (s.toLowerCase().equals("n"))
139                         thermalLoading = false;
140                     else
141                         UTIL.errorMsg("thermalLoading should be"
142                                     + " y/n. Specified: " + s);
143                     break;
144
145                 case includefile:
146                     s = es.next().toLowerCase();
147                     FeScanner R = new FeScanner(s);
148                     readDataFile(R);
149                     break;
150
151                 case end:   return;
152                 }
153             }
154         }
```

Import statements in lines 3–6 make available classes from packages `elem`, `material`, `util` and `Solver`. Constructor `FeModel` sets references to data scanner RD and print writer PR.

Method `readDataFile` reads data describing the finite element model from an ASCII file that is related to the finite element scanner `es` (class `FeScanner`).

Data is read inside the main `while` loop in lines 37–153. This loop continues while there are input items in the scanner `es`. A text data item read in string `varName` (line 38) is changed to lower case `varname`. In the next line we check if this item is symbol #, which is used as a comment sign. In the case of a comment we proceed to the next line of the scanner and try to read the next data item. If this data item is not the comment sign then we try to find the `name` that corresponds to the string `varname` among enumerated `vars`. If `varname` does not correspond to any predetermined value in `vars`, then an error is generated using static method `errorMsg` of class `UTIL`.

If `varname` is found among predetermined values in `vars` then a `switch` statement is executed in line 48 and a particular case statement reads the second part of an input statement that may contain one or more (sometimes many) input items.

In lines 50, 53, and 56 problem parameters nEl (number of elements), nNod (number of nodes), and nDim (problem dimension) are read as integers. It is possible not to specify explicitly the problem dimension nDim since it will be set automatically depending on the stress state parameter stressState (lines 61–70). The parameters physLaw (material physical law) and solver (equation solver) are treated in lines 74–80 and 84–90.

Data input for all elements in the finite element model (line 94) is performed by method readElemConnectivities, shown in lines 158–178.

Statements in lines 98–106 set nodal coordinates of the finite element model. After checking that variables nNod and nDim have nonzero values, an array of nodal coordinates xyz is created and the nodal coordinates are read in this array.

Input of mechanical properties of materials is performed in lines 110–123. Each material is identified by its name. Three parameters should be specified for an elastic material – elasticity modulus e, Poisson's ratio nu, and thermal-expansion coefficient alpha. For the elastic–plastic material, it is necessary to provide three additional parameters – yield stress sY, hardening coefficient km, and hardening power mm. Material objects are stored as hash table materials.

Displacement boundary conditions in the form of constrained displacements are treated in lines 127 and 131 using methods readConstrDisplacements and createBoxConstrDisplacements.

The thermal loading indicator thermalLoading, which can have values Y (yes) and N (no), is treated in lines 135–142.

An important possibility of inserting other files containing data is implemented in lines 146–148 (data statement includeFile fileName). A specified file name is used to create new data scanner R. Reading data from file fileName is achieved by a recursive call of method readDataFile.

Data statement end (line 151) leads to termination of data input for the finite element model.

Input of element connectivities and other element data is performed by method readElemData, which is presented next.

```
156      // Read element type, material and connectivities
157      // for all elements
158      private void readElemData(FeScanner es) {
159
160          if (nEl == 0) UTIL.errorMsg (
161                  "nEl should be defined before elCon");
162          elems = new Element[nEl];
163
164          for (int iel = 0; iel < nEl; iel++) {
165              // Element type
166              String s = es.next().toLowerCase();
167              elems[iel] = Element.newElement(s);
168              // Element material
169              String elMat = es.next();
170              elems[iel].setElemMaterial(elMat);
171              // Element connectivities
172              int nind = elems[iel].ind.length;
173              for (int i = 0; i < nind; i++) {
174                  elCon[i] = es.readInt();
175              }
```

```
176                        elems[iel].setElemConnectivities(elCon,nind);
177              }
178      }
```

This method first confirms that the number of elements nEl already has a nonzero value. Line 162 allocates an array of Element objects. The loop of lines 164–177 reads element data. The element type is read in line 166 as a String. Line 167 constructs element objects using an element name by method byName of the class NewElement (package elem). The element material name and element connectivity numbers are read and assigned to element objects with the help of methods setElemMaterial (line 170) and setElemConnectivities (line 176).

Data for displacement boundary conditions is determined by the following two methods.

```
180      // Read data for specified constrained displacements
181      private void readConstrDisplacements(FeScanner es) {
182          String s = es.next().toLowerCase();
183          int idf = UTIL.direction(s);
184          if (idf == -1) UTIL.errorMsg("constrDispl direction"+
185                  " should be x/y/z. Specified:"+s);
186          if (!es.hasNextDouble())
187              UTIL.errorMsg("constrDispl value is not a double: "
188                      +es.next());
189          double vd = es.nextDouble();
190          it = es.readNumberList(it, idf, nDim, vd);
191      }
192
193      // Create data for constrained displacements
194      // specified inside a box
195      private void createBoxConstrDisplacements(FeScanner es) {
196          String s = es.next().toLowerCase();
197          int idf = UTIL.direction(s);
198          if (idf == -1)
199              UTIL.errorMsg("boxConstrDispl direction should be"
200                      +" x/y/z. Specified:"+s);
201          if (!es.hasNextDouble())
202              UTIL.errorMsg("boxConstrDispl value is not"
203                      + " a double: " + es.next());
204          double vd = es.nextDouble();
205          for (int i = 0; i < 2; i++)
206              for (int j=0; j<nDim; j++)
207                  box[i][j] = es.readDouble();
208          node: for (int i = 0; i < nNod; i++) {
209              for (int j = 0; j < nDim; j++) {
210                  double x = getNodeCoord(i,j);
211                  if (x<box[0][j] || x>box[1][j]) continue node;
212              }
213              it.add(new Dof(nDim *i + idf, vd));
214          }
215      }
216
217 }
```

Method `readConstrDisplacements` just reads a list of node numbers with specified constrained displacements along with a coordinate direction and a displacement value. Line 190 adds constrained degrees of freedom to linked list `defDs` with modification of list iterator `it`.

Method `createBoxConstrDisplacements` uses the coordinates of a box specified by its two diagonally opposite vertices. Lines 205–207 input box coordinates. Constraints are generated for all nodes located inside the box in lines 208–214. Constraints are added to linked list `defDs` using list iterator `it`. Specification of displacement boundary conditions inside the box is convenient for flat surfaces of the finite element model.

6.3 Adding New Data Item

During further development of the presented finite element program we may need to add new data items to the finite element model. Suppose we want to include a new data item `dataItem` of the type `double`. It can be done using the following steps.

1. Place variable declaration in class `FeModelData`
   ```
   public double dataItem;
   ```
   We declare the variable with `public` attribute. Usually, in object-oriented programming it is recommended to hide data and to program methods for getting and setting its value. We do not follow this approach because it considerably inflates the source code.

2. Add member `dataitem` in enumerated `enum vars` (lines 48–53 of class `FeModelData`). In principle, it is possible to use another name in the `vars` list. However, it is more natural to use the same name in lower case.

3. Place the following statements
   ```
   case dataitem: dataItem = es.readDouble();
   break;
   ```
   inside case structure `switch (name)` that starts at line 48 of class `FeModel`.

After data item input, the value of variable `dataItem` can be accessed as a public field of the `FeModel` object.

Problems

6.1. Explain how a linked list data structure is organized. What elements can be placed in a Java `LinkedList` object? Why do we use a `LinkedList` object for storing constrained displacements data?

6.2. Explain what a hash table data structure is and why it is used for storing material properties. Propose another approach to storing material data and to accessing such data.

6.3. We need to input integer array `M[nM]` together with its length nM. Modify classes `FeModelData` and `FeModel` to perform such data input.

6.4. Suppose it is required to input an array of type double with unknown length. Make the necessary modifications to classes `FeModelData` and `FeModel`.

Chapter 7
Elastic Material

Abstract Hooke's law for elastic material in two- and three-dimensional cases is introduced. Class `Material`, which is a parent class for material models, is introduced. Constitutive equations for an elastic material in three-dimensional, plane strain, and plane stress cases are implemented in class `ElasticMaterial`.

7.1 Hooke's Law

In the finite element method, the main variables (and main unknowns) are displacements at nodal points. Using displacement differentiation with the help of shape functions, it is possible to obtain strains ε_{ij} at any point of a finite element. The application of Hooke's law yields stresses σ_{ij}. For a linear elastic material, a convenient form of Hooke's law has the following appearance:

$$\sigma_{ij} = \lambda \theta \delta_{ij} + 2\mu \varepsilon_{ij}^{e},$$

$$\theta = \varepsilon_{ii}^{e}, \tag{7.1}$$

$$\varepsilon_{ij}^{e} = \varepsilon_{ij} - \varepsilon_{ij}^{t}.$$

Here, ε_{ij}^{e} are elastic fractions of strains, ε_{ij}^{t} are thermal fractions of strains, θ is the elastic volume change, and λ and μ are elastic Lame constants. Thermal expansion affects only elongations:

$$\varepsilon_{ij}^{t} = \alpha T \delta_{ij}, \tag{7.2}$$

where α is the thermal expansion coefficient and T is temperature. Lame constants are expressed through the elasticity modulus E and Poisson's ratio v:

$$\lambda = \begin{cases} \dfrac{vE}{(1+v)(1-2v)} & \text{for three-dimensional and plane strain cases,} \\[2mm] \dfrac{vE}{1-v^2} & \text{for plane stress,} \end{cases} \tag{7.3}$$

$$\mu = \frac{E}{2(1+v)}.$$

(7.4)

In some cases it is more convenient to use a matrix form of Hooke's law (as introduced in (3.6)),

$$\{\sigma\} = [E]\{\varepsilon^e\} = [E](\{\varepsilon\} - \{\varepsilon^t\}).$$

(7.5)

For *three-dimensional* problems, stress, strain, and thermal strain vectors contain six components:

$$\{\sigma\} = \{\sigma_x \ \sigma_y \ \sigma_z \ \tau_{xy} \ \tau_{yz} \ \tau_{zx}\},$$

$$\{\varepsilon\} = \{\varepsilon_x \ \varepsilon_y \ \varepsilon_z \ \gamma_{xy} \ \gamma_{yz} \ \gamma_{zx}\},$$

(7.6)

$$\{\varepsilon^t\} = \{\alpha T \ \alpha T \ \alpha T \ 0 \ 0 \ 0\},$$

and the elasticity matrix $[E]$ has the appearance:

$$[E] = \begin{bmatrix} \lambda+2\mu & \lambda & \lambda & 0 & 0 & 0 \\ \lambda & \lambda+2\mu & \lambda & 0 & 0 & 0 \\ \lambda & \lambda & \lambda+2\mu & 0 & 0 & 0 \\ 0 & 0 & 0 & \mu & 0 & 0 \\ 0 & 0 & 0 & 0 & \mu & 0 \\ 0 & 0 & 0 & 0 & 0 & \mu \end{bmatrix}.$$

(7.7)

In *two-dimensional* problems it is sufficient to have four components in stress and strain vectors and to use a four by four elasticity matrix:

$$\{\sigma\} = \{\sigma_x \ \sigma_y \ \tau_{xy} \ \sigma_z\},$$

$$\{\varepsilon\} = \{\varepsilon_x \ \varepsilon_y \ \gamma_{xy} \ \varepsilon_z\},$$

(7.8)

$$\{\varepsilon^t\} = \{\alpha T \ \alpha T \ 0 \ \alpha T\},$$

$$[E] = \begin{bmatrix} \lambda+2\mu & \lambda & 0 & \lambda \\ \lambda & \lambda+2\mu & 0 & \lambda \\ 0 & 0 & \mu & 0 \\ \lambda & \lambda & 0 & \lambda+2\mu \end{bmatrix}.$$

(7.9)

Under plane stress conditions, stress σ_z is equal to zero. For plane strain, strain ε_z has zero value.

7.2 Class for a Material

We expect that both elastic and elastic–plastic materials will be present in our finite element program. To treat both material models as objects of the same type, let us introduce a parent Java^{TM} class with the name `Material`. A listing of this class is shown below.

```java
 1 package material;
 2
 3 import elem.Element;
 4
 5 // Material constitutive relations
 6 public class Material {
 7
 8     // StressContainer state (plstrain/plstress/axisym/threed)
 9     String stressState;
10     // Elasticity modulus
11     double e;
12     // Poisson's ratio
13     double nu;
14     // Thermal expansion
15     double alpha;
16     // Yield stress
17     double sY;
18     // Hardening coefficient
19     double km;
20     // Hardening power
21     double mm;
22
23     public static Material newMaterial (String matPhysLaw,
24                                         String stressState) {
25         if (matPhysLaw.equals("elastic"))
26                 return new ElasticMaterial(stressState);
27         else  return new ElasticPlasticMaterial(stressState);
28     }
29
30     // Given strain increment at integration point ip
31     // element elm, compute stress dsig increment
32     public void strainToStress(Element elm, int ip) {  }
33
34     // Set elastic properties
35     public void setElasticProp(double e, double nu,
36                                double alpha){
37         this.e = e;
38         this.nu = nu;
39         this.alpha = alpha;
40     }
41
42     // Set plastic properties
43     public void setPlasticProp(double sY, double km,
44                                double mm) {
45         this.sY = sY;
46         this.km = km;
47         this.mm = mm;
48     }
49
50     // Returns Lame constant lambda
51     public double getLambda() {
```

```
52          return (stressState.equals("plstress")) ?
53              e*nu/((1+nu)*(1-nu)) : e*nu/((1+nu)*(1-2*nu));
54      }
55
56      // Returns shear modulus
57      public double getMu() { return 0.5*e/(1 + nu); }
58
59      // Returns Poisson's ratio
60      public double getNu() { return nu;   }
61
62      // Returns thermal expansion coefficient
63      public double getAlpha() { return alpha; }
64
65      // Compute elasticity matrix emat
66      public void elasticityMatrix(double[][] emat) {   }
67
68 }
```

Class `Material` belongs to package `material` and imports class `Element` from package `elem`.

In lines 10–21 all the material properties are declared:

e – elasticity modulus;

nu – Poisson's ratio;

`alpha` – thermal expansion coefficient;

`sY` – yield stress;

km – hardening coefficient;

mm – hardening power.

The first three parameters are related to elastic problems and are set by method `setElasticProp`; the final three parameters characterize a deformation curve in an elastic–plastic region, and are set by method `setPlasticProp`.

The class constructor presented in lines 23–28 returns an `ElasticMaterial` or `ElasticPlasticMaterial` object depending on parameter `matPhysLaw`, specified in input data.

The principal purpose of material constitutive relations is to compute the stress increment as a function of strain increment. This function is performed by method `strainToStress`, declared in line 32 and is implemented in subclasses of `Material` class. Stress increment computing is done for an integration point `ip` of a finite element `elm` that are specified as method parameters.

Methods `getLambda` and `getMu` provide values of the Lame parameters for a current material; methods `getNu` and `getAlpha` return values of Poisson's ratio and the thermal-expansion coefficient. Method `elasticityMatrix` computes the elasticity matrix `emat`. It is implemented in class `ElasticMaterial`.

7.3 Class for Elastic Material

Class `ElasticMaterial` inherits from class `Material` and implements constitutive relations for linear elastic material in three- and two-dimensional cases. Two-dimensional problems include plane strain and plane stress conditions and axisymmetric deformation.

```java
1  package material;
2
3  import elem.Element;
4
5  // Constitutive relations for elastic material
6  public class ElasticMaterial extends Material {
7
8      static double[] deps = new double[6];
9      static double[] dsig = new double[6];
10     // Length of strain and stress vectors
11     static int lv;
12
13     public ElasticMaterial(String stressState) {
14
15         this.stressState = stressState;
16         lv = (stressState.equals("threed"))? 6:4;
17     }
18
19     // Hooke's law: increment of stress due to
20     // increment of strain
21     public void strainToStress(Element elm, int ip) {
22
23         deps = elm.getStrainsAtIntPoint(ip);
24         double temp = elm.getTemperatureAtIntPoint(ip);
25
26         double mu = 0.5*e/(1 + nu);
27         double lambda = (stressState.equals("plstress")) ?
28                 e*nu/(1-nu*nu) : e*nu/((1+nu)*(1-2*nu));
29         double beta = lambda + 2.0*mu;
30         double at = alpha*temp;
31
32         if (stressState.equals("threed")) {
33             deps[0] -= at;
34             deps[1] -= at;
35             deps[2] -= at;
36             dsig[0] = beta*deps[0] + lambda*(deps[1]+deps[2]);
37             dsig[1] = beta*deps[1] + lambda*(deps[0]+deps[2]);
38             dsig[2] = beta*deps[2] + lambda*(deps[0]+deps[1]);
39             dsig[3] = mu*deps[3];
40             dsig[4] = mu*deps[4];
41             dsig[5] = mu*deps[5];
42         } else {
43             deps[0] -= at;
44             deps[1] -= at;
45             if (!stressState.equals("plstress")) deps[3] -= at;
46             dsig[0] = beta*deps[0] + lambda*(deps[1]+deps[3]);
```

```
47             dsig[1] = beta*deps[1] + lambda*(deps[0]+deps[3]);
48             dsig[2] = mu*deps[2];
49             dsig[3] = 0.0;
50             if (stressState.equals("plstrain"))
51                 dsig[3] = nu*(dsig[0]+dsig[1]) - e*at;
52             if (stressState.equals("axisym"))
53                 dsig[3] = beta*deps[3]+lambda*(deps[0]+deps[1]);
54         }
55       for (int i=0; i<lv; i++)
56             elm.str[ip].dStress[i] = dsig[i];
57     }
58
59     // Compute elasticity matrix emat
60     public void elasticityMatrix(double[][] emat) {
61         if (stressState.equals("threed"))
62             elasticityMatrix3D(emat);
63         else
64             elasticityMatrix2D(emat);
65     }
66
67     // Elasticity 3D matrix emat [6][6]
68     public void elasticityMatrix3D(double[][] emat) {
69
70         double mu = getMu();
71         double lambda = getLambda();
72         double beta = lambda + 2*mu;
73
74         for (int i=0; i<6; i++)
75             for (int j=0; j<6; j++)
76                 emat[i][j]=0;
77
78         emat[0][0] = emat[1][1] = emat[2][2] = beta;
79         emat[0][1] = emat[1][0] = emat[0][2] = emat[2][0] =
80                     emat[1][2] = emat[2][1] = lambda;
81         emat[3][3] = emat[4][4] = emat[5][5] = mu;
82     }
83
84     // Elasticity 2D matrix emat [4][4]
85     public void elasticityMatrix2D(double[][] emat) {
86
87         double mu = getMu();
88         double lambda = getLambda();
89         double beta = lambda + 2*mu;
90
91         emat[0][0] = emat[3][3] = emat[1][1] = beta;
92         emat[0][1] = emat[1][0] = emat[0][3] = emat[3][0] =
93                     emat[1][3] = emat[3][1] = lambda;
94         emat[2][2] = mu;
95         emat[0][2] = emat[2][0] = emat[1][2] = emat[2][1] =
96                     emat[2][3] = emat[3][2] = 0.;
97     }
98
99 }
```

For simplicity, working arrays for strain `deps` and stress `dsig` increments are allocated with length 6 and a parameter `lv` that keeps the length of the strain–stress vector it introduces. In line 16 `lv` is assigned a value of 6 for three-dimensional problems, otherwise it is set to 4. The stress increment due to the strain increment at integration point `ip` of element `elm` is computed by method `strainToStress`.

Increments of strain `deps` and temperature `temp` are obtained using methods `getStrainsAtIntPoint` and `getTemperatureAtIntPoint` of a class corresponding to a particular element. Computing of stress increment `dsig` according to Hooke's law (7.1) is done in lines 33–41 for the three-dimensional case and in lines 43–53 for the two-dimensional cases of plane stress, plane strain, and axisymmetric deformation. At the end of the method calculated stress increment `dsig` is placed in element object `str`, which is used for storing stresses and strains at integration points.

Method `elasticityMatrix` generates an elasticity matrix `emat` for the three-dimensional case according to Equation 7.7 and for the two-dimensional case according to Equation 7.9. These cases are implemented in lines 68–82 by method `elasticityMatrix3D` and by method `elasticityMatrix2D` (lines 85–97).

Problems

7.1. Express elasticity matrix $[E]$ for a plane stress case using elasticity modulus E and Poisson's ratio v.

7.2. Show the equivalence of Hooke's law in tensor form (7.1) and in matrix form (7.5) for the three-dimensional case of stress calculation.

7.3. Develop a Java method that computes the stress vector as a function of the elastic strain vector with the following interface.

```
// Compute stress vector due to elastic strain vector.
// elStrain [] vector of elastic strains,
// returns stress vector.
public double[] strainToStress(double[] elStrain) {

    return stress;
}
```

The method should work for the three-dimensional case and for two-dimensional cases of plane strain and plane stress. It is supposed that the method belongs to class `ElasticMaterial`, and that all fields and methods of this class are available to the developed method.

7.4. Create a main method for class `ElasticMaterial` that performs any usage of some methods of the class. For example, it could be a simple test of method `elasticityMatrix`.

Chapter 8
Elements

Abstract Finite elements are considered as main objects in both mathematical and programmatical senses. In order to implement the main methods for a finite element, abstract class `Element` is designed. The class holds element data, methods common to all element types, and empty methods specific to particular element types. Overriding of the parent methods allows one to create new element types using standard procedures.

8.1 Element Methods

Finite elements are the main objects (in both mathematical and programmatical senses) of the finite element method. Commercial and scientific finite element programs include large libraries of finite elements. They are able to solve various problems due to using various types of finite elements, so it is important to construct finite element routines in such a way that it is straightforward to add new types of finite elements with minimal changes to already existing code.

Let us restrict ourselves to the problems under consideration involving surface loading and temperature influence, and introduce principal functions, which should be performed by classes implementing finite elements.

According to element equations introduced in Section 3.2, the principal methods of a finite element class include:

Element stiffness matrix $[k]$ evaluation

$$[k] = \int_V [B]^{\mathrm{T}} [E][B] dV. \tag{8.1}$$

Estimation of nodal equivalent $\{p\}$ of a distributed surface load

$$\{p\} = \int_S [N]^{\mathrm{T}} \{p^S\} dS. \tag{8.2}$$

Element thermal vector $\{h\}$ evaluation

$$\{h\} = \int_V [B]^{\mathrm{T}}[E]\{\varepsilon^{\mathrm{t}}\}dV. \tag{8.3}$$

Vector of nodal forces equivalent to stress distribution $\{p_{\mathrm{eq}}\}$

$$\{p_{\mathrm{eq}}\} = \int_V [B]^{\mathrm{T}}\{\sigma\}dV. \tag{8.4}$$

When we implement a particular element type, the element class should contain methods that realize the above functions and other element functions like calculation of strains, extrapolation of stresses to nodes, etc.

8.2 Abstract Class Element

Our finite element code should be able to include any number of different element types with the same functions. Because of this it is natural to introduce an abstract class `Element` that contains data and methods common to all elements, and declares abstract methods that are implemented in classes for particular elements.

8.2.1 Element Data

In the beginning of Java^{TM} class `Element`, static data common to all element objects and data specific to each `Element` object are placed.

```
 1  package elem;
 2
 3  import model.*;
 4  import material.*;
 5  import fea.FE;
 6  import util.UTIL;
 7
 8  // Finite element
 9  public abstract class Element {
10
11      // Finite element model
12      public static FeModel fem;
13      // Finite element load
14      public static FeLoad load;
15      // Material of current element
16      static Material mat;
17      // Element stiffness matrix
18      public static double kmat[][] = new double[60][60];
19      // Element vector
20      public static double evec[] = new double[60];
```

```
21        // Element nodal coordinates
22        static double xy[][] = new double[20][3];
23        // Element nodal temperatures
24        static double dtn[] = new double[20];
25        // Strain vector
26        static double dstrain[] = new double[6];
27
28        // Element name
29        public String name;
30        // Element material name
31        public String matName;
32        // Element connectivities
33        public int ind[];
34        // Stress-strain storage
35        public StressContainer[] str;
```

Static data includes references to the finite element model `fem`, to the load `load`, and to the material for current element `mat`.

The following arrays are also declared as static:

`kmat[60][60]` – element stiffness matrix;

`evec[60]` – element working vector;

`xy[20][3]` – element nodal coordinates;

`dtn[20]` – element nodal temperatures;

`dstrain[6]` – strain vector.

We elected to declare arrays of constant size. It is assumed that the maximum number of nodes in an element is twenty, corresponding to a three-dimensional hexagonal quadratic element. Each coefficient of the element stiffness matrix is identified by four indexes – two nodal indexes and two indexes corresponding to degrees of freedom. Formally, the stiffness matrix is a four-dimensional array. However, in the finite element method it is usually considered as a matrix, so we declare the element stiffness matrix as a two-dimensional array. In order to have similar assembly algorithms for the stiffness matrix and for element vectors, vector `evec` is declared as a one-dimensional array. The array of nodal coordinates `xy` is two-dimensional since it is not involved in assembly operations.

8.2.2 Element Constructor

The next part of `Element` class includes a description of element types and element constructor.

```
37        // Implemented element types
38        static enum elements {
39            quad8 {Element create() {return new ElementQuad2D();}},
40            hex20 {Element create() {return new ElementQuad3D();}};
41
42            abstract Element create();
```

```
43        }
44
45        // Construct new element
46        // name - element name
47        public static Element newElement(String name) {
48            elements el = null;
49            try {
50                el = elements.valueOf(name);
51            } catch (Exception e) {
52                UTIL.errorMsg("Incorrect element type: " + name);
53            }
54            return el.create();
55        }
56
57        // Constructor for an element.
58        // name - element name;
59        // nind - number of nodes;
60        // nstress - number of stress points
61        public Element(String name, int nind, int nstress) {
62            this.name = name;
63            ind = new int[nind];
64            if (FE.main != FE.JMGEN) {
65                str = new StressContainer[nstress];
66                for (int ip=0; ip<nstress; ip++)
67                    str[ip] = new StressContainer(fem.nDim);
68            }
69        }
```

We use the Java enum type to store element types and element constructors (lines
38–43). Here, we placed two element types that we implement:

quad8 – two-dimensional quadrilateral quadratic element with eight nodes; its
constructor is ElementQuad2D;

hex20 – three-dimensional hexahedral quadratic element with twenty nodes
(constructor ElementQuad3D).

Each element record, besides element type, implements method create that calls
the corresponding element constructor. If we want to add new element in the finite
element processor then it is necessary to create a class for a new element and to add
its record in enum elements.

Method newElements (lines 47–55) serves as a constructor when we create
new element objects. It takes string name as a parameter, looks for a particular element in enum elements, and calls the appropriate method create. The method
returns Element object.

In lines 61–69, a constructor of Element object is presented. The constructor
sets element name and allocates memory for element connectivities ind and for
stresses str. Memory for stresses is not allocated if Element object is used for
mesh generation. Line 64 checks if the main method is JMGEN and if so memory–
allocation statements are avoided. Stresses and equivalent plastic strains are placed
in objects of class StressContainer. The length of array str is equal to the

number of reduced integration points where stresses have the highest accuracy. Constructor `Element` should be called by constructors of particular elements using the `super` statement.

8.2.3 *Methods of Particular Elements*

If an element class of a specific type is created then the following element methods should be implemented:

`stiffnessMatrix` – compute the element stiffness matrix and put it into array `kmat` (Equation 8.1);

`thermalVector` – compute the element thermal vector due to nodal temperatures `dtn` and put it into array `evec` (Equation 8.3);

`equivFaceLoad` – compute the element nodal equivalent of the distributed face load and put it into array `evec` (Equation 8.2);

`equivVector` – compute the element nodal equivalent of element stress field `evec` (Equation 8.4);

`getElemFaces` – return the two-dimensional integer array of local node numbers for element faces;

`getStrainsAtIntPoint` – return the double array of strains at requested integration point;

`extrapolateToNodes` – return the two-dimensional double array of extrapolated values at nodes using values at reduced integration points.

The declaration of these methods is illustrated below.

```
71      // Compute element stiffness matrix kmat[][]
72      public void stiffnessMatrix() { }
73
74      // Compute element thermal vector (evec[])
75      public void thermalVector() { }
76
77      // Element nodal equivalent of distributed face load
78      // (evec[])
79      public int equivFaceLoad(ElemFaceLoad surLd) {
80          return -1;
81      }
82
83      // Nodal vector equivalent to stresses (evec[])
84      public void equivStressVector() { }
85
86      // Get local node numbers for element faces
87      // returns elementFaces[nFaces][nNodesOnFace]
88      public int[][] getElemFaces() {
89          return new int[][] {{0},{0}};
90      }
```

```
 91
 92        // Get strains at integration point (stress)
 93        // intPoint - integration point number (stress);
 94        // returns   strain vector [2*ndim]
 95        public double[] getStrainsAtIntPoint(int intPoint) {
 96            return new double[] {0,0};
 97        }
 98
 99        // Get temperature at integration point (stress)
100        // intPoint - integration point number (stress);
101        // returns   temperature
102        public double getTemperatureAtIntPoint(int intPoint) {
103            return 0.0;
104        }
105
106        // Extrapolate quantity from integration points to nodes
107        // fip [nInt][2*nDim] - values at integration points;
108        // fn [nind][2*nDim] - values at nodes (out)
109        public void extrapolateToNodes(double[][] fip,
110                                       double[][] fn) {
111        }
```

8.2.4 Methods Common to All Elements

Finally, class `Element` provides convenience methods common to all element
types, as shown below.

```
113        // Set element connectivities
114        // indel - connectivity numbers
115        // nind - number of element nodes
116        public void setElemConnectivities(int[] indel, int nind) {
117            System.arraycopy(indel, 0, ind, 0, nind);
118        }
119
120        // Set element connectivities
121        // indel - connectivity numbers
122        public void setElemConnectivities(int[] indel) {
123            System.arraycopy(indel, 0, ind, 0, indel.length);
124        }
125
126        // Set element material name
127        // mat - material name
128        public void setElemMaterial(String mat) {
129            matName = mat;
130        }
131
132        // Set element nodal coordinates xy[nind][nDim]
133        public void setElemXy() {
134            for (int i = 0; i < ind.length; i++) {
135                int indw = ind[i] - 1;
136                if (indw >= 0) {
```

```
137                        xy[i] = fem.getNodeCoords(indw);
138                 }
139             }
140     }
141
142     // Set nodal coordinates xy[nind][nDim] and
143     //      temperatures dtn[nind]
144     public void setElemXyT() {
145         for (int i = 0; i < ind.length; i++) {
146             int indw = ind[i] - 1;
147             if (indw >= 0) {
148                 if (fem.thermalLoading)
149                     dtn[i] = FeLoad.dtemp[indw];
150                 xy[i] = fem.getNodeCoords(indw);
151             }
152         }
153     }
154
155     // Assemble element vector.
156     // elVector - element vector;
157     // glVector - global vector (in/out)
158     public void assembleElemVector(double[] elVector,
159                                    double[] glVector) {
160         for (int i = 0; i < ind.length; i++) {
161             int indw = ind[i] - 1;
162             if (indw >= 0) {
163                 int adr = indw*fem.nDim;
164                 for (int j = 0; j < fem.nDim; j++)
165                     glVector[adr+j] += elVector[i*fem.nDim +j];
166             }
167         }
168     }
169
170     // Disassemble element vector (result in evec[]).
171     // glVector - global vector
172     public void disAssembleElemVector(double[] glVector) {
173         for (int i = 0; i < ind.length; i++) {
174             int indw = ind[i] - 1;
175             if (indw >= 0) {
176                 int adr = indw*fem.nDim;
177                 for (int j = 0; j < fem.nDim; j++)
178                     evec[i*fem.nDim +j] = glVector[adr+j];
179             }
180         }
181     }
182
183     // Returns element connectivities
184     public int[] getElemConnectivities() {
185         int indE[] = new int[ind.length];
186         System.arraycopy(ind, 0, indE, 0, ind.length);
187         return indE;
188     }
189
190     //  Accumulate stresses and equivalent plastic strain
```

```
191      public void accumulateStress() {
192          for (int ip=0; ip<str.length; ip++)
193              for (int i = 0; i < 2*fem.nDim; i++)
194                  str[ip].sStress[i] += str[ip].dStress[i];
195              for (int ip=0; ip<str.length; ip++)
196                  str[ip].sEpi += str[ip].dEpi;
197      }
198
199  }
```

Two methods, `setElemConnectivities` in lines 116–118 and 122–124, help to set element connectivities. Method `setElemMaterial` sets the material name for an element. Methods `setElemXy` (lines 133–140) and `setElemXyT` (lines 144–153) provide means to set element nodal coordinates in the first method and both coordinates and nodal temperatures in the second method. Note that if the node index (connectivity number) is zero then coordinates and temperatures are not set for this node. This opens up the possibility to create elements with a variable number of nodes (midside nodes of higher-order elements can be present or not).

Method `assembleElemVector` (lines 158–168) assembles element vector `elVector` into global vector `glVector`. Locations of element entries in global vector are determined by element connectivities.

Method `disAssembleElemVector` shown in lines 172–181 disassembles element entries from the global vector into the element vector according to element connectivities.

Method `getElemConnectivities` returns the connectivity numbers for this finite element.

Accumulation of stresses and equivalent plastic strains are performed by the method `accumulateStress` shown in lines 191–197. This method adds an increment of stress vector to the accumulated stress vector and an increment of equivalent plastic strain to its accumulated value.

8.2.5 Container for Stresses

In elastic–plastic problems, it is necessary to trace the history of stresses and equivalent plastic strains. Both stresses and strains are expressed through derivatives of displacements. Displacement derivatives have the highest accuracy at reduced integration points inside finite elements. Because of this it is natural to store stresses and strains inside element objects. To simplify storage and retrieval procedures let us introduce the special class `StressContainer`, shown below.

```
1  package elem;
2
3  // Stresses and equivalent strains at integration point
4  public class StressContainer {
5
6      // Accumulated stress
7      public double sStress[];
```

```
 8        // Stress increment
 9        public double dStress[];
10        // Accumulated equivalent plastic strain
11        public double sEpi;
12        // Equivalent plastic strain increment
13        public double dEpi;
14
15        StressContainer(int nDim) {
16            sStress = new double[2*nDim];
17            dStress = new double[2*nDim];
18        }
19
20 }
```

The following data is contained in `StressContainer`:

`sStress` – vector of accumulated stress;

`dStress` – vector of stress increment;

`sEpi` – accumulated equivalent plastic strain;

`dEpi` – accumulated plastic strain increment.

Constructor of `StressContainer` allocates arrays for two stress vectors with length 4 for two-dimensional problems and with length 6 for three-dimensional problems.

8.3 Adding New Element Type

Adding new element types is one of the main ways to extend the finite element program. To implement a new element type with name `elname`, number of nodes `nElNodes`, and number of stress points `nElStress`, the following steps should be performed.

1. In class `Element`, modify the enumerated description `elements` (lines 38–43). Insert the following statement:

   ```
   elname {Element create()
       {return new ElClassName();}
   }
   ```

   Here, `ElClassName` is a class name for this element type.

2. Create new class `ElClassName` with the constructor

   ```
   public ElClassName() {
       super ("elname", nElNodes, nElStress);
       // ...
   }
   ```

The first statement of the constructor should be a call to the superclass constructor with specification of element name, number of element nodes, and number of points for storing of element stresses. In quadrilateral isoparametric elements, stresses are usually stored at reduced integration points since they have the highest precision there.

3. Implement element methods described in Section 8.2.3:

> `stiffnessMatrix` – element stiffness matrix;
>
> `thermalVector` – element thermal vector;
>
> `equivFaceLoad` – element nodal equivalent of distributed face load;
>
> `equivVector` – element nodal equivalent of element stress field;
>
> `getElemFaces` – local node numbers for element faces;
>
> `getStrainsAtIntPoint` – strains at requested integration point;
>
> `extrapolateToNodes` – extrapolated values at nodes using values at reduced integration points.

A particular element can be implemented in several classes. A major element class can construct other objects and use methods from other classes related to that element.

Problems

8.1. Explain why class `Element` is declared abstract? What is the difference between an abstract class and an interface in Java language?

8.2. Suppose that a three-dimensional element has element connectivities

```
int[] ind = {21, 23, 17, 15};
```

Create an array containing all degrees of freedom for this element.

8.3. Analyze the algorithm of method `disAssembleElemVector`, which disassembles a global vector into element vectors by selection within the global vector according to element connectivities. For the following data

```
double[] glVector = {1, 2, 3, ... 100};
int[] ind = {21, 23, 17, 15};
```

and `fem.nDim = 2`, find contents of array `evec` as a result of calling the method.

Chapter 9
Numerical Integration

Abstract Different types of finite elements usually use numerical integration for estimating element matrices and vectors. Gauss integration rules of different order are considered for one-, two-, and three-dimensional integration domains. A special 14-point integration rule is used for twenty-node hexagonal finite elements. Java^{TM} class `GaussRule` implements different integration rules for two- and three-dimensional elements.

9.1 Gauss Integration Rule

Integration of expressions for element stiffness matrices and load vectors can not be performed analytically for the general case of finite elements. Instead, stiffness matrices and load vectors are evaluated numerically using some integration rule.

In the finite element method, the Gauss integration rule is usually used because of its high accuracy. It can be applied in cases when an integrated function can be evaluated at arbitrary points inside the integration interval. Since there is no difficulty in fulfilling such a requirement in the finite element algorithms, then the Gauss rule is a suitable integration tool for element matrices and vectors.

Let us derive the Gauss rule for a simple case when two integration points are used. A two-term formula will contain four parameters (the two abscissas and the two weights) and should integrate precisely a polynomial of third degree. To determine the four unknown parameters let us consider the integral on the standard integration interval $[-1, 1]$

$$I = \int_{-1}^{1} f(\xi)d\xi \tag{9.1}$$

and write an integration formula using two points as

$$I = f(\xi_1)w_1 + f(\xi_2)w_2, \tag{9.2}$$

where w_k are weights and ξ_k are undetermined points.

Our formula should be valid for any polynomial of degree 3. Hence, it will work if $f(\xi) = 1$, $f(\xi) = \xi$, $f(\xi) = \xi^2$ and $f(\xi) = \xi^3$:

$$
\begin{aligned}
\int_{-1}^{1} d\xi &= 2 = w_1 + w_2, \\[2mm]
\int_{-1}^{1} \xi\, d\xi &= 0 = \xi_1 w_1 + \xi_2 w_2, \\[2mm]
\int_{-1}^{1} \xi^2\, d\xi &= \frac{2}{3} = \xi_1^2 w_1 + \xi_2^2 w_2, \\[2mm]
\int_{-1}^{1} \xi^3\, d\xi &= 0 = \xi_1^3 w_1 + \xi_2^3 w_2.
\end{aligned}
\tag{9.3}
$$

The limits of integration are symmetric about $\xi = 0$, so we require that points be located symmetrically and set $\xi_2 = -\xi_1$. From the first and second equations above, we get

$$
w_1 = w_2 = 1. \tag{9.4}
$$

With these values, the fourth equation is automatically satisfied. The third equation becomes

$$
\frac{1}{3} = \xi_1^2,
$$

which yields

$$
\xi_1 = \frac{1}{\sqrt{3}} = 0.577350269. \tag{9.5}
$$

The integration formula derived above is the simplest member of the Gauss quadrature rules.

In the general one-dimensional case, the Gauss quadrature rule is expressed as

$$
I = \int_{-1}^{1} f(\xi)\, d\xi = \sum_{i=1}^{n} f(\xi_i) w_i, \tag{9.6}
$$

where n is the number of integration points, ξ_i are abscissas, and w_i are the weights of integration. Abscissas and weights of Gauss quadrature for $n = 1,2,3$ are given in Table 9.1. Since the Gauss integration rule uses $2n$ constants (n abscissas and n weights) it integrates exactly polynomials of order $2n - 1$.

The Gauss quadrature formula for the integral in the two-dimensional case is of the form

$$
I = \int_{-1}^{1}\int_{-1}^{1} f(\xi, \eta)\, d\xi\, d\eta = \sum_{i=1}^{n}\sum_{j=1}^{n} f(\xi_i, \eta_j) w_i w_j, \tag{9.7}
$$

where ξ_i, η_j are abscissas and w_i are the weighting coefficients of the Gauss integration rule.

Gauss integration in the three-dimensional case can be performed by applying the one-dimensional formula thrice:

Table 9.1 Abscissas and weights of Gauss quadrature

n	ξ_i	w_i
1	0	2
2	$-1/\sqrt{3}$	1
	$1/\sqrt{3}$	1
3	$-\sqrt{3/5}$	5/9
	0	8/9
	$\sqrt{3/5}$	5/9

$$
I = \int_{-1}^{1} \int_{-1}^{1} \int_{-1}^{1} f(\xi, \eta, \zeta) \, d\xi \, d\eta \, d\zeta
$$
$$
= \sum_{i=1}^{n} \sum_{j=1}^{n} \sum_{k=1}^{n} f(\xi_i, \eta_j, \zeta_k) w_i w_j w_k. \tag{9.8}
$$

Here, ξ_i, η_j, ζ_k are abscissas and w_i are weights.

Instead of applying one-dimensional formulas twice or thrice for integration over two- or three-dimensional domains, it is possible to derive special integration rules of the Gauss type. One such rule, particularly useful for a three-dimensional twenty-node hexagonal element, is the 14-point integration rule [14, 17] for three-dimensional domains. Since points of this integration rule are not located on a rectangular grid then integration is performed inside a single loop:

$$
I = \int_{-1}^{1} \int_{-1}^{1} \int_{-1}^{1} f(\xi, \eta, \zeta) \, d\xi \, d\eta \, d\zeta = \sum_{i=1}^{14} f(\xi_i, \eta_i, \zeta_i) w_i, \tag{9.9}
$$

Abscissas and weights of 14-point integration rule are listed in Table 9.2, where numerical constants have the following values:

$$
a = 0.7587869106393281, \qquad b = 0.7958224257542215;
$$
$$
W_a = 0.3351800554016621, \qquad W_b = 0.8864265927977839.
$$

9.2 Implementation of Numerical Integration

In our finite element program, which is designed to solve two- and three-dimensional problems, we need to perform integration over one-, two- and three-dimensional domains. Instead of integrating inside single, double, and triple loops, let us implement any integration as a single-loop process similar to (9.9):

Table 9.2 Abscissas and weights of 14-point integration rule

i	ξ_i	η_i	ζ_i	W_i
1	$-a$	$-a$	$-a$	W_a
2	a	$-a$	$-a$	W_a
3	$-a$	a	$-a$	W_a
4	$-a$	$-a$	a	W_a
5	a	a	$-a$	W_a
6	$-a$	a	a	W_a
7	a	$-a$	a	W_a
8	a	a	a	W_a
9	$-b$	0	0	W_b
10	b	0	0	W_b
11	0	$-b$	0	W_b
12	0	b	0	W_b
13	0	0	$-b$	W_b
14	0	0	b	W_b

$$I = \sum_{i=1}^{N} f(\mathbf{P}_i)W_i. \tag{9.10}$$

Here, I is an integral value over one-, two-, or three-dimensional domains, $\mathbf{P}_i$ are integration points having a corresponding number of coordinates, and W_i are integration weights. Integration points for one-, two-, and three-dimensional cases are represented as vectors of corresponding length:

$$\mathbf{P}_i^{1\mathrm{D}} = \xi_i,$$

$$\mathbf{P}_i^{2\mathrm{D}} = \{\xi_i\ \eta_i\}, \tag{9.11}$$

$$\mathbf{P}_i^{3\mathrm{D}} = \{\xi_i\ \eta_i\ \zeta_i\}.$$

Abscissas and weights for numerical integration using (9.10) are created in class GaussRule, which is a member of the package util. A listing of class GaussRule is shown below.

```
1  package util;
2
3  // Gauss integration rule
4  public class GaussRule {
5
6      // Abscissas of the Gauss rule
7      public double[] xii, eti, zei;
8      // Integration weights
9      public double[] wi;
10     // Total number of integration poins
11     public int nIntPoints;
12
13     // Abscissas and weights for 1, 2 and 3-point rules
```

```java
14      private static final double[][] X = {{0.0},
15              {-1.0/Math.sqrt(3.0), 1.0/Math.sqrt(3.0)},
16              {-Math.sqrt(0.6), 0.0, Math.sqrt(0.6)}};
17      private static final double[][] W = {{2.0},
18              {1.0, 1.0},
19              {5.0/9.0, 8.0/9.0, 5.0/9.0}};
20      // Abscissas and weights for 14-point rule (3D)
21      private static final double a = 0.7587869106393281,
22              b = 0.7958224257542215;
23      private static final double[] X14 =
24              {-a, a, -a, -a, a, -a, a, a, -b, b, 0, 0, 0, 0};
25      private static final double[] Y14 =
26              {-a, -a, a, -a, a, a, -a, a, 0, 0, -b, b, 0, 0};
27      private static final double[] Z14 =
28              {-a, -a, -a, a, -a, a, a, a, 0, 0, 0, 0, -b, b};
29      private static final double Wa = 0.3351800554016621,
30              Wb = 0.8864265927977839;
31
32      // Construct Gauss integration rule.
33      // nGauss - number of Gauss points in each direction
34      // (excluding 14-point rule),
35      // nDim - number of dimensions
36      public GaussRule(int nGauss, int nDim) {
37
38          if (!((nGauss>=1 && nGauss<=3) || nGauss==14))
39              UTIL.errorMsg("nGauss has forbidden value: "
40                      + nGauss);
41          if (!(nDim>=1 && nDim<=3)) UTIL.errorMsg(
42                  "GaussRule: nDim has forbidden value: "
43                      + nDim);
44
45          if (nGauss == 14) nIntPoints = 14;
46          else {
47              nIntPoints = 1;
48              for (int i = 0; i < nDim; i++)
49                  nIntPoints *= nGauss;
50          }
51
52          xii = new double[nIntPoints];
53          wi = new double[nIntPoints];
54          if (nDim > 1) eti = new double[nIntPoints];
55          if (nDim > 2) zei = new double[nIntPoints];
56
57          if (nGauss == 14) {
58              for (int i = 0; i < nGauss; i++) {
59                  xii[i] = X14[i];
60                  eti[i] = Y14[i];
61                  zei[i] = Z14[i];
62                  wi[i] = (i < 8) ? Wa : Wb;
63              }
64          }
65          else {
66              int ip = 0;
67              int n = nGauss - 1;
```

```
68                    switch (nDim) {
69                    case 1:
70                        for (int i = 0; i < nGauss; i++) {
71                            xii[ip] = X[n][i];
72                            wi[ip++] = W[n][i];
73                        }
74                        break;
75
76                    case 2:
77                        for (int i = 0; i < nGauss; i++) {
78                            for (int j = 0; j < nGauss; j++) {
79                                xii[ip] = X[n][i];
80                                eti[ip] = X[n][j];
81                                wi[ip++] = W[n][i]*W[n][j];
82                            }
83                        }
84                        break;
85
86                    case 3:
87                        for (int i = 0; i < nGauss; i++) {
88                            for (int j = 0; j < nGauss; j++) {
89                                for (int k = 0; k < nGauss; k++) {
90                                    xii[ip] = X[n][i];
91                                    eti[ip] = X[n][j];
92                                    zei[ip] = X[n][k];
93                                    wi[ip++] =
94                                        W[n][i]*W[n][j]*W[n][k];
95                                }
96                            }
97                        }
98                        break;
99                    }
100               }
101           }
102
103   }
```

Public members of the class declared in lines 7, 9, and 11 are set by a class constructor and later are used for numerical integration. Arrays xii, eti, and zei are integration abscissas ξ_i, η_i, and ζ_i and array wi contains integration weights W_i. Scalar nIntPoints specifies the total number of integration points.

Lines 38–40 check that the number of Gauss points in each coordinate direction is from 1 to 3 or it is the special 14-point rule for integration in the three-dimensional case. Lines 41–43 confirm that the number of dimensions is one, two or three.

The total number of integration points is determined in lines 45–50. Abscissas and weights for the requested Gauss integration rule are set in lines 57–100.

In order to perform numerical integration it is necessary to construct an integration rule object GaussRule with specification of nGauss – number of integration points in each dimension of the coordinate system (except the 14-point rule), and nDim – number of dimensions. The 14-point integration rule is used only for the three-dimensional case `nDim` $= 3$. Integration is performed using the created abscissas `xii`, `eti` and `zei` and weights `wi` inside the single integration loop.

The following snippet shows how to integrate function $f(\xi, \eta)$ over the two-dimensional domain $-1 \leq \xi, \eta \leq 1$ using a 3 by 3 integration rule.

```
GaussRule g = new GaussRule(3, 2);
double I = 0;
for (int ip = 0; ip < g.nIntPoints; ip++) {
    I += f(g.xii[ip],g.eti[ip])*g.wi[ip];
}
```

It can be seen that using class `GaussRule` makes numerical integration a relatively simple task.

Problems

9.1. Consider analytical and numerical calculation of the following integral:

$$I = \int_{-1}^{1} (x^3 - 2x^2 + 3x - 1)dx.$$

Calculate the integral value using Gauss rules with one, two, and three integration points. Compare the integral values with each other and with the analytical value.

9.2. Compute the following integral

$$I = \int_{0}^{1} \sin(\pi\xi)d\xi$$

using the three-point Gauss rule. Compare the numerical result with the analytical value.

9.3. Write a Java code fragment that performs numerical integration of the function

$$f(\xi, \eta, \zeta) = \xi^2 + \eta^2 + \zeta^2$$

over the three-dimensional domain $-1 \leq \xi, \eta, \zeta \leq 1$ using the $2 \times 2 \times 2$ Gauss rule.

Chapter 10
Two-dimensional Isoparametric Elements

Abstract The mathematical foundations of two-dimensional isoparametric finite elements are considered. The concept of shape functions for interpolation of unknown fields and for description of element shape is introduced. Use of a Jacobi matrix for creation of a displacement differentiation matrix is explained. It is shown how to compute element matrices and vectors using numerical integration. Extrapolation of stress values from reduced integration points to element nodes is demonstrated.

10.1 Shape Functions

Isoparametric finite elements are based on the parametric definition of both coordinate and displacement functions. The same shape functions are used for specification of the element shape and for interpolation of the displacement field.

Linear and quadratic quadrilateral two-dimensional isoparametric finite elements are presented in Figure 10.1. Shape functions N_i are defined in local coordinates ξ, η ($-1 \leq \xi$, $\eta \leq 1$). Interpolations of displacements and coordinates are performed in the following way:

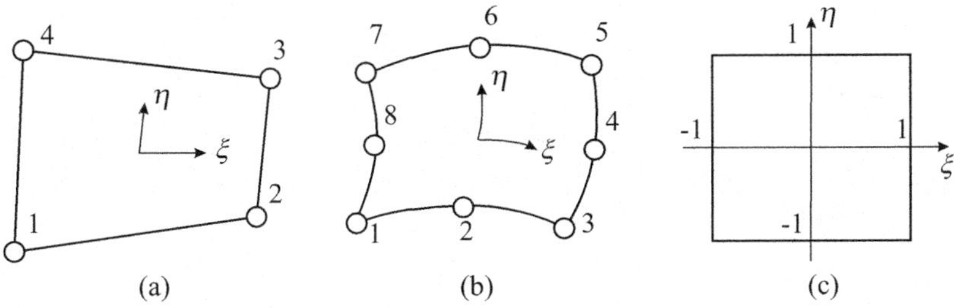

Fig. 10.1 Linear (a) and quadratic (b) quadrilateral finite elements. Both elements are mapped to a square $-1 \leq \xi$, $\eta \leq 1$ in the local coordinate system (c)

$$u = \sum N_i u_i, \quad v = \sum N_i v_i, \tag{10.1}$$

$$x = \sum N_i x_i, \quad y = \sum N_i y_i, \tag{10.2}$$

where u, v are displacement components at the point with local coordinates (ξ, η); u_i, v_i are displacement values at the nodes of the finite element; x, y are point coordinates and x_i, y_i are coordinates of element nodes. The matrix form of the relations for displacement interpolations is as follows:

$$\{u\} = [N]\{q\},$$

$$\{u\} = \{u \ v\}, \tag{10.3}$$

$$\{q\} = \{u_1 \ v_1 \ u_2 \ v_2 \ ...\},$$

where $\{u\}$ is a displacement vector at a point inside an element, $\{q\}$ is an element displacement vector including displacements at all element nodes and the interpolation matrix (matrix of shape functions) is:

$$[N] = \begin{bmatrix} N_1 & 0 & N_2 & 0 & ... \\ 0 & N_1 & 0 & N_2 & ... \end{bmatrix}. \tag{10.4}$$

Interpolation of coordinates $\{x\}$ from their nodal values $\{x^e\}$ is performed in a similar way:

$$\{x\} = [N]\{x^e\},$$

$$\{x\} = \{x \ y\}, \tag{10.5}$$

$$\{x^e\} = \{x_1 \ y_1 \ x_2 \ y_2 \ ...\}.$$

It appears that the element shape is determined with the help of interpolation functions N_i. This explains why interpolation functions are called *shape functions* in the finite element method.

Shape functions for the linear two-dimensional isoparametric elements with four nodes are given by

$$N_i = \frac{1}{4}(1 + \xi_0)(1 + \eta_0). \tag{10.6}$$

In the above relations the following notation is used: $\xi_0 = \xi \xi_i$, $\eta_0 = \eta \eta_i$, where ξ_i, η_i are the values of local coordinates ξ, η at nodes. The shape function for node 1 is shown in Figure 10.2 as a three-dimensional surface over the element plane.

Shape functions for the quadratic isoparametric element with eight nodes have the following appearance:

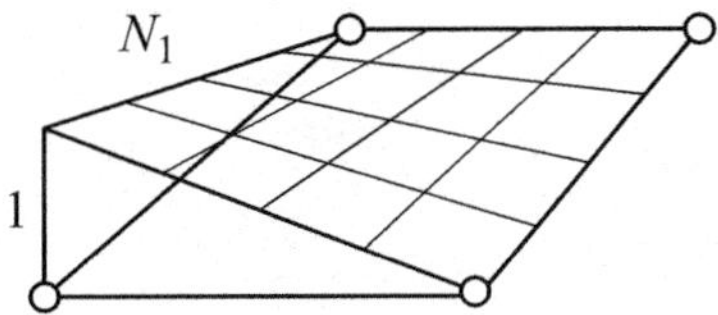

Fig. 10.2 Shape function for a linear finite element

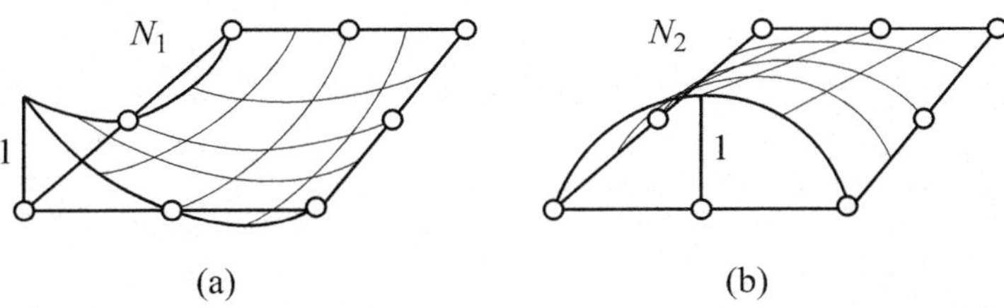

Fig. 10.3 Shape functions of a corner node N_1 (a) and of a midside node N_2 (b) for a quadratic finite element

$$N_i = \frac{1}{4}(1+\xi_0)(1+\eta_0) - \frac{1}{4}(1-\xi^2)(1+\eta_0)$$

$$- \frac{1}{4}(1+\xi_0)(1-\eta^2), \quad i = 1, 3, 5, 7,$$

$$N_i = \frac{1}{2}(1-\xi^2)(1+\eta_0), \quad i = 2, 6,$$

$$N_i = \frac{1}{2}(1+\xi_0)(1-\eta^2), \quad i = 4, 8.$$

(10.7)

As previously, we denote $\xi_0 = \xi\xi_i$, $\eta_0 = \eta\eta_i$ and ξ_i, η_i are nodal values of local co-ordinates ξ, η. Examples of shape functions for a corner node N_1 and for a midside node $N2$ are depicted in Figure 10.3.

It can be seen that shape functions for corner nodes of the quadratic element are combinations of shape functions for the linear element and shape functions for midside nodes of the quadratic element. This allows construction of isoparametric elements with a variable number of nodes changing from four to eight.

Shape functions of such elements are computed as follows. Shape functions for midside nodes are equal to:

$$N_i = \frac{1}{2}\left(1-\xi^2\right)(1+\eta_0)\,k_i\,, \quad \xi_i = 0,$$

$$N_i = \frac{1}{2}(1+\xi_0)\left(1-\eta^2\right)k_i\,, \quad \eta_i = 0,$$

(10.8)

where $k_i = 1$ if the ith midside node is present, otherwise $k_i = 0$. Shape functions for corner nodes are expressed by the relations:

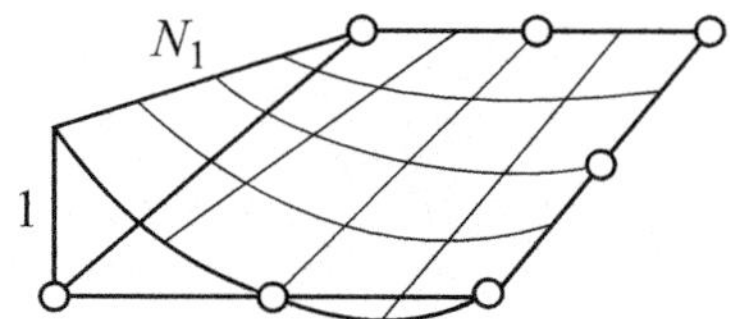

Fig. 10.4 Corner shape function for an element with a missing midside node

$$N_i = \frac{1}{4}\left(1 + \xi_0\right)\left(1 + \eta_0\right) - \frac{1}{2}\left(N_{i-1} + N_{i+1}\right). \tag{10.9}$$

Here, N_{i-1} and N_{i+1} are the shape functions of neighboring midside nodes. An example of a corner shape function for an element with a missing midside node is shown in Figure 10.4

A useful possibility for creation of finite element meshes is element degeneration. Placement of nodes belonging to one element side to the same position and assignment to them of the same connectivity number transforms a quadrilateral element into a triangular one. A linear quadrilateral element degenerated into a triangle (Figure 10.5a) has the same shape functions as a normal element.

However, degeneration of the quadratic quadrilateral element shown in Figure 10.5b requires modifying three shape functions for nodes located at a side opposite to the degenerated side. In the case of degeneration with coincident nodes 1, 2 and 3 the shape functions N_5, N_6 and N_7 should be modified in the following way [32]:

$$N_5' = N_5 + \Delta,$$

$$N_6' = N_6 - 2\Delta,$$

$$N_7' = N_7 + \Delta, \tag{10.10}$$

$$\Delta = \frac{1}{8}\left(1 - \xi^2\right)\left(1 - \eta^2\right).$$

10.2 Strain–Displacement Matrix

The displacement differentiation matrix $[B(\xi, \eta)]$ () is used to compute strains $\{\varepsilon\}$ at any point inside the element using a vector of nodal displacements $\{q\}$:

$$\{\varepsilon\} = [B]\{q\}. \tag{10.11}$$

Matrix $[B]$ can be presented in a block form

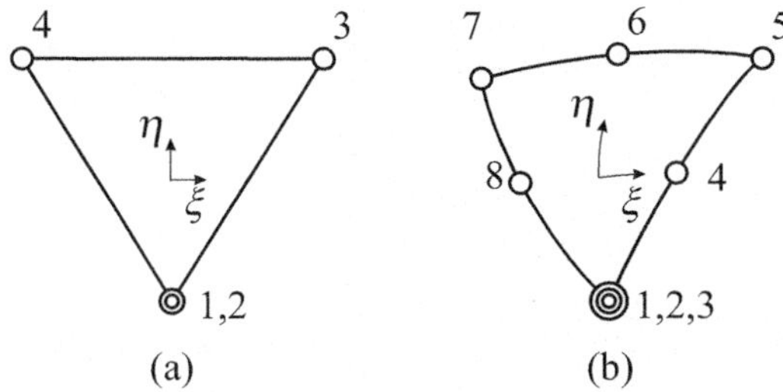

Fig. 10.5 Degenerated linear (a) and quadratic (b) elements. Nodes 1 and 2 in a linear element and nodes 1, 2 and 3 in a quadratic element have the same position and connectivity number

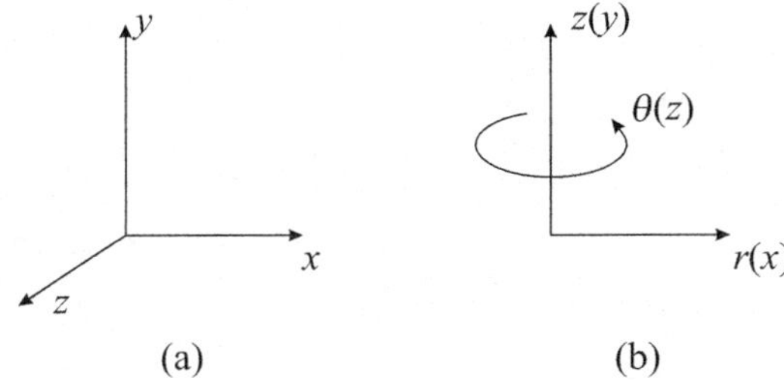

Fig. 10.6 Cartesian (a) and axisymmetric (b) coordinate systems

$$[B] = [B_1 \ B_2 \ ...].$$ (10.12)

Each block corresponds to displacements of one node. For plane problems in x, y, z coordinates shown in Figure 10.6a the strain vector contains three components:

$$\{\varepsilon\} = \{\varepsilon_x \ \varepsilon_y \ \gamma_{xy}\} = \left\{ \frac{\partial u}{\partial x} \quad \frac{\partial v}{\partial y} \quad \frac{\partial v}{\partial x} + \frac{\partial u}{\partial y} \right\}.$$ (10.13)

A block of the strain–displacement matrix that corresponds to the ith node has the appearance:

$$[B_i] = \begin{bmatrix} \dfrac{\partial N_i}{\partial x} & 0 \\ 0 & \dfrac{\partial N_i}{\partial y} \\ \dfrac{\partial N_i}{\partial y} & \dfrac{\partial N_i}{\partial x} \end{bmatrix}.$$ (10.14)

It is possible also to include in our considerations axisymmetric problems since such problems can be treated as two-dimensional. Using coordinates r (radial), z (along the axis of symmetry) and θ (angular) depicted in Figure 10.6b the strain vector can be presented as:

$$\{\varepsilon\} = \{\varepsilon_r \ \varepsilon_z \ \gamma_{rz} \ \varepsilon_\theta\} = \left\{ \frac{\partial u}{\partial r} \ \frac{\partial w}{\partial z} \ \frac{\partial w}{\partial r} + \frac{\partial u}{\partial z} \ \frac{u}{r} \right\}. \tag{10.15}$$

The strain–displacement matrix block for axisymmetric problems is as follows:

$$[B_i] = \begin{bmatrix} \dfrac{\partial N_i}{\partial r} & 0 \\ 0 & \dfrac{\partial N_i}{\partial z} \\ \dfrac{\partial N_i}{\partial z} & \dfrac{\partial N_i}{\partial r} \\ 0 & \dfrac{N_i}{r} \end{bmatrix}. \tag{10.16}$$

In order to unify notation for Java^{TM} code development it is possible to use coordinate axes x, y and z instead of r, z and θ, as shown in Figure 10.6b.

While shape functions are expressed through local coordinates ξ, η, the strain–displacement matrix contains derivatives with respect to the global coordinates x and y. Derivatives can be transformed from one coordinate system to the other by means of the chain rule of partial differentiation:

$$\left\{ \begin{array}{c} \dfrac{\partial N_i}{\partial \xi} \\ \dfrac{\partial N_i}{\partial \eta} \end{array} \right\} = \left\{ \begin{array}{cc} \dfrac{\partial x}{\partial \xi} & \dfrac{\partial y}{\partial \xi} \\ \dfrac{\partial x}{\partial \eta} & \dfrac{\partial y}{\partial \eta} \end{array} \right\} \left\{ \begin{array}{c} \dfrac{\partial N_i}{\partial x} \\ \dfrac{\partial N_i}{\partial y} \end{array} \right\} = [J] \left\{ \begin{array}{c} \dfrac{\partial N_i}{\partial x} \\ \dfrac{\partial N_i}{\partial y} \end{array} \right\}, \tag{10.17}$$

where $[J]$ is the Jacobian matrix. The derivatives with respect to the global coordinates are computed with the use of the inverse of the Jacobian matrix:

$$\left\{ \begin{array}{c} \dfrac{\partial N_i}{\partial x} \\ \dfrac{\partial N_i}{\partial y} \end{array} \right\} = [J]^{-1} \left\{ \begin{array}{c} \dfrac{\partial N_i}{\partial \xi} \\ \dfrac{\partial N_i}{\partial \eta} \end{array} \right\}. \tag{10.18}$$

The components of the Jacobian matrix are calculated using derivatives of shape functions N_i with respect to the local coordinates ξ, η and global coordinates of element nodes x_i, y_i:

$$\frac{\partial x}{\partial \xi} = \sum \frac{\partial N_i}{\partial \xi} x_i, \quad \frac{\partial x}{\partial \eta} = \sum \frac{\partial N_i}{\partial \eta} x_i,$$

$$\frac{\partial y}{\partial \xi} = \sum \frac{\partial N_i}{\partial \xi} y_i, \quad \frac{\partial y}{\partial \eta} = \sum \frac{\partial N_i}{\partial \eta} y_i. \tag{10.19}$$

The determinant of the Jacobian matrix $|J|$ is used for the transformation of integrals from the global coordinate system to the local coordinate system. Assuming unit thickness in plane problems, it is possible to represent an elementary volume as:

$$dV = dxdy = |J|d\xi d\eta. \tag{10.20}$$

For axisymmetric problems an elementary volume includes the length of a circle with a current radius:

$$dV = 2\pi r dr dz = 2\pi r |J|d\xi d\eta. \tag{10.21}$$

10.3 Element Properties

Element matrices and vectors are calculated as follows:

stiffness matrix

$$[k] = \int_{-1}^{1}\int_{-1}^{1} [B]^{\mathrm{T}}[E][B]t\,|J|d\xi d\eta, \tag{10.22}$$

thermal vector (fictitious forces to simulate thermal expansion)

$$\{h\} = \int_{-1}^{1}\int_{-1}^{1} [B]^{\mathrm{T}}[E]\{\varepsilon^{t}\}t\,|J|d\xi d\eta \tag{10.23}$$

force vector (surface load)

$$\{p\} = \int_{-1}^{1} [N]^{\mathrm{T}}\{p^{S}\}t\frac{ds}{d\xi}d\xi, \tag{10.24}$$

equivalent stress vector (with negative sign)

$$\{p_\sigma\} = -\int_{-1}^{1}\int_{-1}^{1} [B]^{\mathrm{T}}\{\sigma\}t\,|J|d\xi d\eta. \tag{10.25}$$

The elasticity matrix $[E]$ is given by (7.9). A "thickness" t for the current position is introduced in all integrals:

$$t = \begin{cases} 1 & \text{for plane problems,} \\ 2\pi r & \text{for axisymmetric problems.} \end{cases} \tag{10.26}$$

Integration of expressions for stiffness matrices and load vectors can not be performed analytically for the general case of isoparametric elements. Instead, stiffness matrices and load vectors are evaluated numerically using Gauss quadrature over quadrilateral regions. The Gauss quadrature formula for the volume integral in the two-dimensional case is of the form:

$$I = \int_{-1}^{1}\int_{-1}^{1} f(\xi,\eta)d\xi d\eta = \sum_{i=1}^{n}\sum_{j=1}^{n} f(\xi_i,\eta_j)w_i w_j$$
$$= \sum_{k=1}^{N} f(\xi_k,\eta_k)W_k, \tag{10.27}$$

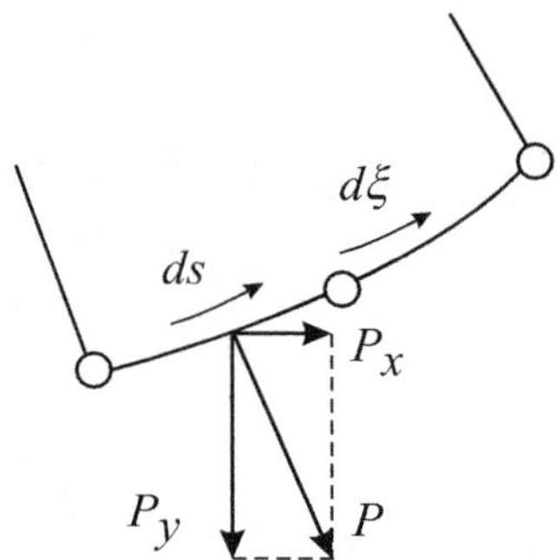

Fig. 10.7 Distributed normal load on a side of a quadratic element

where ξ_i, η_j are abscissas, w_i are weighting coefficients of the one-dimensional Gauss integration rule, $N = n \times n$ and W_k are products of pairs of one-dimensional integration weights.

10.4 Nodal Equivalent of the Surface Load

To compute the nodal equivalent of the surface load, the surface integral is replaced by the line integration along an element side. The fraction of the surface load is evaluated as:

$$\{p\} = \int_{-1}^{1} [N]^{\mathrm{T}} \left\{ \begin{matrix} p_x^s \\ p_y^s \end{matrix} \right\} t \frac{ds}{d\xi} d\xi, \tag{10.28}$$

$$\frac{ds}{d\xi} = \sqrt{\left(\frac{dx}{d\xi}\right)^2 + \left(\frac{dy}{d\xi}\right)^2}. \tag{10.29}$$

Here, s is a global coordinate along the element side and ξ is a local coordinate along the element side.

If the distributed load with intensity p is applied along the normal to the element side as shown in Figure 10.7 then its components along global coordinate axes are;

$$p_x^s = p\frac{dy}{ds}, \quad p_y^s = -p\frac{dx}{ds} \tag{10.30}$$

and the nodal equivalent of such a load is:

$$\{p\} = \int_S [N]^{\mathrm{T}} p \left\{ \begin{matrix} \dfrac{dy}{ds} \\ -\dfrac{dx}{ds} \end{matrix} \right\} t \frac{ds}{d\xi} d\xi = \int_{-1}^{1} [N]^{\mathrm{T}} p \left\{ \begin{matrix} \dfrac{dy}{d\xi} \\ -\dfrac{dx}{d\xi} \end{matrix} \right\} t d\xi. \tag{10.31}$$

10.5 Example: Computing Nodal Equivalents of a Distributed Load

Calculate the nodal equivalents of a distributed load with constant intensity applied to the edge of a two-dimensional quadratic element depicted in Figure 10.8.

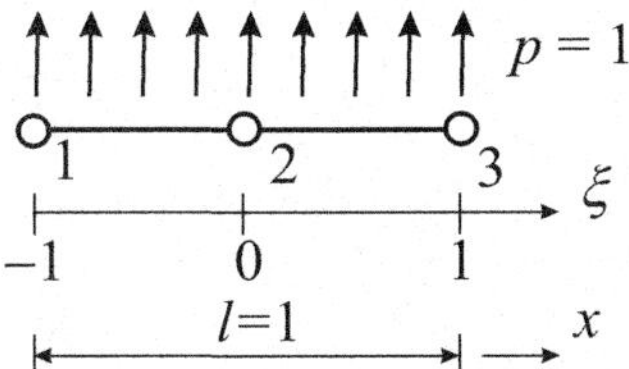

Fig. 10.8 Distributed load on an edge of a quadratic element

Solution

The nodal equivalent of the distributed load is calculated as:

$$\{p\} = \int_{-1}^{1} [N]^{\mathrm{T}} p \frac{dx}{d\xi} d\xi$$

or

$$\{p\} = \left\{ \begin{array}{c} p_1 \\ p_2 \\ p_3 \end{array} \right\} = \int_{-1}^{1} \left\{ \begin{array}{c} N_1 \\ N_2 \\ N_3 \end{array} \right\} p \frac{dx}{d\xi} d\xi, \quad \frac{dx}{d\xi} = \frac{1}{2}.$$

The shape functions for the edge of a quadratic element are:

$$N_1 = -\frac{1}{2}\xi(1-\xi), \quad N_2 = 1-\xi^2, \quad N_3 = \frac{1}{2}\xi(1+\xi).$$

The values of nodal forces at nodes 1, 2 and 3 are defined by integration:

$$p_1 = -\int_{-1}^{1} \frac{1}{2}\xi(1-\xi)\frac{1}{2}d\xi = \frac{1}{6},$$

$$p_2 = \int_{-1}^{1} (1-\xi^2)\frac{1}{2}d\xi = \frac{2}{3},$$

$$p_3 = \int_{-1}^{1} \frac{1}{2}\xi(1+\xi)\frac{1}{2}d\xi = \frac{1}{6}.$$

The example shows that a physical approach with proportional load distribution cannot be used for the estimation of nodal equivalents of a surface load. It works for linear elements. However, it does not work for higher-order elements because they have nonlinear shape functions.

10.6 Calculation of Strains and Stresses

Strains at any point of an element are determined using the Cauchy relations (10.13) with the use of the displacement differentiation matrix (10.14) for plane problems or (10.16) for axisymmetric problems. Stresses are calculated with Hooke's law.

The precision of strains and stresses is significantly dependent on the point location where they are computed. The highest precision for displacement gradients is at the geometric center for the linear element and at reduced integration points 2×2 for the quadratic quadrilateral element.

For quadratic elements with eight nodes, strains and stresses have the best precision at 2×2 integration points with local coordinates ξ, $\eta = \pm 1/\sqrt{3}$. A possible way to create a continuous stress field with reasonable accuracy consists of: 1) extrapolation of stresses from reduced integration points to nodes; 2) averaging contributions from finite elements at all nodes of the finite element model. Later, stresses can be interpolated from nodes using quadratic shape functions.

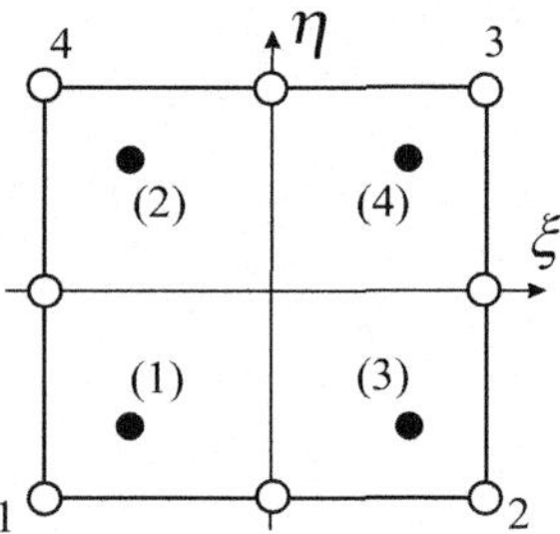

Fig. 10.9 Numbering of integration points and vertices for the eight-node isoparametric element

Let us consider a quadratic element in the local coordinate system ξ, η as shown in Figure 10.9 where integration points are numbered as (1)–(4). The order of integration points is determined by class `GaussRule` presented in Chapter 9. Corner nodes are numbered 1–4 in anticlockwise order.

Let us first interpolate values at corner nodes 1–4 to the reduced integration points (1)–(4) using linear shape functions $N_i^{\mathrm{L}}(\xi, \eta)$:

$$\{f_{(m)}\} = \left[N_{i(m)}^{\mathrm{L}}\right] \{f_i\}, \tag{10.32}$$

where $\{f_{(m)}\}$ is the vector of values at integration points, $\left[N_{i(m)}^{\mathrm{L}}\right]$ is the matrix of values of shape functions at integration points, and $\{f_i\}$ is the vector of nodal values.

After finding the inverse of the interpolation matrix

$$\left[L_{i(m)}\right] = \left[N_{i(m)}^{\mathrm{L}}\right]^{-1}$$

the extrapolation relation can be presented as follows:

$$\{f_i\} = \left[L_{i(m)}\right]\{f_{(m)}\}, \tag{10.33}$$

$$\left[L_{i(m)}\right] = \begin{bmatrix} A & B & B & C \\ B & C & A & B \\ C & B & B & A \\ B & A & C & B \end{bmatrix},$$

$$A = 1 + \frac{\sqrt{3}}{2}, \tag{10.34}$$

$$B = -\frac{1}{2},$$

$$C = 1 - \frac{\sqrt{3}}{2}.$$

The above relation extrapolates known values at integration points $f_{(m)}$ to element corners giving values f_i at nodes 1–4.

Problems

10.1. Using Equation 10.6 write down explicit expressions for the shape functions of the linear quadrilateral element.

10.2. Write expressions for shape functions N_2 and N_3 of the quadratic element. Use the general relations (10.7).

10.3. Evaluate the Jacobian matrix $[J]$ and its determinant $|J|$ for the four-node element shown below.

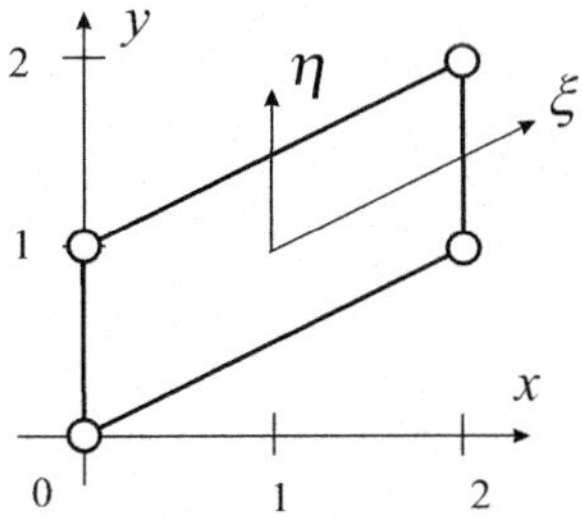

Use relations (10.19) for computing elements of the Jacobian matrix. Estimate the same partial derivatives by the ratios of coordinate increments (for example, $\partial x/\partial \xi = \Delta x/\Delta \xi$) and compare the results.

10.4. Estimate the nodal equivalents of a distributed load with linearly varying intensity applied to the edge of the two-dimensional quadratic element below.

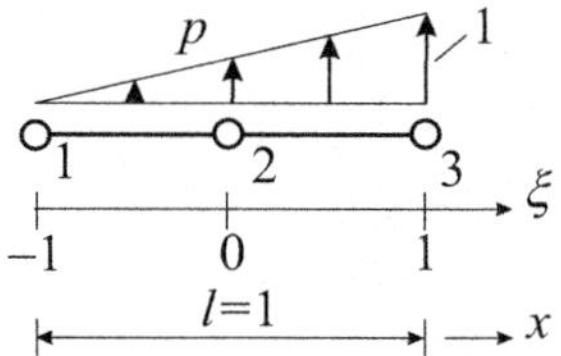

10.5. The result values at four reduced integration points of the quadratic element shown in Figure 10.9 are

$$\{f_{(m)}\} = \{0.0 \quad 1.0 \quad 2.0 \quad 3.0\}.$$

Find the result values at the corner nodes of the element using extrapolation (10.33).

10.6. Determine the Jacobian matrix $[J]$ for the following four-node finite element.

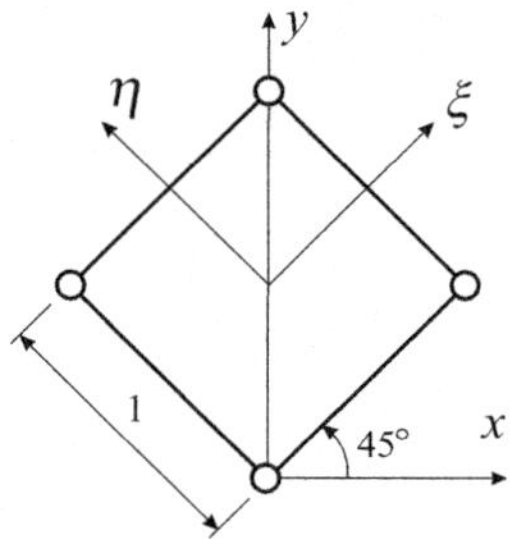

The element has a square shape with unit edge length and is inclined at $45°$ angle.

Chapter 11
Implementation of Two-dimensional Quadratic Element

Abstract Two-dimensional isoparametric quadratic elements have quadrilateral shape and eight nodes. Implementation of this element type is done through development of two classes. Class ShapeQuad2D provides means for calculating the shape functions and derivatives of shape functions for elements with the number of nodes from 4 to 8. Class ElementQuad2D contains methods for calculating the element stiffness matrix, displacement differentiation matrix, thermal vector, distributed load vector, and equivalent stress vector. Extrapolation of stresses from reduced integration points to element nodes is also performed.

11.1 Class for Shape Functions and Their Derivatives

Java^{TM} class ShapeQuad2D is included in package elem. It provides the means for calculating shape functions for two-dimensional quadratic isoparametric elements with four to eight nodes. Element nodes are numbered in an anticlockwise direction starting from any corner node. The local numbers of nodes are shown in Figure 11.1.

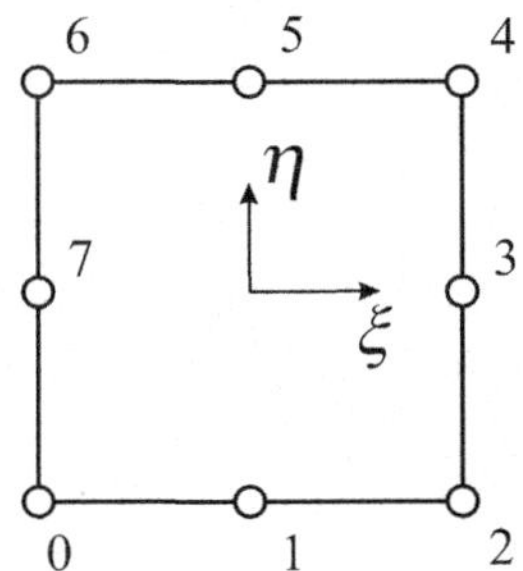

Fig. 11.1 Local numbering of nodes for the quadratic isoparametric element

Corner nodes should always be present in an element. Intermediate midside nodes can be absent in any order. Absence of a midside node is coded by a zero connectivity number.

11.1.1 Element Degeneration

A quadrilateral element can have a triangular shape. Such an element is created by degeneration of a side into a point. If a linear element with four nodes becomes degenerate then two neighboring corner nodes have the same connectivity numbers and the shape functions need no change. For a degenerate quadratic element three connectivity numbers belonging to one side should have the same connectivity numbers. Shape functions of the degenerate element have to be modified.

Below is the first fragment of the class `ShapeQuad2D` containing the method responsible for discovering degenerate quadratic elements.

```
 1 package elem;
 2
 3 import util.UTIL;
 4
 5 // Quadratic 2D shape functions and their derivatives
 6 public class ShapeQuad2D {
 7
 8     // Degeneration check.
 9     // If element is triangular then the method returns
10     // a local number (starting from 0) of the midside node
11     // opposite to degenerated side.
12     // ind - connectivity numbers
13     static int degeneration(int[] ind) {
14         int deg = 0;
15         for (int i = 0; i < 7; i += 2) {
16             if (ind[i] == ind[i + 1]) {
17                 deg = (i + 5) % 8;
18                 break;
19             }
20         }
21         return deg;
22     }
```

Method `degeneration` checks if a corner connectivity number is equal to the next midside node. If so, then the method returns the local number of a midside node opposite to the degenerated side.

11.1.2 Shape Functions

Method `shape` computes element shape functions an for specified local coordinates xi (ξ) and et (η). Connectivity numbers ind are used as information on the existence of midside nodes.

```
24        // Shape functions.
25        // xi, et - local coordinates;
26        // ind[8] - element connectivities;
27        // an[8] - shape functions (out)
28        static void shape(double xi, double et, int[] ind,
29                          double[] an) {
30
31            // Shape functions of midside nodes
32            an[1] = an[3] = an[5] = an[7] = 0;
33            if (ind[1] > 0) an[1] = 0.5*(1 - xi*xi)*(1 - et);
34            if (ind[3] > 0) an[3] = 0.5*(1 - et*et)*(1 + xi);
35            if (ind[5] > 0) an[5] = 0.5*(1 - xi*xi)*(1 + et);
36            if (ind[7] > 0) an[7] = 0.5*(1 - et*et)*(1 - xi);
37
38            // Shape functions of corner nodes
39            an[0] = 0.25*(1 - xi)*(1 - et) - 0.5*(an[7] + an[1]);
40            an[2] = 0.25*(1 + xi)*(1 - et) - 0.5*(an[1] + an[3]);
41            an[4] = 0.25*(1 + xi)*(1 + et) - 0.5*(an[3] + an[5]);
42            an[6] = 0.25*(1 - xi)*(1 + et) - 0.5*(an[5] + an[7]);
43
44            // Modification of functions due to degeneration
45            int deg = degeneration(ind);
46            if (deg > 0 && ind[1] > 0 && ind[3] > 0 &&
47                    ind[5] > 0 && ind[7] > 0) {
48                double delta = 0.125*(1 - xi*xi)*(1 - et*et);
49                an[deg - 1] += delta;
50                an[deg] -= 2.*delta;
51                an[(deg + 1)%8] += delta;
52            }
53        }
```

Shape functions of midside nodes are calculated in lines 32–36. If the corresponding connectivity number is nonzero (positive) then shape function an[i], i = 1, 3, 5, 7 is set according to its expression for the quadratic element, otherwise it remains zero. Shape functions of corner nodes are computed in lines 39–42 as combinations of linear shape functions and quadratic shape functions of the neighboring midside nodes. Modifications of three shape functions for a triangular degenerate element are done in lines 45–52.

11.1.3 *Derivatives of Shape Functions*

Derivatives of shape functions with respect to global coordinates dnxy are given by method deriv. The method parameters are: xi, et – local coordinates ξ and η, ind – element connectivities, xy – array of nodal coordinates.

```
55      // Derivatives of shape functions
56      // with respect to global coordinates x and y.
57      // xi, et - local coordinates;
58      // ind[8] - element connectivities;
59      // xy[8][2] - nodal coordinates;
60      // dnxy[8][2] - derivatives of shape functions (out);
61      // returns  determinant of the Jacobian matrrix
62      static double deriv(double xi, double et, int[] ind,
63                          double[][] xy, double[][] dnxy) {
64          // Derivatives in local coords dN/dXi, dN/dEta
65          // Midside nodes
66          double[][] dnxe = new double[8][2];
67          dnxe[1][0] = dnxe[1][1] = dnxe[3][0] = dnxe[3][1] =
68          dnxe[5][0] = dnxe[5][1] = dnxe[7][0] = dnxe[7][1] = 0;
69          if (ind[1] > 0) {
70              dnxe[1][0] = -xi*(1-et);
71              dnxe[1][1] = -0.5*(1-xi*xi);
72          }
73          if (ind[3] > 0) {
74              dnxe[3][0] = 0.5*(1-et*et);
75              dnxe[3][1] = -et*(1+xi);
76          }
77          if (ind[5] > 0) {
78              dnxe[5][0] = -xi*(1+et);
79              dnxe[5][1] = 0.5*(1-xi*xi);
80          }
81          if (ind[7] > 0) {
82              dnxe[7][0] = -0.5*(1-et*et);
83              dnxe[7][1] = -et*(1-xi);
84          }
85          // Corner nodes
86          dnxe[0][0] = -0.25*(1-et)-0.5*(dnxe[7][0]+dnxe[1][0]);
87          dnxe[0][1] = -0.25*(1-xi)-0.5*(dnxe[7][1]+dnxe[1][1]);
88          dnxe[2][0] =  0.25*(1-et)-0.5*(dnxe[1][0]+dnxe[3][0]);
89          dnxe[2][1] = -0.25*(1+xi)-0.5*(dnxe[1][1]+dnxe[3][1]);
90          dnxe[4][0] =  0.25*(1+et)-0.5*(dnxe[3][0]+dnxe[5][0]);
91          dnxe[4][1] =  0.25*(1+xi)-0.5*(dnxe[3][1]+dnxe[5][1]);
92          dnxe[6][0] = -0.25*(1+et)-0.5*(dnxe[5][0]+dnxe[7][0]);
93          dnxe[6][1] =  0.25*(1-xi)-0.5*(dnxe[5][1]+dnxe[7][1]);
94
95          // Modification of derivatives due to degeneration
96          int deg = degeneration(ind);
97          if (deg > 0 && ind[1] > 0 && ind[3] > 0 &&
98                  ind[5] > 0 && ind[7] > 0) {
99              double z = -0.25*xi*(1-et*et);
100             double t = -0.25*(1-xi*xi)*et;
101             int j = (deg + 1)%8;
```

```
102                  dnxe[deg - 1][0] = dnxe[deg - 1][0] + z;
103                  dnxe[deg - 1][1] = dnxe[deg - 1][1] + t;
104                  dnxe[deg][0] = dnxe[deg][0] - 2*z;
105                  dnxe[deg][1] = dnxe[deg][1] - 2*t;
106                  dnxe[j][0] = dnxe[j][0] + z;
107                  dnxe[j][1] = dnxe[j][1] + t;
108              }
109
110          // Jacobian matrix
111          double[][] aj = new double[2][2];
112          for (int j = 0; j < 2; j++) {
113              for (int i = 0; i < 2; i++) {
114                  aj[i][j] = 0.0;
115                  for (int k = 0; k < 8; k++)
116                      aj[i][j] += dnxe[k][j]*xy[k][i];
117              }
118          }
119          double det = aj[0][0]*aj[1][1] - aj[0][1]*aj[1][0];
120          // Zero or negative determinant
121          if (det<=0) UTIL.errorMsg("Negative/zero Jacobian "+
122              "determinant for 8N element "+(float)det);
123          // Jacobian inverse
124          double aj00 = aj[1][1]/det;
125          aj[1][1] =   aj[0][0]/det;
126          aj[0][0] =   aj00;
127          aj[1][0] = -aj[1][0]/det;
128          aj[0][1] = -aj[0][1]/det;
129
130          // Derivatives in global coordinates dN/dx, dN/dy
131          for (int k = 0; k < 8; k++)
132              for (int i = 0; i < 2; i++)
133                  dnxy[k][i] = aj[0][i]*dnxe[k][0]
134                              + aj[1][i]*dnxe[k][1];
135          return det;
136      }
```

The statements in lines 66–93 evaluate derivatives of shape functions with respect to the local coordinates. Analogously to evaluation of shape functions, we start with computation of derivatives for midside nodes. A shape function derivative is nonzero if a midside node exists. Then the derivatives for corner nodes are calculated using the derivatives of neighboring midside nodes. For a degenerate element, the derivatives of three nodes are modified.

Calculation of the Jacobian matrix aj according to Equation 10.19 is done in lines 111–118. If the Jacobian determinant is not positive, which means an error in the element data, the error message is generated in lines 121–122. The Jacobian matrix inverse is performed in lines 124–128. Derivatives of shape functions with respect to global coordinates x, y are obtained by multiplication of the Jacobian matrix and derivatives with respect to local coordinates ξ, η (lines 131–134). The method returns a determinant of the Jacobian matrix. This is done in line 135 for the normal method completion.

11.1.4 One-dimensional Shape Functions and Their Derivatives

One-dimensional shape functions and derivatives of shape functions with respect to local coordinates are necessary for estimation of the nodal equivalent of a surface load on the element edge (face) of the quadratic element according to Equation 10.28. Method shapeDerivFace calculates three shape functions an and their derivatives dndxi with respect to the local coordinate ξ. Other method parameters are: xi – local coordinate ξ along the side and parameter kmid – connectivity number of a midside node at this edge (zero connectivity number means absence of the midside node).

```
138        // One-dimensional quadratic shape functions and
139        //     their derivatives in local coordinates
140        // xi - local coordinate;
141        // kmid - index of midside node (=0 no midside node);
142        // an[3] - shape functions (out);
143        // dndxi[3] - derivatives of shape functions (out)
144        public static void shapeDerivFace(double xi, int kmid,
145                                     double[] an, double[] dndxi) {
146            double x1 = 1 - xi;
147            double x2 = 1 + xi;
148            if (kmid > 0) {
149                an[1] =   x1*x2;
150                dndxi[1] = -2*xi;
151            }
152            an[0]   =  0.5*x1 - 0.5*an[1];
153            an[2]   =  0.5*x2 - 0.5*an[1];
154            dndxi[0] = -0.5 - 0.5*dndxi[1];
155            dndxi[2] =  0.5 - 0.5*dndxi[1];
156        }
157
158 }
```

11.2 Class for Eight-node Element

Class ElementQuad2D extends class Element and implements methods for computing element matrices and vectors. Below are given the class fields and the class constructor.

```
1 package elem;
2
3 import model.*;
4 import material.*;
5 import util.*;
6
7 // 2D quadratic isoparametric element (4-8 nodes)
8 class ElementQuad2D extends Element {
9     // Element edges (local numbers)
10        private static int[][] faceInd =
```

```
11                    {{0,1,2},{2,3,4},{4,5,6},{6,7,0}};
12        // Shape functions
13        private static double[] an = new double[8];
14        // Derivatives of shape functions
15        private static double[][] dnxy = new double[8][2];
16        // Displacements differentiation matrix
17        private static double[][] bmat = new double[4][16];
18        // Elasticity matrix
19        private static double[][] emat = new double[4][4];
20        // Thermal strains
21        private static double[] ept = new double[4];
22        // Radius in the axisymmetric problem
23        private static double r;
24        // Gauss rules for stiffness matrix, thermal vector,
25        // surface load and stress integration
26        private static GaussRule gk = new GaussRule(3,2);
27        private static GaussRule gh = new GaussRule(3,2);
28        private static GaussRule gf = new GaussRule(3,1);
29        private static GaussRule gs = new GaussRule(2,2);
30
31        // Constructor for 2D quadratic element
32        public ElementQuad2D() {
33            super ("quad8", 8, 4);
34        }
```

Lines 10–11 contain specification of element sides (faces). Local node numbering from 0 to 7 is used. In lines 13, 15, 17, 19 and 21 the following arrays are declared:

an – shape functions;

dnxy – derivatives of shape functions with respect to global coordinates x, y (first index is related to node number, second index to x and y);

bmat – displacement differentiation matrix, emat – elasticity matrix,

ept – vector of thermal strains.

Lines 26–29 create GaussRule objects for integration of the element stiffness matrix, thermal vector, surface load, and equivalent stress vector.

Constructor ElementQuad2D calls the constructor of parent class Element and passes to it the element name quad8, the number of element nodes (8) and the number of points for storing stresses and strains.

11.2.1 Stiffness Matrix

Computation of the element stiffness matrix according to Equation 10.22 is performed by method stiffnessMatrix. The pseudocode given below illustrates integration of the element stiffness matrix with the use of the Gauss rule.

Clean stiffness matrix $[k] = 0$
Check if element is degenerate

Set elasticity matrix $[E]$
do loop over integration points i
    Set displacement differentiation matrix $[B]$ at ξ_i, η_i
    Numerical integration $[k] = [k] + [B]^T[E][B]dV \cdot W_i$
end do

Here, W_i is the combined integration weight for current integration point. Implementation of the stiffness matrix computation is as follows.

```
36        // Compute stiffness matrix
37        public void stiffnessMatrix() {
38
39            // Zeros to stiffness matrix kmat
40            for (int i = 0; i < 16; i++)
41                for (int j = 0; j < 16; j++)
42                            kmat[i][j] = 0;
43
44            // ld = length of strain/stress vector (3 or 4)
45            int ld = (FeModel.stressState
46                        == FeModel.StrStates.axisym ) ? 4 : 3;
47            // Material mat
48            mat = (Material)fem.materials.get(matName);
49            if (mat == null) UTIL.errorMsg(
50                    "Element material name: " + matName);
51            mat.elasticityMatrix(emat);
52
53            // Gauss integration loop
54            for (int ip = 0; ip < gk.nIntPoints; ip++) {
55                // Set displacement differentiation matrix bmat
56                double det = setBmatrix(gk.xii[ip], gk.eti[ip]);
57                double dv = det*gk.wi[ip];
58                if (FeModel.stressState==FeModel.StrStates.axisym)
59                    dv *= 2*Math.PI*r;
60                // Upper symmetrical part of the stiffness matrix
61                for (int i = 0; i < 16; i++) {
62                    for (int j = i; j < 16; j++) {
63                        double s = 0;
64                        for (int k = 0; k < ld; k++) {
65                            for (int l = 0; l < ld; l++) {
66                                s +=
67                                bmat[l][i]*emat[l][k]*bmat[k][j];
68                            }
69                        }
70                        kmat[i][j] += s*dv;
71                    }
72                }
73            }
74        }
```

The stiffness matrix `kmat` is set to zero in lines 40–42. Integer variable `ld` (line 45) represents a length of the strain or stress vector. For plane problems `ld` is equal to 3. In axisymmetric problems strain and stress vectors have size 4.

In line 48 `Material` object `mat` is extracted from the hash table `materials` using the material name. The elasticity matrix `emat` is set in line 49.

Numerical integration of the element stiffness matrix is performed in a loop with a parameter `ip` denoting integration point number (lines 54–73). Integration is performed inside a single loop since the constructor of class `GaussRule` always produces abscissas and weights, which are placed in one-dimensional arrays. In the case of the stiffness matrix we use the integration rule object `gk` that contains abscissas `xii` and `eti`, weights `wi` and the number of integration points `nIntPoints`. In line 56 method `setBmatrix` sets the displacement differentiation matrix `bmat` and returns a determinant of the Jacobian matrix `det`. Variable `dv` includes an integration weight and a circle length in the case of the axisymmetric problem.

Note that the loop parameter `j` (line 62) starts from `i`, which is a parameter of the outer loop. In doing this we compute just the upper symmetric part of the stiffness matrix thus economizing computations. An incomplete stiffness matrix structure should be taken into account during assembly of the global stiffness matrix from element contributions. Integration of stiffness matrix coefficients is performed by summation of a triple product in line 70.

11.2.2 Displacement Differentiation Matrix

Method `setBmatrix` performs computation of a displacement differentiation matrix `bmat` for specified local coordinates `xi` and `et` and returns the determinant of the Jacobian matrix.

```
76        // Set displacement differentiation matrix bmat.
77        // xi, et - local coordinates,
78        // returns  determinant of Jacobian matrix
79        private double setBmatrix(double xi, double et) {
80
81            // Derivatives of shape functions
82            double det = ShapeQuad2D.deriv(xi, et, ind, xy, dnxy);
83            if (det <= 0) UTIL.errorMsg(
84                    "Negative/zero 8N element area");
85            if (FeModel.stressState == FeModel.StrStates.axisym) {
86                ShapeQuad2D.shape(xi, et, ind, an);
87                r = 0;
88                for (int i = 0; i < 8; i++) r += an[i] * xy[i][0];
89            }
90            // Eight blocks of the displacement differentiation
91            //    matrix
92            for (int ib = 0; ib < 8; ib++) {
93                bmat[0][2*ib]   = dnxy[ib][0];
94                bmat[0][2*ib+1] = 0.;
95                bmat[1][2*ib]   = 0.;
96                bmat[1][2*ib+1] = dnxy[ib][1];
97                bmat[2][2*ib]   = dnxy[ib][1];
98                bmat[2][2*ib+1] = dnxy[ib][0];
99                if(FeModel.stressState==FeModel.StrStates.axisym) {
```

```
100                      bmat[3][2*ib]   = an[ib] / r;
101                      bmat[3][2*ib+1] = 0.;
102                  }
103              }
104          return det;
105      }
```

Derivatives of shape functions dnxy with respect to global coordinates x, y are computed in line 82 using method deriv. A value of the determinant of the Jacobi matrix det is checked in line 83. A negative determinant means an error in the element shape, thus leading to an error message. A possible reason for such an error can be related to incorrect order in numbering element nodes (incorrect order in element connectivities). In the case of the axisymmetric problem element shape functions an are used for evaluation of the current radius r (lines 85–89).

Eight blocks of the displacement differentiation matrix bmat are created in a loop that starts in line 92. The first three rows of each block are common for all two-dimensional problems, the fourth row (lines 100–101) is set only for axisymmetric problems. The method returns the value of the Jacobian matrix determinant.

11.2.3 Thermal Vector

Method thermalVector sets an element thermal vector to array evec according to Equation 10.23. An algorithm for computation of a thermal vector is similar to the algorithm for the element stiffness matrix.

```
107      // Compute thermal vector
108      public void thermalVector() {
109
110          // Zeros to thermal vector evec
111          for (int i = 0; i < 16; i++) evec[i] = 0.;
112          int ld = (FeModel.stressState
113                      == FeModel.StrStates.axisym) ? 4 : 3;
114          // Material mat
115          mat = (Material)fem.materials.get(matName);
116          mat.elasticityMatrix(emat);
117          double alpha = mat.getAlpha();
118          double nu = mat.getNu();
119
120          // Gauss integration loop
121          for (int ip = 0; ip < gh.nIntPoints; ip++) {
122              // Set displacement differentiation matrix bmat
123              double det = setBmatrix(gh.xii[ip], gh.eti[ip]);
124              // Shape functions an
125              ShapeQuad2D.shape(gh.xii[ip], gh.eti[ip], ind, an);
126              double t = 0;
127              for (int i = 0; i < 8; i++) t += an[i]*dtn[i];
128              double dv = det*gh.wi[ip];
129              if (FeModel.stressState==FeModel.StrStates.axisym)
130                  dv *= 2*Math.PI*r;
```

```
131                         ept[0] = alpha*t;
132                         if(FeModel.stressState==FeModel.StrStates.plstrain)
133                                   ept[0] *= (1 + nu);
134                         ept[1] = ept[0];
135                         ept[2] = 0.;
136                         ept[3] = ept[0];
137
138                         for (int i = 0; i < 16; i++) {
139                             double s = 0;
140                             for (int j = 0; j < ld; j++) {
141                                 for (int k = 0; k < ld; k++) {
142                                     s += bmat[k][i]*emat[j][k]*ept[j];
143                                 }
144                             }
145                             evec[i] += s*dv;
146                         }
147                     }
148                 }
```

First, array `evec` is set to zero. Method `elasticityMatrix` sets the elasticity matrix `emat`. Thermal vector integration is performed in a loop with parameter `ip` – integration point number. Displacements differentiation matrix `bmat` is set in line 123. Computation of shape functions an (line 125) is necessary for estimation of the temperature `t` at the integration point. A vector of thermal strains is formed in lines 132–136. Integration of thermal vector coefficients is done in line 145.

11.2.4 Nodal Equivalent of a Distributed Load

A load distributed along an element side is transformed into a nodal vector according to the algorithm described in Section 10.4. Method `equivFaceLoad` computing a nodal equivalent of surface load is presented below.

```
150     // Set nodal equivalent of distributed face load to evec.
151     // surLd - object describing element face load;
152     // returns loaded element face
153     // or -1 (loaded nodes do not match elem face)
154     public int equivFaceLoad(ElemFaceLoad surLd) {
155         // Shape functons
156         double an[] = new double[3];
157         // Derivatives of shape functions
158         double xin[] = new double[3];
159
160         for (int i=0; i<16; i++) evec[i] = 0.;
161         int loadedFace = surLd.rearrange(faceInd, ind);
162         if (loadedFace == -1) return -1;
163
164         // Gauss integration loop
165         for (int ip=0; ip<gf.nIntPoints; ip++){
166             ShapeQuad2D.shapeDerivFace(gf.xii[ip],
167                     surLd.faceNodes[1], an, xin);
```

```
168              double p = r = 0;
169              double xs = 0;
170              double ys = 0;
171              for (int i=0; i<3; i++){
172                  p  += an[i]*surLd.forceAtNodes[i];
173                  int j = faceInd[loadedFace][i];
174                  r  += an[i]*xy[j][0];
175                  xs += xin[i]*xy[j][0];
176                  ys += xin[i]*xy[j][1];
177              }
178              double dl = Math.sqrt(xs*xs+ys*ys);
179              double ds = dl;
180              if (FeModel.stressState==FeModel.StrStates.axisym)
181                  ds *= 2*Math.PI*r;
182              double p1, p2;
183              // direction=0 - normal load, =1,2 - along axes x,y
184              if (surLd.direction == 0 && ds > 0.0) {
185                  p1 = p*ys/dl;
186                  p2 = -p*xs/dl;
187              }
188              else if (surLd.direction == 1) { p1 = p; p2 = 0; }
189              else                           { p1 = 0; p2 = p; }
190
191              for (int i=0; i<3; i++){
192                  int j = faceInd[loadedFace][i];
193                  evec[2*j    ]  += an[i]*p1*ds*gf.wi[ip];
194                  evec[2*j + 1]  += an[i]*p2*ds*gf.wi[ip];
195              }
196          }
197      return loadedFace;
198   }
```

Description of a distributed load applied to an element side is contained in
surLd, an instance of class ElemFaceLoad (this class will be presented later in
relation to finite element load).

Array evec is first set to zero in line 160. Method rearrange is used to put
load information in order, corresponding to local element numbering (line 161).
Two arrays are employed for this purpose: faceInd containing local numbers for
element faces and ind – element connectivities.

The nodal equivalent of surface load is found by numerical integration using
one-dimensional integration rule gf with three Gauss points. The integration loop
is started in line 165. Method shapeDerivFace provides one-dimensional shape
functions an and derivatives of shape functions xin with respect to local coordinate
ξ changing along the considered element side. Shape functions are used to find
load intensity p and radial coordinate r at integration point ip (lines 171–174).
Derivatives $dx/d\xi$ (xs) and $dy/d\xi$ (ys) are found with the use of shape function
derivatives in lines 175–176. Variable ds in line 179 is a derivative of arc length
with respect to local coordinate $ds/d\xi$.

Object surLd contains the direction of surface load coded as an integer parameter. In two-dimensional problems the direction parameter can take the values: 0 –
load normal to the element side, 1– load along x, 2 – load along y. Load compo-

nents p1 along x and p2 along y are calculated in lines 188–189. Computation of the nodal equivalent of the distributed load and its placement in array evec is done in lines 191–195. In the case of normal completion the method returns a loaded face number, otherwise -1.

11.2.5 Equivalent Stress Vector

Method equivStressVector is designed for computation of a nodal force vector, which is equivalent to element stress field according to Equation 10.25. A negative sign is assigned to the resulting vector in order to estimate the difference between the global force vector and the global equivalent stress vector using an ordinary assembly algorithm.

```java
200     // Compute equivalent stress vector (with negative sign)
201     public void equivStressVector() {
202
203         for (int i = 0; i < 16; i++) evec[i] = 0.;
204         int ld = (FeModel.stressState
205                     == FeModel.StrStates.axisym) ? 4 : 3;
206
207         for (int ip = 0; ip < gs.nIntPoints; ip++) {
208             // Accumulated stress
209             double[] s = new double[4];
210             for (int i=0; i<4; i++)
211                 s[i] = str[ip].sStress[i] + str[ip].dStress[i];
212             // Set displacement differentiation matrix bmat
213             double det = setBmatrix(gs.xii[ip], gs.eti[ip]);
214             double dv = det*gs.wi[ip];
215             if (FeModel.stressState==FeModel.StrStates.axisym)
216                 dv *= 2*Math.PI*r;
217
218             for (int i=0; i<16; i++) {
219                 double a = 0;
220                 for (int j=0; j<ld; j++) a += bmat[j][i]*s[j];
221                 evec[i] -= a*dv;
222             }
223         }
224     }
```

The resulting equivalent stress vector is placed in vector evec, which is first set to zero in line 203. Integration of a product of the displacement differentiation matrix and the stress vector is performed with the use of the Gauss rule gs (2 by 2 integration points). In lines 210–211 the current level of stress s is calculated as a sum of the accumulated stress (before the beginning of this step) and the stress increment. Line 213 sets the displacement differentiation matrix at the current integration point with local coordinates gs.xii[ip] and gs.eti[ip]. A loop in lines 218–222 contains multiplication of the displacements differentiation matrix by the stress vector and its numerical integration with assignment of a negative sign.

11.2.6 Extrapolation from Integration Points to Nodes

Extrapolation of stresses from reduced integration points to element nodes is necessary for creation of a continuous stress field for the finite element model and its subsequent visualization. Stress extrapolation from integration points to nodes is done by method `extrapolateToNodes` according to Equation 10.33. The parameters of the method are stresses at integration points `fip[4][4]` (the first index is the integration point number) and stresses at nodes `fn[8][4]` (the first index is the local node number). The second indices of both parameters correspond to stress components. The results of extrapolation are placed in array `fn`.

```
226     // Extrapolate values from integration points to nodes.
227     // fip [4][4] - values at integration points;
228     // fn [8][4] - values at nodes (out)
229     public void extrapolateToNodes (double[][] fip,
230                                     double[][] fn) {
231         final double A = 1 + 0.5*Math.sqrt(3.),
232                      B = -0.5,
233                      C = 1 - 0.5*Math.sqrt(3.);
234         // Extrapolation matrix
235         final double lim[][] = {{A, B, B, C},
236                                 {B, C, A, B},
237                                 {C, B, B, A},
238                                 {B, A, C, B}};
239
240         for (int i = 1; i < 8; i+=2)
241             for (int j = 0; j < 4; j++) fn[i][j] = 0;
242
243         for (int corner = 0; corner < 4; corner++) {
244             int n = (corner==0) ? 7 : 2*corner-1;
245             for (int k = 0; k < 4; k++) {
246                 double c = 0.0;
247                 for (int ip = 0 ; ip < 4; ip++)
248                     c += lim[corner][ip]*fip[ip][k];
249                 fn[2*corner][k] = c;   // corner node
250                 fn[n][k] += 0.5*c;
251                 fn[2*corner+1][k] += 0.5*c;
252             }
253         }
254     }
```

The extrapolation matrix given by (10.34) is specified in lines 233–236 as a two-dimensional array `lim`. Lines 240–241 assign zeros to the resulting values at midside nodes. The loop in lines 243–253 contains multiplication of the extrapolation matrix `lim` by stress values at integration points `fip` according to Equation 10.33. Stresses at midside nodes are computed as averages of two neighboring corner node values (lines 250–251).

11.2.7 Other Methods

Below are source codes of the following element methods: getElemFaces, getStrainsAtIntPoint and getTemperatureAtIntPoint.

```
256        // Get local node numbers for element faces.
257        // returns elementFaces[nFaces][nNodesOnFace]
258        public int[][] getElemFaces() {
259            return faceInd;
260        }
261
262        // Get strains at integration point.
263        // ip - integration point number (stress);
264        // returns  strain vector (ex, ey, gxy, ez)
265        public double[] getStrainsAtIntPoint(int ip) {
266
267            // Set displacement differentiation matrix bmat
268            setBmatrix(gs.xii[ip], gs.eti[ip]);
269            double strain[] = new double[4];
270            for (int i=0; i<4; i++) {
271                strain[i] = 0;
272                for (int j=0; j<16; j++)
273                    strain[i] += bmat[i][j]*evec[j];
274            }
275            return strain;
276        }
277
278        // Get temperature at integration point (stress)
279        public double getTemperatureAtIntPoint(int ip) {
280            ShapeQuad2D.shape(gs.xii[ip], gs.eti[ip], ind, an);
281            double t = 0;
282            for (int i=0; i<8; i++) t += an[i]*dtn[i];
283            return t;
284        }
285
286 }
```

Method getElemFaces simply returns local numbers for four element sides specified in array faceInd (lines 10 and 11).

Method getStrainsAtIntPoint returns strains at the requested integration point ip. Strains are calculated for subsequent estimation of stresses, thus the displacement differentiation matrix is set at reduced integration points specified by the Gauss rule used for stresses (line 268). The resulting strains are obtained with the use of vector evec. Before using this method, element displacements should be placed in array evec.

Method getTemperatureAtIntPoint returns temperature at the reduced integration point ip, which is interpolated from nodal points using shape functions.

Problems

11.1. Study method `degeneration` in Section 11.1.1. What value will the method return if the following connectivity array is passed to it

    ind = {34, 35, 36, 46, 56, 56, 56, 45 }

as an input parameter? To what global node number does the returned value correspond?

11.2. Propose how to check the correctness of shape functions determined in method `shape` of class `ShapeQuad2D`. Use the properties of the shape functions. Write a Java code fragment for checking the shape functions that can be placed at the end of method `shape`.

11.3. Write the main method in class `ShapeQuad2D` that evaluates an area of the quadratic finite element shown below.

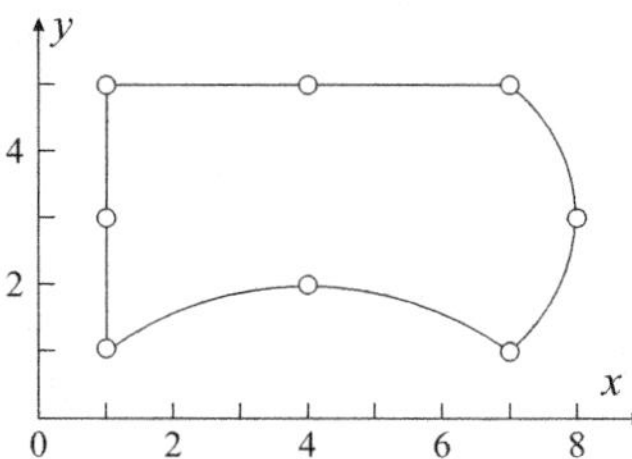

Use method `ShapeQuad2D.deriv` for calculation of the determinants of the Jacobian matrix. For area evaluation, perform 3 by 3 numerical integration using the class `GaussRule` described in Chapter 9.

11.4. Read and understand the source code of method `stiffnessMatrix` that computes the stiffness matrix for the two-dimensional isoparametric quadratic element. The double loop of lines 61–62 is such that the method computes the upper symmetrical part of the stiffness matrix. Modify the code in such a way that: 1) the full stiffness matrix is computed; 2) the lower symmetrical part of the stiffness matrix is computed.

11.5. Develop a main method for class `ElementQuad2D` that extrapolates values from reduced integration points to nodes.

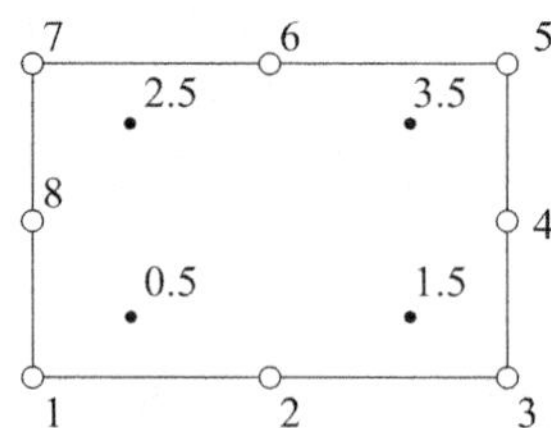

Use method `extrapolateToNodes`. Values at reduced integration points $\xi, \eta = \pm 1/\sqrt{3}$ are shown as numbers near the black dots. Results should be obtained at nodes 1–8.

Chapter 12
Three-dimensional Isoparametric Elements

Abstract Three-dimensional isoparametric elements are considered. Shape functions for hexahedral elements with eight nodes and twenty nodes are given. A method for computing derivatives of shape functions in the global coordinate system is presented. Numerical integration is used for estimation of element matrices and vectors. It is explained how to organize efficient computation of an element stiffness matrix. Algorithms for calculating nodal equivalents of surface loads and stresses at integration points and nodes are demonstrated.

12.1 Shape Functions

Hexahedral (or brick-type) linear eight-node and quadratic twenty-node three-dimensional isoparametric elements are depicted in Figure 12.1. The term "isoparametric" means that the geometry and displacement fields are specified in parametric form and are interpolated with the same functions. The shape functions used for interpolation are polynomials of the local coordinates ξ, η and ζ ($-1 \leq \xi, \eta, \zeta \leq 1$). Displacements are interpolated according to the following relations:

$$\{u\} = [N]\{q\},$$

$$\{u\} = \{u \ v \ w\},$$

$$\{q\} = \{u_1 \ v_1 \ w_1 \ u_2 \ v_2 \ w_2 \ ...\},$$

(12.1)

where $\{u\}$ is a displacement vector at a point with local coordinates ξ, η and ζ; u, v, w are displacement components along global coordinate axes x, y and z; $\{q\}$ is a nodal displacement vector with entries u_i, v_i, w_i (displacement values at nodes); $[N]$ is a matrix of shape functions.

Element shape is determined by coordinate interpolation using the matrix of shape functions $[N]$:

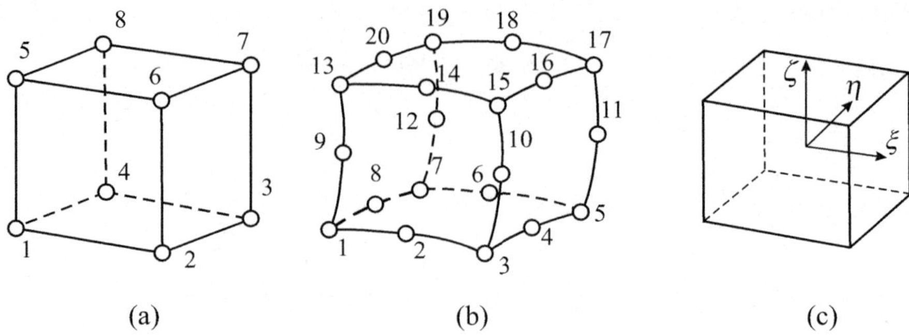

Fig. 12.1 Linear (a) and quadratic (b) three-dimensional finite elements and their representation in the local coordinate system (c)

$${x} = [N]{x^e},$$

$${x} = {x \ y \ z},$$ (12.2)

$${x^e} = {x_1 \ y_1 \ z_1 \ x_2 \ y_2 \ z_2 \ ...}.$$

Here, x, y, z are point coordinates and x_i, y_i, z_i are coordinates of nodes. The matrix of shape functions is defined as:

$$[N] = \begin{bmatrix} N_1 & 0 & 0 & N_2 & 0 & 0 & ... \\ 0 & N_1 & 0 & 0 & N_2 & 0 & ... \\ 0 & 0 & N_1 & 0 & 0 & N_2 & ... \end{bmatrix}.$$ (12.3)

The shape functions of the three-dimensional linear element are:

$$N_i = \frac{1}{8}(1 + \xi_0)(1 + \eta_0)(1 + \zeta_0),$$

$$\xi_0 = \xi \xi_i,$$ (12.4)

$$\eta_0 = \eta \eta_i,$$

$$\zeta_0 = \zeta \zeta_i.$$

A linear hexahedral element can be degenerated into a triangular prism (Figure 12.2a) by shrinking an element face into a straight line. Linear shape functions are not affected by this degeneration.

For the quadratic element with twenty nodes, the shape functions can be written in the following form:

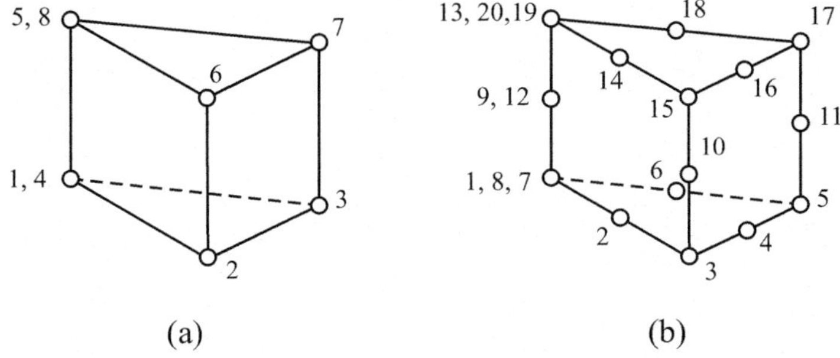

Fig. 12.2 Linear (a) and quadratic (b) elements degenerated into triangular prisms

$$N_i = \frac{1}{8}(1+\xi_0)(1+\eta_0)(1+\zeta_0)(\xi_0+\eta_0+\zeta_0-2) \quad \text{at vertices,}$$

$$N_i = \frac{1}{4}(1-\xi^2)(1+\eta_0)(1+\zeta_0)\,, \quad i=2,\,6,\,14,\,18,$$

$$N_i = \frac{1}{4}(1-\eta^2)(1+\xi_0)(1+\zeta_0)\,, \quad i=4,\,8,\,16,\,20,$$

$$N_i = \frac{1}{4}(1-\zeta^2)(1+\xi_0)(1+\eta_0)\,, \quad i=9,\,10,\,11,\,12. \tag{12.5}$$

In the above relations, ξ_i, η_i, ζ_i are values of local coordinates ξ, η, ζ at nodes.

For the degenerated quadratic hexahedral element shown in Figure 12.2b six shape functions for nodes located at a face opposite to the degenerated face should be modified. In the case of the degeneration depicted in Figure 12.2b the shape functions are modified in the following way [32]:

$$N_i' = N_i + \Delta, \quad i = 3,5,15,17,$$

$$N_i' = N_i - 2\Delta, \quad i = 4,16,$$

$$\Delta = \frac{1}{16}\left(1-\xi^2\right)\left(1-\eta^2\right)\left(1+\zeta\zeta_i\right). \tag{12.6}$$

12.2 Strain–Displacement Matrix

The strain vector $\{\varepsilon\}$ contains six different components of the strain tensor:

$$\{\varepsilon\} = \{\varepsilon_x \ \varepsilon_y \ \varepsilon_z \ \gamma_{xy} \ \gamma_{yz} \ \gamma_{zx}\}. \tag{12.7}$$

The strain–displacement matrix for three-dimensional elements is:

$$[B] = [D][N] = [B_1 \ B_2 \ B_3 \ ...], \tag{12.8}$$

$$[B_i] = \begin{bmatrix} \dfrac{\partial N_i}{\partial x} & 0 & 0 \\[2mm] 0 & \dfrac{\partial N_i}{\partial y} & 0 \\[2mm] 0 & 0 & \dfrac{\partial N_i}{\partial z} \\[2mm] \dfrac{\partial N_i}{\partial y} & \dfrac{\partial N_i}{\partial x} & 0 \\[2mm] 0 & \dfrac{\partial N_i}{\partial z} & \dfrac{\partial N_i}{\partial y} \\[2mm] \dfrac{\partial N_i}{\partial z} & 0 & \dfrac{\partial N_i}{\partial x} \end{bmatrix}. \tag{12.9}$$

The derivatives of three-dimensional shape functions with respect to global coordinates are obtained as follows:

$$\left\{ \begin{array}{c} \dfrac{\partial N_i}{\partial x} \\[2mm] \dfrac{\partial N_i}{\partial y} \\[2mm] \dfrac{\partial N_i}{\partial z} \end{array} \right\} = [J]^{-1} \left\{ \begin{array}{c} \dfrac{\partial N_i}{\partial \xi} \\[2mm] \dfrac{\partial N_i}{\partial \eta} \\[2mm] \dfrac{\partial N_i}{\partial \zeta} \end{array} \right\}, \tag{12.10}$$

where the Jacobian matrix has the appearance

$$[J] = \begin{bmatrix} \dfrac{\partial x}{\partial \xi} & \dfrac{\partial y}{\partial \xi} & \dfrac{\partial z}{\partial \xi} \\[2mm] \dfrac{\partial x}{\partial \eta} & \dfrac{\partial y}{\partial \eta} & \dfrac{\partial z}{\partial \eta} \\[2mm] \dfrac{\partial x}{\partial \zeta} & \dfrac{\partial y}{\partial \zeta} & \dfrac{\partial z}{\partial \zeta} \end{bmatrix}. \tag{12.11}$$

The partial derivatives of x, y, z with respect to ξ, η, ζ are found by differentiation of displacements expressed through shape functions and nodal displacement values:

$$\frac{\partial x}{\partial \xi} = \sum \frac{\partial N_i}{\partial \xi} x_i, \quad \frac{\partial x}{\partial \eta} = \sum \frac{\partial N_i}{\partial \eta} x_i, \quad \frac{\partial x}{\partial \zeta} = \sum \frac{\partial N_i}{\partial \zeta} x_i,$$

$$\frac{\partial y}{\partial \xi} = \sum \frac{\partial N_i}{\partial \xi} y_i, \quad \frac{\partial y}{\partial \eta} = \sum \frac{\partial N_i}{\partial \eta} y_i, \quad \frac{\partial y}{\partial \zeta} = \sum \frac{\partial N_i}{\partial \zeta} y_i, \tag{12.12}$$

$$\frac{\partial z}{\partial \xi} = \sum \frac{\partial N_i}{\partial \xi} z_i, \quad \frac{\partial z}{\partial \eta} = \sum \frac{\partial N_i}{\partial \eta} z_i, \quad \frac{\partial z}{\partial \zeta} = \sum \frac{\partial N_i}{\partial \zeta} z_i.$$

The transformation of integrals from the global coordinate system to the local coordinate system is performed with the use of the determinant of the Jacobian matrix:

$$dV = dx\,dy\,dz = |J|\,d\xi\,d\eta\,d\zeta. \tag{12.13}$$

12.3 Element Properties

Element matrices and vectors for the three-dimensional hexagonal element are given
by the following expressions:

stiffness matrix

$$[k] = \int_{-1}^{1} \int_{-1}^{1} \int_{-1}^{1} [B]^{\mathrm{T}}[E][B]|J|d\xi d\eta d\zeta, \tag{12.14}$$

thermal vector (fictitious forces to simulate thermal expansion)

$$\{h\} = \int_{-1}^{1} \int_{-1}^{1} \int_{-1}^{1} [B]^{\mathrm{T}}[E]\{\varepsilon^{t}\}|J|d\xi d\eta d\zeta, \tag{12.15}$$

force vector due to distributed face load

$$\{p\} = \int_{-1}^{1} \int_{-1}^{1} [N]^{\mathrm{T}}\{p^{S}\}\frac{ds}{d\xi}d\xi d\eta, \tag{12.16}$$

equivalent stress vector (with negative sign)

$$\{p_{\sigma}\} = -\int_{-1}^{1} \int_{-1}^{1} \int_{-1}^{1} [B]^{\mathrm{T}}\{\sigma\}|J|d\xi d\eta d\zeta. \tag{12.17}$$

In the above relations $[E]$ is the elasticity matrix of the material, $\{\varepsilon^{t}\}$ is the vector
of thermal strains and $\{\sigma\}$ is the current stress vector. Three-time application of the
one-dimensional Gauss quadrature rule leads to the following numerical integration
procedure:

$$\begin{aligned}
I &= \int_{-1}^{1} \int_{-1}^{1} \int_{-1}^{1} f(\xi, \eta, \zeta)d\xi d\eta d\zeta \\
&= \sum_{i=1}^{n} \sum_{j=1}^{n} \sum_{k=1}^{n} f(\xi_{i}, \eta_{j}, \zeta_{k})w_{i}w_{j}w_{k} \\
&= \sum_{m=1}^{N} f(\xi_{m}, \eta_{m}, \zeta_{m})W_{m}.
\end{aligned} \tag{12.18}$$

Here, $\xi_{i}, \eta_{j}, \zeta_{k}$ are the abscissas of numerical integrations; w_{i}, w_{j}, w_{k} are the corresponding weights. The third line of the above equation is a practical implementation
of three-dimensional integration rules as a single loop. In this case, $N = n \times n \times n$,
and W_{m} are triple products of the Gauss weights.

Usually, $2 \times 2 \times 2$ integration is used for three-dimensional linear elements and
$3 \times 3 \times 3$ integration is applied to evaluation of the stiffness matrix for quadratic elements. For more efficient integration, a special 14-point Gauss-type rule [14, 17] is
employed, which provides sufficient precision of integration for three-dimensional
quadratic elements.

12.4 Efficient Evaluation of Element Matrices and Vectors

Calculation of the element stiffness matrix by multiplication of three matrices involves many arithmetic operations with zeros. In addition, the three-dimensional case does not contain any additional variants of the elasticity matrix and displacement differentiation matrix as in the two-dimensional case. Because of this, a simple and efficient three-dimensional algorithm for the stiffness matrix can be formulated by performing matrix multiplications in closed form. Then, the coefficients of the element stiffness matrix $[k]$ are expressed as follows:

$$k_{ii}^{mn} = \int_V \left[\beta \frac{\partial N_m}{\partial x_i} \frac{\partial N_n}{\partial x_i} + \mu \left(\frac{\partial N_m}{\partial x_{i+1}} \frac{\partial N_m}{\partial x_{i+1}} + \frac{\partial N_m}{\partial x_{i+2}} \frac{\partial N_m}{\partial x_{i+2}} \right) \right] dV,$$

$$k_{ij}^{mn} = \int_V \left(\lambda \frac{\partial N_m}{\partial x_i} \frac{\partial N_n}{\partial x_j} + \mu \frac{\partial N_m}{\partial x_j} \frac{\partial N_n}{\partial x_i} \right) dV, \tag{12.19}$$

$$\beta = \lambda + 2\mu.$$

Here, λ and μ are elastic Lame constants; m, n are local node numbers; i, j are indices related to coordinate axes (x_1, x_2, x_3). The cyclic rule is employed in the above equation if the coordinate indices exceed 3.

The analogous expression for the element thermal vector is

$$h_i^m = \int_V (3\lambda + 2\mu)\, \alpha T \frac{\partial N_m}{\partial x_i} dV, \tag{12.20}$$

where m is node number, i is coordinate index, α is the thermal-expansion coefficient, and T is temperature.

A convenient expression for estimation of the nodal vector equivalent to the stress distribution is

$$(p_\sigma)_i^m = - \int_V \frac{\partial N_m}{\partial x_j} \sigma_{ij} dV. \tag{12.21}$$

Here, we suppose summation over repeated index j. The nodal vector is calculated with a negative sign for convenience of assembly.

12.5 Calculation of Nodal Equivalents for External Loads

The nodal equivalent of the external surface load is estimated according to Equation 3.20

$$\{p\} = \int_S [N]^T \{p^S\} dS. \tag{12.22}$$

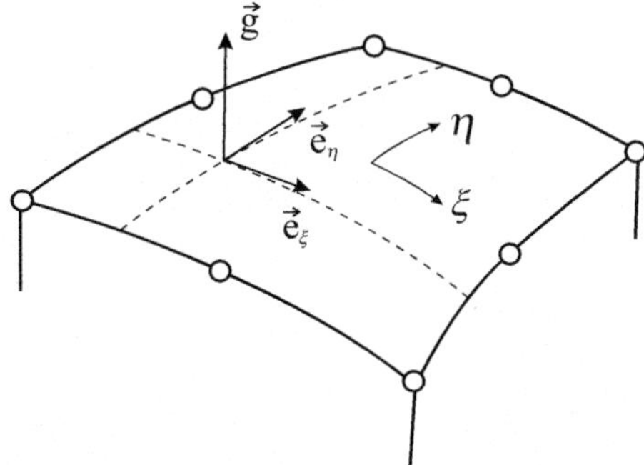

Fig. 12.3 Determination of a normal vector $\bar{g}$ on the curved face of the finite element

Integration of the nodal equivalent of the surface load (12.16) requires explanation related to expressing an area element dS through local coordinates ξ, η.

For integration of the surface load over the finite element face, it is necessary to determine the outward normal vector at any point of the face. Vectors $\mathbf{e}_\xi$ and $\mathbf{e}_\eta$ tangent to the local coordinates ξ and η have the following components in the global coordinate system (Figure 12.3):

$$\mathbf{e}_\xi = \left\{ \frac{\partial x}{\partial \xi} \; \frac{\partial y}{\partial \xi} \; \frac{\partial z}{\partial \xi} \right\},$$

$$\mathbf{e}_\eta = \left\{ \frac{\partial x}{\partial \eta} \; \frac{\partial y}{\partial \eta} \; \frac{\partial z}{\partial \eta} \right\}. \tag{12.23}$$

Derivatives of the global coordinates with respect to local coordinates can be estimated using two-dimensional shape functions:

$$\frac{\partial x}{\partial \xi} = \sum \frac{\partial N_m}{\partial \xi} x_m, \quad \ldots$$

$$\frac{\partial x}{\partial \eta} = \sum \frac{\partial N_m}{\partial \eta} x_m, \quad \ldots. \tag{12.24}$$

The normal vector $\mathbf{g}$ is equal to the vector product of the tangential vectors $\mathbf{e}_\xi$ and $\mathbf{e}_\eta$:

$$\mathbf{g} = \mathbf{e}_\xi \times \mathbf{e}_\eta. \tag{12.25}$$

The components of the outward normal vector $\mathbf{g}$ can be conveniently computed through the following expressions:

$$g_x = \frac{\partial y}{\partial \xi}\frac{\partial z}{\partial \eta} - \frac{\partial z}{\partial \xi}\frac{\partial y}{\partial \eta},$$

$$g_y = \frac{\partial z}{\partial \xi}\frac{\partial x}{\partial \eta} - \frac{\partial x}{\partial \xi}\frac{\partial z}{\partial \eta}, \qquad (12.26)$$

$$g_z = \frac{\partial x}{\partial \xi}\frac{\partial y}{\partial \eta} - \frac{\partial y}{\partial \xi}\frac{\partial x}{\partial \eta}.$$

Since integration of the surface load is performed in the local coordinate system ξ, η, the Jacobian of transformation between the physical surface coordinates and the local coordinates should be determined:

$$|J(\xi,\eta)| = |\mathbf{g}| = \sqrt{g_x^2 + g_y^2 + g_z^2}. \qquad (12.27)$$

The element of area dS is expressed through local coordinates ξ, η as

$$dS = |J(\xi,\eta)|\, d\xi d\eta, \qquad (12.28)$$

and the nodal equivalent for the surface load applied to a finite element face is calculated in the form

$$[p] = \int_{-1}^{1}\int_{-1}^{1} [N(\xi,\eta)]^{\mathrm{T}}\{p^S(\xi,\eta)\}\, |J(\xi,\eta)|\, d\xi d\eta. \qquad (12.29)$$

If the surface load is directed along the surface normal, the components of the unit normal vector (normalized vector $\mathbf{g}$) are used to produce components of the surface load at integration points.

12.6 Example: Nodal Equivalents of a Distributed Load

Estimate the nodal equivalents of a distributed load with constant intensity $p^s = 1$ applied to the flat square face with size $L = 1$ of the three-dimensional quadratic element shown in Figure 12.4.

Solution

The nodal equivalent of the distributed load can be calculated according to Equation 12.29. The calculation procedure is simplified if we take into account that the element face is a flat square with sides of length $L = 1$ parallel to coordinate axes x and y. In this particular case

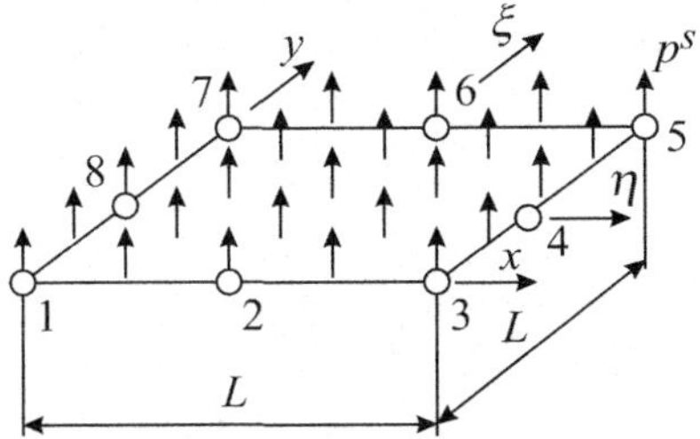

Fig. 12.4 Distributed load on a face of a three-dimensional quadratic element

$$\frac{dx}{d\xi} = \frac{1}{2}, \qquad \frac{dy}{d\eta} = \frac{1}{2}$$

and integration in (12.29) can be performed in a simpler way:

$$[p] = \int_{-1}^{1}\int_{-1}^{1} [N]^{\mathrm{T}}\{p^{S}\}\,\frac{1}{4}d\xi\,d\eta.$$

From symmetry properties, the equivalent nodal forces have identical values at all corner nodes and at all midside nodes. It is sufficient to determine the nodal equivalents at nodes 1 and 2.

Using two-dimensional shape functions $[N_1]$ and $[N_2]$ (10.7) nodal equivalents for corner node 1 and midside node 2 are determined as:

$$p_1 = -\int_{-1}^{1}\int_{-1}^{1}\frac{1}{4}(1-\xi)(1-\eta)(1+\xi+\eta)\frac{1}{4}d\xi\,d\eta = -\frac{1}{12},$$

$$p_2 = \int_{-1}^{1}\int_{-1}^{1}\frac{1}{2}(1-\xi^2)(1-\eta)\frac{1}{4}d\xi\,d\eta = \frac{1}{3}.$$

Nodal equivalents of the distributed load with constant intensity for the face of the three-dimensional quadratic element are shown in Figure 12.5.

At first glance it seems unusual to observe nodal equivalents at corner nodes directed in the opposite direction of the applied load. After counting the total force

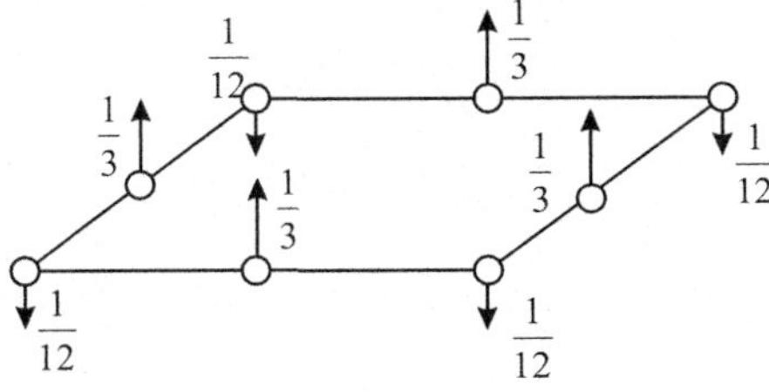

Fig. 12.5 Distributed load on a face of a three-dimensional quadratic element

across the face, we find that it is equal to unity confirming our calculations. Negative directions of corner nodal forces are related to nonlinearity of shape functions.

12.7 Calculation of Strains and Stresses

After computing element matrices and vectors, the assembly process is used to compose the global equation system. Solution of the global equation system provides displacements at nodes of the finite element model. Using disassembly, nodal displacement for each element can be obtained.

Strains inside an element are determined with the use of the displacement differentiation matrix:

$$\{\varepsilon\} = [B]\{q\}. \tag{12.30}$$

Stresses are calculated with Hooke's law:

$$\{\sigma\} = [E]\{\varepsilon^e\} = [E](\{\varepsilon\} - \{\varepsilon^t\}), \tag{12.31}$$

where $\{\varepsilon^t\}$ is the vector of free thermal expansion:

$$\{\varepsilon^t\} = \{\alpha T \quad \alpha T \quad \alpha T \quad 0 \quad 0 \quad 0\}. \tag{12.32}$$

It should be noted that displacement gradients (and hence strains and stresses) have quite difference precision at different points inside finite elements. The highest precisions for displacement gradients are at the geometric center for the linear element and at reduced integration points $2 \times 2 \times 2$ for the quadratic hexagonal element.

12.8 Extrapolation of Strains and Stresses

For quadratic elements, displacement derivatives have the best precision at $2 \times 2 \times 2$ integration points with local coordinates ξ, η, $\zeta = \pm 1/\sqrt{3}$. In order to build a continuous field of strains or stresses, it is necessary to extrapolate result values from $2 \times 2 \times 2$ integration points to the vertices of a twenty-node element (numbering of integration points and vertices is shown in Figure 12.6).

Results are calculated at 8 integration points, and a trilinear extrapolation in the local coordinate system ξ, η, ζ is used:

$$f_i = L_{i(m)} f_{(m)}, \tag{12.33}$$

where $f_{(m)}$ are known function values at reduced integration points, f_i are function values at vertex nodes, and $L_{i(m)}$ is the extrapolation matrix:

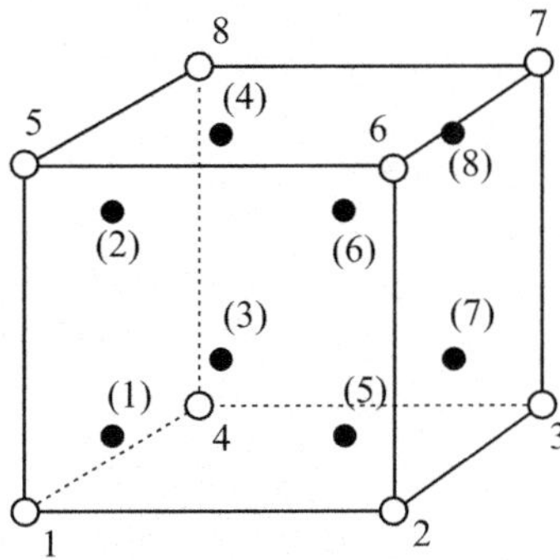

Fig. 12.6 Numbering of integration points and vertices for the twenty-node isoparametric element

$$
L_{i(m)} =
\begin{bmatrix}
A & B & B & C & B & C & C & D \\
B & C & C & D & A & B & B & C \\
C & D & B & C & B & C & A & B \\
B & C & A & B & C & D & B & C \\
B & A & C & B & C & B & D & C \\
C & B & D & C & B & A & C & B \\
D & C & C & B & C & B & B & A \\
C & B & B & A & D & C & C & B
\end{bmatrix},
\tag{12.34}
$$

$$
A = \frac{5 + \sqrt{3}}{4}, \quad B = -\frac{\sqrt{3} + 1}{4},
$$

$$
C = \frac{\sqrt{3} - 1}{4}, \quad D = \frac{5 - \sqrt{3}}{4}.
$$

Stresses are extrapolated from integration points to all nodes of elements. Values for midside nodes can be calculated as an average between values for two vertex nodal values. Then averaging of contributions from the neighboring finite elements is performed for all nodes of the finite element model. Averaging produces a continuous field of secondary results specified at nodes of the model with quadratic variation inside finite elements. Later, the results can be interpolated to any point inside an element or on its surface using quadratic shape functions.

Problems

12.1. Derive explicit expressions for shape functions N_6 and N_{15} for the three-dimensional quadratic element. Use the general relations (12.5).

12.2. Show that three-dimensional linear shape functions (12.4) satisfy the equality

$$\sum_{i=1}^{8} N_i = 1$$

at any point ξ, η, ζ.

12.3. Obtain the derivatives of corner node shape functions with respect to local coordinates ξ, η, and ζ for the three-dimensional quadratic element. Relations for the shape functions are given by (12.5).

12.4. Estimate the Jacobian matrix $[J]$ and its determinant $|J|$ for the following eight-node element.

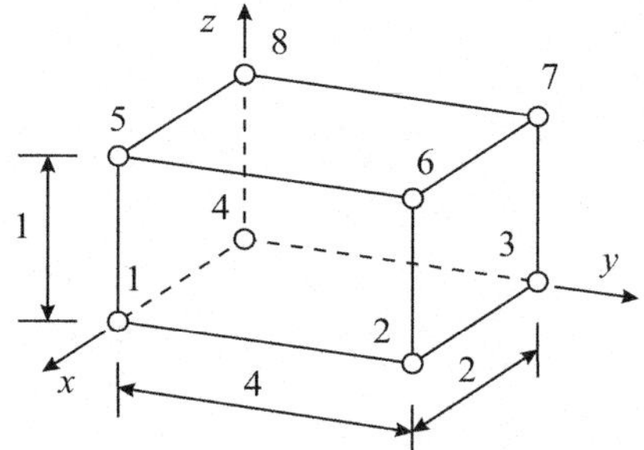

The element has dimensions 2, 4, and 1 along coordinate axes x, y, and z, respectively.

12.5. Show that the index form equation for evaluation of thermal vector components h_i^m (12.20) in the three-dimensional case follows from matrix Equation 12.15.

Chapter 13
Implementation of Three-dimensional Quadratic Element

Abstract Java^{TM} classes implementing a three-dimensional hexagonal element with twenty nodes are presented. Class ShapeQuad3D computes shape functions and their derivatives. Class ElementQuad3D provides methods for computation of element matrices and vectors including element stiffness matrix, thermal vector, nodal equivalent of distributed load, and equivalent stress vector.

13.1 Class for Shape Functions and Their Derivatives

Class ShapeQuad3D includes methods for calculating shape functions and their derivatives for three-dimensional isoparametric elements of hexahedral shapes. The maximum number of nodes in this element is twenty. Any node located at a center of an element edge can be absent. Thus, it is possible to construct elements with number of nodes from ten to twenty. When the element has twenty nodes, interpolation of displacements and coordinates is quadratic. For eight-node elements, interpolation is linear. Local node numbers for a twenty-node element are shown in Figure 13.1. If a midside node is absent, the corresponding connectivity number is set to zero.

13.1.1 Element Degeneration

A hexahedral element can have many types of degeneration. In order to keep the length and complexity of the code reasonable, we consider just one degeneration, presented in Figure 13.2. Such degeneration is achieved when the following three groups of nodes

 0, 7, 6,
 8, 11, and
 12, 19, 18

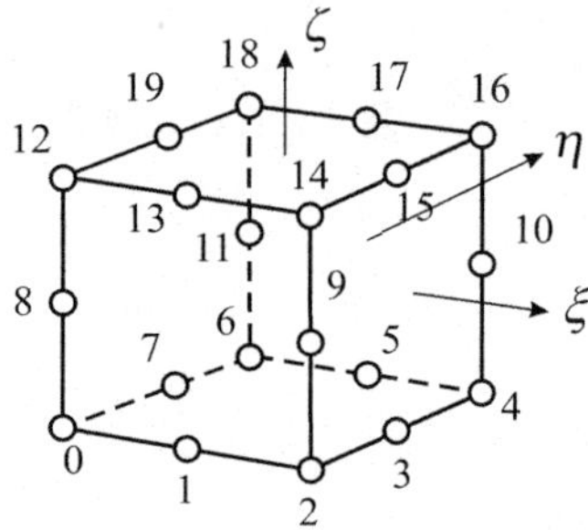

Fig. 13.1 Local numbering of nodes for the hexahedral quadratic isoparametric element

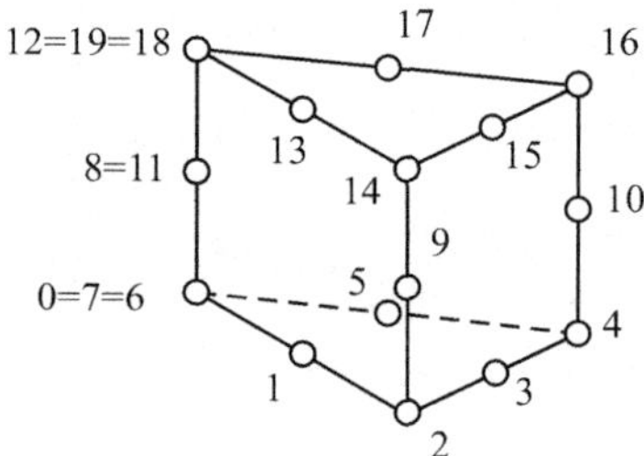

Fig. 13.2 Triangular prism obtained by degeneration of hexahedral element

have the same connectivity numbers, and each group its own connectivity number.

Method degeneration below, just checks the coincidence of connectivity numbers (array ind) for three groups of nodes. It is supposed that an element has all midside nodes and connectivity numbers do not create other degenerations except the above listed. The method returns 1 if the allowed degeneration is discovered, otherwise it returns 0.

```
 1 package elem;
 2
 3 import util.UTIL;
 4
 5 // Quadratic 3D shape functions and their derivatives
 6 public class ShapeQuad3D {
 7
 8      // Degeneration check.
 9      // The only degeneration is: 0=7=6,8=11,12=19=18
10      static int degeneration(int[] ind) {
11          // Element should be quadratic
12          if ((ind[0] == ind[7] && ind[7] == ind[6]) &&
13              (ind[8] == ind[11]) &&
14              (ind[12] == ind[19] && ind[19] == ind[18]))
15              return 1;
16          else return 0;
17      }
```

13.1.2 Shape Functions

Shape functions are computed by method shape. The method obtains values of local coordinates xi (ξ), et (η) and ze (ζ) and element connectivities ind. Connectivities allow detection of the absence of midside nodes and determination of degenerate collapse of the element into a triangular prism. Array an passed as a parameter contains results for shape functions after a call to method shape.

```
19        // Shape functions.
20        // xi, et, ze - local coordinates;
21        // ind - element connectivities;
22        // n - shape functions (out)
23        static void shape(double xi, double et, double ze,
24                          int[] ind, double[] n) {
25            double s0 = 1 + xi;
26            double t0 = 1 + et;
27            double d0 = 1 + ze;
28            double s1 = 1 - xi;
29            double t1 = 1 - et;
30            double d1 = 1 - ze;
31            double s2 = 1 - xi*xi;
32            double t2 = 1 - et*et;
33            double d2 = 1 - ze*ze;
34            // Midside nodes
35            n[1] = n[3] = n[5] = n[7] = n[8] = n[9] = n[10] =
36                    n[11] = n[13] = n[15] = n[17] = n[19] = 0;
37            if (ind[ 1] > 0) n[1]  = 0.25*s2*t1*d1;
38            if (ind[ 5] > 0) n[5]  = 0.25*s2*t0*d1;
39            if (ind[17] > 0) n[17] = 0.25*s2*t0*d0;
40            if (ind[13] > 0) n[13] = 0.25*s2*t1*d0;
41            if (ind[ 7] > 0) n[7]  = 0.25*t2*s1*d1;
42            if (ind[ 3] > 0) n[3]  = 0.25*t2*s0*d1;
43            if (ind[15] > 0) n[15] = 0.25*t2*s0*d0;
44            if (ind[19] > 0) n[19] = 0.25*t2*s1*d0;
45            if (ind[ 8] > 0) n[8]  = 0.25*d2*s1*t1;
46            if (ind[ 9] > 0) n[9]  = 0.25*d2*s0*t1;
47            if (ind[10] > 0) n[10] = 0.25*d2*s0*t0;
48            if (ind[11] > 0) n[11] = 0.25*d2*s1*t0;
49            // Vertex nodes
50            n[ 0] = 0.125*s1*t1*d1 - 0.5*(n[ 1]+n[ 7]+n[ 8]);
51            n[ 2] = 0.125*s0*t1*d1 - 0.5*(n[ 1]+n[ 3]+n[ 9]);
52            n[ 4] = 0.125*s0*t0*d1 - 0.5*(n[ 3]+n[ 5]+n[10]);
53            n[ 6] = 0.125*s1*t0*d1 - 0.5*(n[ 5]+n[ 7]+n[11]);
54            n[12] = 0.125*s1*t1*d0 - 0.5*(n[ 8]+n[13]+n[19]);
55            n[14] = 0.125*s0*t1*d0 - 0.5*(n[ 9]+n[13]+n[15]);
56            n[16] = 0.125*s0*t0*d0 - 0.5*(n[10]+n[15]+n[17]);
57            n[18] = 0.125*s1*t0*d0 - 0.5*(n[11]+n[17]+n[19]);
58            // Modification of functions due to degeneration
59            if (degeneration(ind) == 1) {
60                double dn1 = 0.0625*s2*t2*d1;
61                double dn2 = 0.0625*s2*t2*d0;
62                n[2] += dn1;
63                n[3] -= 2*dn1;
```

```
64                 n[4]  += dn1;
65                 n[14] += dn2;
66                 n[15] -= 2*dn2;
67                 n[16] += dn2;
68             }
69         }
```

Shape functions for midside nodes are first initialized to zero. A shape function for a midside node is evaluated if the connectivity number for this node is nonzero (lines 37–48). Shape functions for vertex nodes are computed in lines 50–57 as combinations of linear shape functions and quadratic shape functions of three neighboring midside nodes. Possible degeneration of the element is taken into account in lines 59–68.

13.1.3 Derivatives of Shape Functions

Method `deriv` computes the derivatives of shape functions with respect to global coordinates `dnxy` at a point with local coordinates `xi`, `et`, and `ze` (ξ, η and ζ). The method also obtains `ind` – element connectivities, `xy` – array of nodal coordinates.

```
71      // Derivatives of shape functions
72      //    with respect to global coordinates xy.
73      // xi, et, ze - local coordinates;
74      // ind - element connectivities;
75      // xy - nodal coordinates;
76      // dnxy - derivatives of shape functions (out);
77      // returns  determinant of the Jacobian matrrix
78      static double deriv(double xi, double et, double ze,
79                     int[] ind, double[][] xy, double[][] dnxy) {
80          // Derivatives with respect to local coordinates d
81          double[][] d = new double[20][3];
82          double s0 = 1 + xi;
83          double t0 = 1 + et;
84          double d0 = 1 + ze;
85          double s1 = 1 - xi;
86          double t1 = 1 - et;
87          double d1 = 1 - ze;
88          double s2 = 1 - xi*xi;
89          double t2 = 1 - et*et;
90          double d2 = 1 - ze*ze;
91          // Midside nodes
92          if (ind[1] > 0) { d[1][0] = -0.5*xi*t1*d1;
93                            d[1][1] = -0.25*s2*d1;
94                            d[1][2] = -0.25*s2*t1;
95          } else { d[0][1] = d[1][1] = d[2][1] = 0;   }
96
97          if (ind[5] > 0) { d[5][0] = -0.5*xi*t0*d1;
98                            d[5][1] =  0.25*s2*d1;
99                            d[5][2] = -0.25*s2*t0;
100         } else { d[5][0] = d[5][1] = d[5][2] = 0;   }
```

```
101
102     if (ind[17] > 0) { d[17][0] = -0.5*xi*t0*d0;
103                        d[17][1] =  0.25*s2*d0;
104                        d[17][2] =  0.25*s2*t0;
105     } else {  d[17][0] = d[17][1] = d[17][2] = 0;   }
106
107     if (ind[13] > 0) { d[13][0] = -0.5*xi*t1*d0;
108                        d[13][1] = -0.25*s2*d0;
109                        d[13][2] =  0.25*s2*t1;
110     } else { d[13][0] = d[13][1] = d[13][2] = 0;   }
111
112     if (ind[7] > 0) {  d[7][0] = -0.25*t2*d1;
113                        d[7][1] = -0.5*et*s1*d1;
114                        d[7][2] = -0.25*t2*s1;
115     } else { d[7][0] = d[7][1] = d[7][2] = 0;   }
116
117     if (ind[3] > 0) {  d[3][0] =  0.25*t2*d1;
118                        d[3][1] = -0.5*et*s0*d1;
119                        d[3][2] = -0.25*t2*s0;
120     } else { d[3][0] = d[3][1] = d[3][2] = 0;   }
121
122     if (ind[15] > 0) { d[15][0] =  0.25*t2*d0;
123                        d[15][1] = -0.5*et*s0*d0;
124                        d[15][2] =  0.25*t2*s0;
125     } else { d[15][0] = d[15][1] = d[15][2] = 0;   }
126
127     if (ind[19] > 0) { d[19][0] = -0.25*t2*d0;
128                        d[19][1] = -0.5*et*s1*d0;
129                        d[19][2] =  0.25*t2*s1;
130     } else { d[19][0] = d[19][1] = d[19][2] = 0;   }
131
132     if (ind[8] > 0) {  d[8][0] = -0.25*d2*t1;
133                        d[8][1] = -0.25*d2*s1;
134                        d[8][2] = -0.5*ze*s1*t1;
135     } else { d[8][0] = d[8][1] = d[8][2] = 0;   }
136
137     if (ind[9] > 0) {  d[9][0] =  0.25*d2*t1;
138                        d[9][1] = -0.25*d2*s0;
139                        d[9][2] = -0.5*ze*s0*t1;
140     } else { d[9][0] = d[9][1] = d[9][2] = 0;   }
141
142     if (ind[10] > 0) { d[10][0] =  0.25*d2*t0;
143                        d[10][1] =  0.25*d2*s0;
144                        d[10][2] = -0.5*ze*s0*t0;
145     } else { d[10][0] = d[10][1] = d[10][2] = 0; }
146
147     if (ind[11] > 0) { d[11][0] = -0.25*d2*t0;
148                        d[11][1] =  0.25*d2*s1;
149                        d[11][2] = -0.5*ze*s1*t0;
150     } else { d[11][0] = d[11][1] = d[11][2] = 0; }
151     // Vertex nodes
152     d[0][0]=-0.125*t1*d1-0.5*(d[1][0]+d[7][0]+d[8][0]);
153     d[0][1]=-0.125*s1*d1-0.5*(d[1][1]+d[7][1]+d[8][1]);
154     d[0][2]=-0.125*s1*t1-0.5*(d[1][2]+d[7][2]+d[8][2]);
```

```
155
156     d[2] [0]= 0.125*t1*d1-0.5*(d[1] [0]+d[3] [0]+d[9] [0]);
157     d[2] [1]=-0.125*s0*d1-0.5*(d[1] [1]+d[3] [1]+d[9] [1]);
158     d[2] [2]=-0.125*s0*t1-0.5*(d[1] [2]+d[3] [2]+d[9] [2]);
159
160     d[4] [0]= 0.125*t0*d1-0.5*(d[3] [0]+d[5] [0]+d[10][0]);
161     d[4] [1]= 0.125*s0*d1-0.5*(d[3] [1]+d[5] [1]+d[10][1]);
162     d[4] [2]=-0.125*s0*t0-0.5*(d[3] [2]+d[5] [2]+d[10][2]);
163
164     d[6] [0]=-0.125*t0*d1-0.5*(d[5] [0]+d[7] [0]+d[11][0]);
165     d[6] [1]= 0.125*s1*d1-0.5*(d[5] [1]+d[7] [1]+d[11][1]);
166     d[6] [2]=-0.125*s1*t0-0.5*(d[5] [2]+d[7] [2]+d[11][2]);
167
168     d[12][0]=-0.125*t1*d0-0.5*(d[8] [0]+d[13][0]+d[19][0]);
169     d[12][1]=-0.125*s1*d0-0.5*(d[8] [1]+d[13][1]+d[19][1]);
170     d[12][2]= 0.125*s1*t1-0.5*(d[8] [2]+d[13][2]+d[19][2]);
171
172     d[14][0]= 0.125*t1*d0-0.5*(d[9] [0]+d[13][0]+d[15][0]);
173     d[14][1]=-0.125*s0*d0-0.5*(d[9] [1]+d[13][1]+d[15][1]);
174     d[14][2]= 0.125*s0*t1-0.5*(d[9] [2]+d[13][2]+d[15][2]);
175
176     d[16][0]= 0.125*t0*d0-0.5*(d[10][0]+d[15][0]+d[17][0]);
177     d[16][1]= 0.125*s0*d0-0.5*(d[10][1]+d[15][1]+d[17][1]);
178     d[16][2]= 0.125*s0*t0-0.5*(d[10][2]+d[15][2]+d[17][2]);
179
180     d[18][0]=-0.125*t0*d0-0.5*(d[11][0]+d[17][0]+d[19][0]);
181     d[18][1]= 0.125*s1*d0-0.5*(d[11][1]+d[17][1]+d[19][1]);
182     d[18][2]= 0.125*s1*t0-0.5*(d[11][2]+d[17][2]+d[19][2]);
183     // Modification of derivatives due to degeneration
184     if (degeneration(ind) == 1) {
185         double dn1[] = new double[3];
186         double dn2[] = new double[3];
187         dn1[0] = -0.125*xi*t2*d1;
188         dn1[1] = -0.125*et*s2*d1;
189         dn1[2] = -0.0625*s2*t2;
190         dn2[0] = -0.125*xi*t2*d0;
191         dn2[1] = -0.125*et*s2*d0;
192         dn2[2] = -dn1[2];
193         for (int i = 0; i < 3; i++) {
194             d[2] [i] += dn1[i];
195             d[3] [i] -= 2*dn1[i];
196             d[4] [i] += dn1[i];
197             d[14][i] += dn2[i];
198             d[15][i] -= 2*dn2[i];
199             d[16][i] += dn2[i];
200         }
201     }
202     // Jacobian matrix ja
203     double[][] ja = new double[3][3];
204     for (int i = 0; i < 3; i++)
205         for (int j = 0; j < 3; j++) {
206             ja[j][i] = 0.;
207             for (int k = 0; k < 20; k++)
208                 ja[j][i] += d[k][i]*xy[k][j];
```

```
209                    }
210            // Determinant of Jacobian matrix det
211            double det =
212                ja[0][0]*(ja[1][1]*ja[2][2] - ja[2][1]*ja[1][2])
213              - ja[1][0]*(ja[0][1]*ja[2][2] - ja[0][2]*ja[2][1])
214              + ja[2][0]*(ja[0][1]*ja[1][2] - ja[0][2]*ja[1][1]);
215            if (det<=0) UTIL.errorMsg("Negative/zero Jacobian "+
216                "determinant for 20N element "+(float)det);
217            // Jacobian inverse ja1
218            double[][] ja1 = new double[3][3];
219            double v = 1.0/det;
220            ja1[0][0] = (ja[1][1]*ja[2][2]-ja[2][1]*ja[1][2])*v;
221            ja1[1][0] = (ja[0][2]*ja[2][1]-ja[0][1]*ja[2][2])*v;
222            ja1[2][0] = (ja[0][1]*ja[1][2]-ja[0][2]*ja[1][1])*v;
223            ja1[0][1] = (ja[1][2]*ja[2][0]-ja[1][0]*ja[2][2])*v;
224            ja1[1][1] = (ja[0][0]*ja[2][2]-ja[2][0]*ja[0][2])*v;
225            ja1[2][1] = (ja[0][2]*ja[1][0]-ja[0][0]*ja[1][2])*v;
226            ja1[0][2] = (ja[2][1]*ja[1][0]-ja[2][0]*ja[1][1])*v;
227            ja1[1][2] = (ja[0][1]*ja[2][0]-ja[0][0]*ja[2][1])*v;
228            ja1[2][2] = (ja[0][0]*ja[1][1]-ja[1][0]*ja[0][1])*v;

230            for (int k = 0; k < 20; k++)
231                for (int i = 0; i < 3; i++) {
232                    dnxy[k][i] = 0.;
233                    for (int j = 0; j < 3; j++)
234                        dnxy[k][i] += ja1[i][j]*d[k][j];
235                }
236            return det;
237        }
```

Array d declared in line 81 contains the derivatives of shape functions with respect to local coordinates ξ, η, ζ. After evaluation and modification (due to possible element degeneration) of derivatives d, Jacobian matrix ja is determined in lines 203–209. A determinant of the Jacobian matrix det is evaluated in lines 211–214. The error message method UTIL.errorMsg is called if the determinant is not positive, indicating that the element data has errors. The inverse of the Jacobian matrix is found in lines 218–228. The array of shape function derivatives with respect to global coordinates dnxy is determined in lines 230–235. The method returns det, the determinant of the Jacobian matrix.

13.1.4 Shape Functions and Their Derivatives for an Element Face

Two-dimensional shape functions and their derivatives are necessary for treating distributed loads applied to faces of three-dimensional isoparametric elements. While methods performing similar computations exist in the class for two-dimensional quadratic element, it is better to implement a similar method here since element classes should be independent of each other.

Computations of two-dimensional quadratic shape functions and their derivatives are performed by method `shapeDerivFace` presented below. It evaluates shape functions an and their derivatives dn with respect to local coordinates for a specified local coordinated ξ, η. As usual, face connectivity numbers are used for detecting midside nodes and for checking face degeneration into a triangle.

```
239    // Two-dimensional shape functions and derivatives
240    //    for a face of 3d 8-20n element.
241    // xi, et - local coordinates;
242    // ind - element connectivities;
243    // an - shape functions (out);
244    // dn  derivatives of shape functions
245    //      with respect to xi and et (out)
246    public static void shapeDerivFace(double xi, double et,
247                        int[] ind, double[] an, double[][] dn) {
248        double x1 = 1 - xi;
249        double x2 = 1 + xi;
250        double e1 = 1 - et;
251        double e2 = 1 + et;
252        // Shape funcrions for midside nodes
253        an[1] = an[3] = an[5] = an[7] = 0;
254        if (ind[1]>0) an[1] =  x1*x2*e1*0.5;
255        if (ind[3]>0) an[3] =  e1*e2*x2*0.5;
256        if (ind[5]>0) an[5] =  x1*x2*e2*0.5;
257        if (ind[7]>0) an[7] =  e1*e2*x1*0.5;
258        // Shape functions for corner nodes
259        an[0] = x1*e1*0.25 - 0.5*(an[7]+an[1]);
260        an[2] = x2*e1*0.25 - 0.5*(an[1]+an[3]);
261        an[4] = x2*e2*0.25 - 0.5*(an[3]+an[5]);
262        an[6] = x1*e2*0.25 - 0.5*(an[5]+an[7]);
263
264        dn[1][0] = dn[1][1] = dn[3][0] = dn[3][1] =
265            dn[5][0] = dn[5][1] = dn[7][0] = dn[7][1] = 0;
266        // Derivatives for midside nodes
267        if (ind[1]>0) {
268            dn[1][0] = -xi*e1; dn[1][1] = -0.5*x1*x2;}
269        if (ind[3]>0) {
270            dn[3][0] = 0.5*e1*e2;  dn[3][1] = -et*x2;}
271        if (ind[5]>0) {
272            dn[5][0] = -xi*e2;  dn[5][1] = 0.5*x1*x2;}
273        if (ind[7]>0) {
274            dn[7][0] = -0.5*e1*e2; dn[7][1] = -et*x1;}
275        // Derivatives for corner nodes
276        dn[0][0] = -0.25*e1 - 0.5*(dn[7][0]+dn[1][0]);
277        dn[0][1] = -0.25*x1 - 0.5*(dn[7][1]+dn[1][1]);
278        dn[2][0] =  0.25*e1 - 0.5*(dn[1][0]+dn[3][0]);
279        dn[2][1] = -0.25*x2 - 0.5*(dn[1][1]+dn[3][1]);
280        dn[4][0] =  0.25*e2 - 0.5*(dn[3][0]+dn[5][0]);
281        dn[4][1] =  0.25*x2 - 0.5*(dn[3][1]+dn[5][1]);
282        dn[6][0] = -0.25*e2 - 0.5*(dn[5][0]+dn[7][0]);
283        dn[6][1] =  0.25*x1 - 0.5*(dn[5][1]+dn[7][1]);
284
285        // Degeneration check
286        int ig = 0;
```

```
287            for (int i = 0; i < 7; i += 2) {
288                if (ind[i] == ind[i + 1]) {
289                    ig = (i + 5) % 8;
290                    break;
291                }
292            }
293            if (ig>0&&ind[1]>0&&ind[3]>0&&ind[5]>0&&ind[7]>0){
294                double delta = 0.125*x1*x2*e1*e2;
295                double z = -0.25*xi*e1*e2;
296                double t = -0.25*x1*x2*et;
297                int j = (ig + 1)%8;
298                an[ig-1] +=  delta;
299                an[ig]   -=  2.*delta;
300                an[j]    +=  delta;
301                dn[ig-1][0] += z;
302                dn[ig-1][1] += t;
303                dn[ig][0]    -= 2*z;
304                dn[ig][1]    -= 2*t;
305                dn[j][0]     += z;
306                dn[j][1]     += t;
307            }
308        }
309
310 }
```

If a midside node is absent, zeros are assigned to the corresponding shape function and its derivatives. Shape functions and derivatives of corner nodes combine linear corner functions and quadratic contributions from neighboring midside nodes.

Element degeneration is checked in lines 286–292. If a connectivity number `ind[i]` for a corner node is equal to a connectivity number `ind[i + 1]` for an intermediate node we consider the side as having collapsed into a point and the element face as having collapsed into a triangle. Modification of shape functions and derivatives due to degeneration is done only for a fully quadratic element face with eight nodes. This condition is checked in line 293.

13.2 Class for Twenty-node Element

Class `ElementQuad3D` contains methods that are necessary for implementing three-dimensional quadratic hexahedral ("brick-type") elements with the number of nodes from eight to twenty.

```
1 package elem;
2
3 import model.*;
4 import material.*;
5 import util.*;
6
7 // 3D 8-20 node isoparametric brick-type element.
8 public class ElementQuad3D extends Element {
9
```

```
10      private static int[][] faceInd = {
11          { 0, 8,12,19,18,11, 6, 7},
12          { 2, 3, 4,10,16,15,14, 9},
13          { 0, 1, 2, 9,14,13,12, 8},
14          { 4, 5, 6,11,18,17,16,10},
15          { 0, 7, 6, 5, 4, 3, 2, 1},
16          {12,13,14,15,16,17,18,19} };
17      private static double[] an = new double[20];
18      private static double[][] dnxy = new double[20][3];
19      // Gauss rules for stiffness matrix, thermal vector,
20      // surface load and stress integration
21      private static GaussRule gk = new GaussRule(14,3);
22      private static GaussRule gh = new GaussRule(3,3);
23      private static GaussRule gf = new GaussRule(3,2);
24      private static GaussRule gs = new GaussRule(2,3);
25
26      // Constructor for 3D 20 node element.
27      public ElementQuad3D() {
28          super ("hex20", 20, 8);
29      }
```

Information on element faces is specified in array `faceInd` (lines 10–16). Each face is defined by eight local node numbers (see Figure 13.1). The order of nodes is such that nodes are listed in an anticlockwise direction looking from the outer normal to the face. The starting node is located at any corner of the face. Arrays `an` and `dnxy` (lines 17–18) allocate memory for storing element shape functions and derivatives of shape functions with respect to global coordinates.

Lines 21–24 construct the Gauss rules for integration of the stiffness matrix (`gk`, 14-point rule), thermal vector (`gh`, $3 \times 3 \times 3$ rule), face load (`gf`, 3×3), and for integration stresses (`gs`, $2 \times 2 \times 2$).

Constructor `ElementQuad3D` contains a call to the constructor of the parent class `Element` with three parameters: element name "hex20", the number of element nodes (20) and the number of points for storing element stresses (8).

13.2.1 Stiffness Matrix

Method `stiffnessMatrix` computes a stiffness matrix for a three-dimensional hexaxedral element. Instead of matrix multiplications done in the two-dimensional case, here we perform direct computation of stiffness matrix coefficients according to Equation 12.19, as shown below.

```
31      // Compute stiffness matrix
32      public void stiffnessMatrix() {
33
34          for (int i = 0; i < 60; i++)
35              for (int j = i; j < 60; j++) kmat[i][j] = 0.;
36          // Material mat
37          mat = (Material)fem.materials.get(matName);
38          if (mat == null) UTIL.errorMsg(
```

```
39                    "Element material name: " + matName);
40            double lambda = mat.getLambda();
41            double mu     = mat.getMu();
42            double beta   = lambda + 2*mu;
43
44            for (int ip = 0; ip < gk.nIntPoints; ip++) {
45                double det = ShapeQuad3D.deriv(gk.xii[ip],
46                        gk.eti[ip], gk.zei[ip], ind, xy, dnxy);
47                double dv = det*gk.wi[ip];
48                // Upper symmetrical part of the matrix by rows
49                for (int i = 0; i < 20; i++) { // i = row
50                    // dNi/dx, dNi/dy, dNi/dz
51                    double dix = dnxy[i][0];
52                    double diy = dnxy[i][1];
53                    double diz = dnxy[i][2];
54                    for (int j = i; j <20; j++) { // j = column
55                        // dNj/dx, dNj/dy, dNj/dz
56                        double djx = dnxy[j][0];
57                        double djy = dnxy[j][1];
58                        double djz = dnxy[j][2];
59
60                        kmat[i*3  ][j*3  ] += (beta*dix*djx
61                            + mu*(diy*djy + diz*djz))*dv;
62                        kmat[i*3  ][j*3+1] += (lambda*dix*djy
63                            + mu*diy*djx)*dv;
64                        kmat[i*3  ][j*3+2] += (lambda*dix*djz
65                            + mu*diz*djx)*dv;
66
67                        if (j > i) kmat[i*3+1][j*3  ]
68                            += (lambda*diy*djx + mu*dix*djy)*dv;
69                        kmat[i*3+1][j*3+1] += (beta*diy*djy
70                            + mu*(diz*djz + dix*djx))*dv;
71                        kmat[i*3+1][j*3+2] += (lambda*diy*djz
72                            + mu*diz*djy)*dv;
73
74                        if (j > i) {
75                            kmat[i*3+2][j*3  ]
76                                += (lambda*diz*djx + mu*dix*djz)*dv;
77                            kmat[i*3+2][j*3+1]
78                                += (lambda*diz*djy + mu*diy*djz)*dv;
79                        }
80                        kmat[i*3+2][j*3+2] += (beta*diz*djz
81                                + mu*(dix*djx + diy*djy))*dv;
82                    }
83                }
84            }
85    }
```

First, we set the upper symmetric part of the stiffness matrix kmat to zero in
lines 34–35. The elastic material properties λ and μ (Lame constants) are obtained
from material object mat in lines 40–42.

The special Gauss rule gk with 14 integration points is used for numerical integration of coefficients of the stiffness matrix. The integration loop starts at line

44. The derivatives of shape functions at integration point `ip` are calculated in lines 45–46. Two loops with parameters `i` in line 49 and `j` in line 54 are employed for treating all node combinations for element nodes. Parameter `j` starts from value `i`, thus computing just the upper symmetric part of the stiffness matrix. The derivatives of shape functions $\partial N_i/\partial x$, $\partial N_i/\partial y$ and $\partial N_i/\partial z$ are denoted as `dix`, `diy diz` (lines 51–53). Identifiers `djx`, `djy djz` are used for derivatives $\partial N_j/\partial x$, $\partial N_j/\partial y$ and $\partial N_j/\partial z$ in lines 55–57.

Block 3 by 3 of the stiffness matrix is calculated in lines 60–81. For numerical integration, each computed coefficient is multiplied by `dv` that is composed of the determinant of the Jacobian matrix and the integration weight.

13.2.2 Thermal Vector

Method `thermalVector` computes an element thermal vector according to Equation 12.20. The algorithm is similar to that for stiffness matrix estimation.

```
87          // Compute thermal vector
88          public void thermalVector() {
89
90              for (int i = 0; i < 60; i++) evec[i] = 0.;
91
92              mat = (Material)fem.materials.get(matName);
93              double alpha = mat.getAlpha();
94              double lambda = mat.getLambda();
95              double mu = mat.getMu();
96              double g = 3*lambda + 2*mu;
97
98              for (int ip = 0; ip < gh.nIntPoints; ip++) {
99                  ShapeQuad3D.shape(gh.xii[ip], gh.eti[ip],
100                         gh.zei[ip], ind, an);
101                  // Temperature at integration point
102                  double t = 0;
103                  for (int i = 0; i < 20; i++) t += an[i]*dtn[i];
104                  double det = ShapeQuad3D.deriv(gh.xii[ip],
105                         gh.eti[ip], gh.zei[ip], ind, xy, dnxy);
106                  double dv = g*alpha*t*det*gh.wi[ip];
107                  for (int i=0; i<20; i++) {
108                      for (int j=0; j<3; j++) {
109                          evec[i*3+j] += dnxy[i][j]*dv;
110                      }
111                  }
112              }
113          }
```

Elastic material properties are set in lines 92–96. Integration is performed with the use of the Gauss rule `gh` inside a loop, which starts in line 98. Shape functions an estimated in lines 99–100 are used for temperature interpolation in lines 102–103. Partial derivatives of shape functions are calculated in lines 104–105 and then used for accumulation of contributions to the thermal vector `evec` (line 109).

13.2.3 Nodal Equivalent of a Distributed Load

Method `equivFaceLoad` performs evaluation of the nodal equivalent of a face distributed load according to the algorithm presented in Section 12.5. The method gets object `ElemFaceLoad surld`, describing application of the distributed load to element face.

```
115      // Set nodal equivalent of distributed face load to evec.
116      // surLd - object describing element face load;
117      // returns loaded element face
118      // or -1 (loaded nodes does not match element face)
119      public int equivFaceLoad(ElemFaceLoad surLd) {
120          // Shape functons
121          double an[] = new double[8];
122          // Derivatives of shape functions
123          double xin[][] = new double[8][2];
124          // Tangent vectors along xi and eta
125          double e[][] = new double[2][3];
126          // Normal vector
127          double g[] = new double[3];
128          double ps[] = new double[3];
129
130          for (int i=0; i<60; i++) evec[i] = 0.;
131
132          int loadedFace = surLd.rearrange(faceInd, ind);
133          if (loadedFace == -1) return -1;
134
135          for (int ip=0; ip<gf.nIntPoints; ip++){
136              ShapeQuad3D.shapeDerivFace(gf.xii[ip], gf.eti[ip],
137                                         ind, an, xin);
138              double p = 0.;
139              for (int i=0; i<8; i++)
140                  p += an[i]*surLd.forceAtNodes[i];
141              // Tangent vectors
142              for (int i=0; i<2; i++) {
143                  for (int j=0; j<3; j++) {
144                      double s = 0;
145                      for (int k=0; k<8; k++)
146                          s += xin[k][i]
147                                  *xy[faceInd[loadedFace][k]][j];
148                      e[i][j] = s;
149                  }
150              }
151              // Normal vector g
152              g[0] = (e[0][1]*e[1][2]-e[1][1]*e[0][2]);
153              g[1] = (e[0][2]*e[1][0]-e[1][2]*e[0][0]);
154              g[2] = (e[0][0]*e[1][1]-e[1][0]*e[0][1]);
155
156              // Element of surface ds
157              double ds = Math.sqrt(g[0]*g[0]
158                              + g[1]*g[1] + g[2]*g[2]);
159              if (ds<=0) UTIL.errorMsg(
160                  "Negative/zero element face");
```

```
161          // Surface load components ps:
162          // direction=0 - normal, x=1, y=2, z=3
163          if (surLd.direction == 0) {
164              for (int i=0; i<3; i++) ps[i] = p*g[i]/ds;
165          }
166          else {
167              for (int i=0; i<3; i++) ps[i] = 0;
168              ps[surLd.direction-1] = p;
169          }
170          for (int i=0; i<8; i++) {
171              int k = faceInd[loadedFace][i];
172              for (int j=0; j<3; j++) {
173                  evec[3*k + j] += an[i]*ps[j]*ds*gf.wi[ip];
174              }
175          }
176      }
177      return loadedFace;
178  }
```

In the beginning of the method, several working arrays are allocated: an – shape functions; xin – derivatives of shape functions with respect to local face coordinates ξ, η; e – a two-dimensional array containing tangent vectors along coordinates ξ and η; g – a vector normal to the element face.

The surface load surLd is rearranged according to the order of face connectivity numbers in line 131. If rearrangement is impossible (which implies errors in input data) then the method returns -1.

The integration rule with 9 points (3 by 3) is used for estimating the nodal equivalent of the surface load (line 135). Two-dimensional shape functions and their derivatives with respect to local coordinates are calculated in lines 136–137 by method shapeDerivFace. Shape functions are employed to interpolate nodal values of distributed load to an integration point. Lines 142–150 determine tangent vectors along local coordinate axes. The normal vector g is calculated through vector product of the two tangent vectors in lines 152–154. The direction of the surface load can be normal to the surface (surLd.direction = 0) or it can be along a coordinate axis (surLd.direction = 1, 2, 3 corresponds to x, y, and z). Integration of the surface load is carried out in line 173. The nodal equivalent of the surface load is accumulated in element vector evec.

13.2.4 Equivalent Stress Vector

A nodal force vector, which is equivalent to the element stress field, is used to estimate equilibrium of the finite element model. This vector is determined according to Equation 12.21 by method equivStressVector.

```
180      // Compute equivalent stress vector (with negative sign)
181      public void equivStressVector() {
182
183          for (int i = 0; i < 60; i++) evec[i] = 0.;
```

```
184
185            for (int ip = 0; ip < gs.nIntPoints; ip++) {
186                // Accumulated stress   s
187                double[] s = new double[6];
188                for (int i=0; i<6; i++)
189                    s[i] = str[ip].sStress[i] + str[ip].dStress[i];
190                double det = ShapeQuad3D.deriv(gs.xii[ip],
191                        gs.eti[ip], gs.zei[ip], ind, xy, dnxy);
192                double dv = det * gs.wi[ip];
193
194                for (int i = 0; i < 20; i++) {
195                    double a0 = dnxy[i][0];
196                    double a1 = dnxy[i][1];
197                    double a2 = dnxy[i][2];
198                    evec[i*3  ] -= (a0*s[0]+a1*s[3]+a2*s[5])*dv;
199                    evec[i*3+1] -= (a1*s[1]+a0*s[3]+a2*s[4])*dv;
200                    evec[i*3+2] -= (a2*s[2]+a1*s[4]+a0*s[5])*dv;
201                }
202            }
203        }
```

The equivalent stress vector is estimated using integration rule gs with $2 \times 2 \times 2$ Gauss points. Stress values are computed and stored at reduced integration points $2 \times 2 \times 2$ since stresses have the highest precision at these points. This dictates the usage of the particular integration rule.

Current stresses s are evaluated as a sum of accumulated stress from the previous load step sStress and stress increment at current load step dStress in lines 188 and 189. Derivatives of shape functions are calculated in lines 190 and 191. Integration of stresses multiplied by derivatives of shape functions is performed in a loop from line 194 to line 201. The resulting equivalent vector is set in array evec.

13.2.5 Extrapolation from Integration Points to Nodes

Method extrapolateToNodes takes nodes at reduced integration points and using trilinear extrapolation computes nodal values of stresses. The parameters of the method are: fip are stresses at eight integration points, fn are the resulting stresses at nodes.

```
205        // Extrapolate stresses from integration points to nodes.
206        // fip [8][6] - stresses at integration points;
207        // fn [20][6] - stresses at nodes (out)
208        public void extrapolateToNodes(double[][] fip,
209                                       double[][] fn) {
210            // Vertices
211            final int vn[] = {0, 2, 4, 6, 12, 14, 16, 18};
212             // Midside nodes
213            final int mn[] = {8, 9, 10, 11, 8, 9, 10, 11};
214            // Extrapolation matrix
215            final double A =   0.25*(5 + 3*Math.sqrt(3.)),
```

```
216                      B =  -0.25*(Math.sqrt(3.) + 1),
217                      C =   0.25*(Math.sqrt(3.) - 1),
218                      D =   0.25*(5 - 3*Math.sqrt(3.));
219         final double lim[][] = {{A, B, B, C, B, C, C, D},
220                                 {B, C, C, D, A, B, B, C},
221                                 {C, D, B, C, B, C, A, B},
222                                 {B, C, A, B, C, D, B, C},
223                                 {B, A, C, B, C, B, D, C},
224                                 {C, B, D, C, B, A, C, B},
225                                 {D, C, C, B, C, B, B, A},
226                                 {C, B, B, A, D, C, C, B}};
227
228         for (int i = 0; i < 20; i++)
229             for (int j = 0; j < 6; j++) fn[i][j] = 0;
230
231         for (int vertex = 0; vertex < 8; vertex++) {
232             int i = vn[vertex];   // node at vertex
233             int im = i - 1;
234             if (i == 0) im = 7;
235             if (i == 12) im = 19;
236             for (int k = 0; k < 6; k++) {
237                 double c = 0.0;
238                 for (int j = 0 ; j < 8; j++)
239                     c += fip[j][k]*lim[vertex][j];
240                 fn[i][k] = c;
241                 fn[im][k] += 0.5*c;
242                 fn[i+1][k] += 0.5*c;
243                 fn[mn[vertex]][k] += 0.5*c;
244             }
245         }
246     }
```

Arrays vn (line 210) and mn (line 213) contain local numbers of vertex nodes and some of the midside nodes, correspondingly. Lines 215–226 specialize matrix lim for trilinear extrapolation from reduced integration points 2×2 to vertex nodes.

A loop over vertex nodes starts in line 231. The integer variable i in line 232 contains local node number for the vertex number. The previous midside node im is set in lines 233–235. The extrapolated value of a stress component is accumulated in variable c (line 239). Then, this value is assigned to vertex i in line 240. Half of the vertex value is added to neighboring midside nodes in the next three lines.

13.2.6 Other Methods

Methods getElemFaces (element face numbers), getStrainsAtIntPoint (strains at reduced integration points) and getTemperatureAtIntPoint (temperature at reduced integration point), also necessary for particular element implementation, are given below.

```
248     // Get local node numbers for element faces.
249     // returns elementFaces[nFaces][nNodesOnFace]
```

```
250        public int[][] getElemFaces() {
251            return faceInd;
252        }
253
254        // Get strains at integration point.
255        // ip - integration point number (stress);
256        // returns  strain vector (ex, ey, ez, gxy, gyz, gzx)
257        public double[] getStrainsAtIntPoint(int ip) {
258
259            // Derivatives of shape functions
260            ShapeQuad3D.deriv(gs.xii[ip], gs.eti[ip], gs.zei[ip],
261                    ind, xy, dnxy);
262
263            // Derivatives of displacements
264            double dux,duy,duz,dvx,dvy,dvz,dwx,dwy,dwz;
265            dux=duy=duz=dvx=dvy=dvz=dwx=dwy=dwz = 0;
266            for (int i = 0; i < 20; i++) {
267                double dnx = dnxy[i][0];
268                double dny = dnxy[i][1];
269                double dnz = dnxy[i][2];
270                double u = evec[3*i  ];
271                double v = evec[3*i+1];
272                double w = evec[3*i+2];
273                dux += dnx*u;   duy += dny*u;   duz += dnz*u;
274                dvx += dnx*v;   dvy += dny*v;   dvz += dnz*v;
275                dwx += dnx*w;   dwy += dny*w;   dwz += dnz*w;
276            }
277            // Strains
278            double strain[] = new double[6];
279            strain[0] = dux;   strain[1] = dvy;   strain[2] = dwz;
280            strain[3] = duy + dvx;
281            strain[4] = dvz + dwy;
282            strain[5] = duz + dwx;
283            return strain;
284        }
285
286        // Returns temperature at integration point (stress)
287        public double getTemperatureAtIntPoint(int ip) {
288
289            ShapeQuad3D.shape(gs.xii[ip], gs.eti[ip], gs.zei[ip],
290                            ind, an);
291            double t = 0;
292            for (int i=0; i<20; i++) t += an[i]*dtn[i];
293            return t;
294        }
295
296 }
```

Method getElemFaces returns local numbers for six element faces, which are assigned to members of array faceInd (lines 10–16).

The strain vector is calculated by method getStrainsAtIntPoint. The method returns an array of six strain components for the requested integration point ip, which belongs to the set of reduced integration points $2 \times 2 \times 2$ (Gauss rule gs). Derivatives of shape functions are determined in lines 260–261. Derivatives of displacements with respect to coordinates x, y, and z are calculated in the loop from line 266 to line 276. Displacements u, v, and w are taken from array `evec`, which should be set previously. Strains in lines 279–282 are determined as combinations of displacement derivatives.

Method `getTemperatureAtIntPoint` returns the temperature at the reduced integration point `ip` using shape functions for interpolation of nodal temperature values.

Problems

13.1. Study method `degeneration` in Section 13.1.1. Suppose that the following local nodes have the same connectivities numbers:

```
0 = 1 = 2,
3 = 7,
4 = 5 = 6.
```

Modify method `degeneration` to discover this element degeneration. What modifications should be done in method `shape` (Section 13.1.2) to correct element shape functions when this kind of element degeneration occurs?

13.2. Write a Java method that calculates and prints shape function values of the twenty-node hexahedral element at point $\xi = 0.2$, $\eta = 0.2$, $\zeta = -0.5$. Use method `shape` for shape-function calculation.

13.3. Write a method in class `ShapeQuad3D` that evaluates a value of the Jacobian determinant at the center $\xi = \eta = \zeta = 0$ of the hexahedral element shown below.

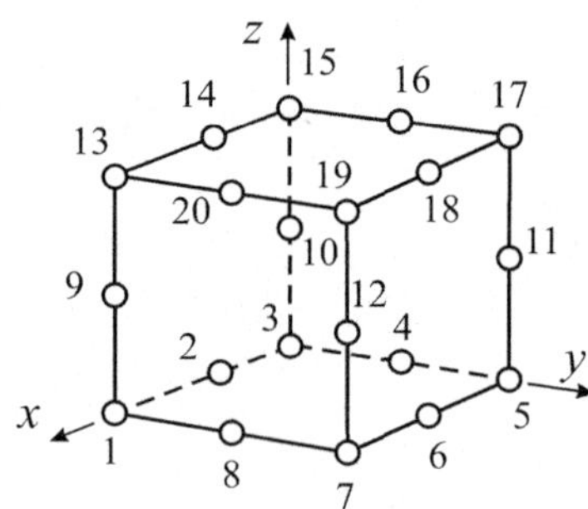

The element is a cube with all edges of unit length and the following connectivities:

```
ind = {1,  2,  3,  4,  5,  6,  7,  8,
       9, 10, 11, 12,.
      13, 14, 15, 16, 17, 18, 19, 20}.
```

Use method `ShapeQuad3D.deriv`. Explain the negative value of the determinant.

13.4. Read and understand the source code of method `stiffnessMatrix` that computes the upper symmetrical part of the stiffness matrix for the three-dimensional isoparametric quadratic element. Note that the method directly computes 3×3 blocks of the stiffness matrix, which correspond to a pair of element nodes. Modify the code in such a way that: 1) the full stiffness matrix is computed; 2) the lower symmetrical part of the stiffness matrix is computed.

Chapter 14
Assembly and Solution

Abstract The chapter presents general approaches to assembly of global vectors and matrices and to solution of finite element equation systems. First, algorithms of vector disassembly and vector assembly are explained. Then, assembly of a global matrix from element matrices is considered for full storage of the global matrix. Two methods of application of displacement boundary conditions are described. Abstract class `Solver` is presented. New finite element solvers can be included by extending this basic class.

14.1 Disassembly and Assembly

The finite element model is composed of finite elements. Mathematically, both the finite element model and each finite element are described by vectors and matrices. Disassembly and assembly operations allow transformation from global vectors and matrices to local (element) vectors and matrices and *vice versa*. Element connectivity information is used to perform disassembly and assembly operations.

14.1.1 Disassembly of Vectors

Disassembly is a selection of element vectors from a given global vector. The disassembly operation is described by the matrix relation (3.36).

Disassembly operation (3.36) includes matrix multiplication with the use of a large matrix $[A]$, which provides correspondence between local and global node enumerations. Matrix $[A]$ is almost completely composed of zeros. It has only one nonzero (unit) entry in each row. Instead of matrix multiplication it is possible just to take vector elements from their global positions and to assign them to appropriate positions in the element vector. Element connectivities determine which entries of the global vector should be assigned to element vectors.

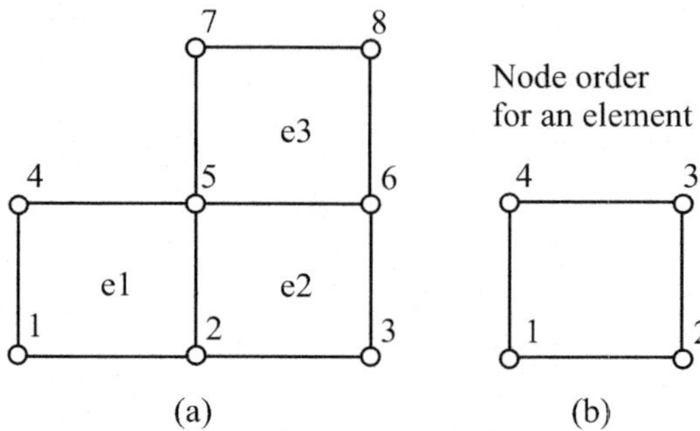

Fig. 14.1 (a) Simple finite element mesh with global node numbering and (b) a typical element with local node numbers

Let us consider an example of a disassembly operation for a mesh shown in Figure 14.1a. The mesh consists of three elements and eight nodes. Local node numbering is shown in Figure 14.1b. For simplicity we assume that each node has one degree of freedom. Then, a global vector, for example, displacement vector Q, contains eight entries Q_1, Q_2, ... Q_8. The element vector includes four entries. Illustration of the disassembly procedure is presented in Figure 14.2. Arrows show copying of entries from the global array to the local array for the second element. Element connectivity numbers determine indices of global array entries that should be placed into an element local array. The algorithm of vector disassembly is given below.

Disassembly of the global vector into element vectors

n = number of degrees of freedom per element
N = total number of degrees of freedom
E = number of elements
$C[E,n]$ = connectivity array
$q[n]$ = element displacement vector
$Q[N]$ = global displacement vector

do $e = 1, E$
   do $i = 1, n$
      $f[i] = F[C[e,i]]$
   end do
   Use element vector f
end do

In the above algorithm we assume that connectivity information is related to degrees of freedom. If the connectivity array contains global node numbers and each node has more than one degree of freedom then instead of one number the block related to the node should be selected. For example, in a three-dimensional elasticity problem each node is associated with three displacements (three degrees of freedom).

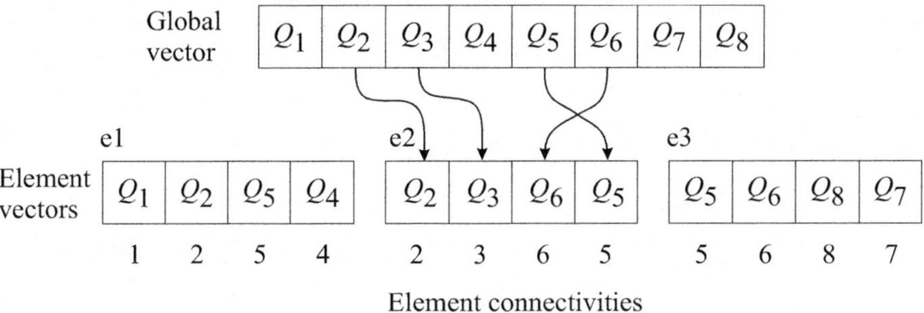

Fig. 14.2 Illustration of disassembly procedure. Assignment of global entries to element vectors is done according to connectivities

14.1.2 Assembly of Vectors

Vector assembly is the operation of joining element vectors in a global vector. Instead of matrix $[A]$, assembly procedures are usually based on direct summation with the use of the element connectivity array.

Suppose that we need to assemble global load vector F using element load vectors f and connectivity array C. The assembly procedure for the mesh of Figure 14.1 is graphically illustrated in Figure 14.3. Element entries go to global addresses that correspond to connectivity element numbers.

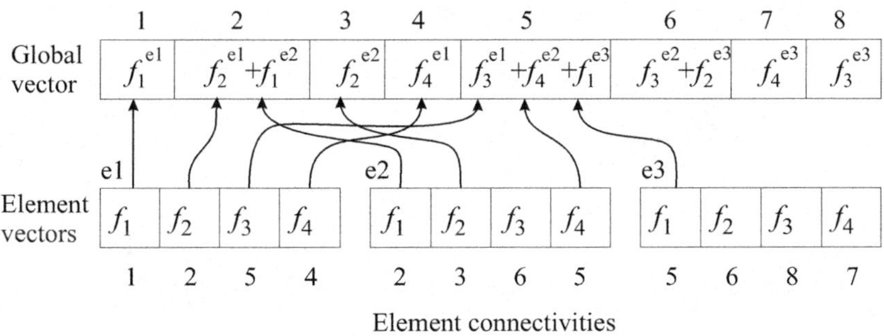

Fig. 14.3 Illustration of assembly procedure for vectors

During assembly more than one element vector can contribute to the same entry of the global vector. In Figure 14.3, we can see that the fifth entry of the global vector combines coefficients from three elements of the mesh. Because of such possibilities the assembly algorithm starts by assigning zeros to all entries of the global vector.

Elementary assembly operation is an addition of a local entry to a corresponding global entry. A pseudocode description of the assembly algorithm is as follows:

Assembly of the global vector

n = number of degrees of freedom per element
N = total number of degrees of freedom
E = number of elements
$C[E,n]$ = connectivity array
$f[n]$ = element load vector
$F[N]$ = global load vector

do $i = 1, N$
  $F[i] = 0$
end do
do $e = 1, E$
  Generate f
  do $i = 1, n$
     $F[C[e,i]] = F[C[e,i]] + f[i]$
  end do
end do

Element load vectors are generated when they are necessary for assembly. It can be seen that the connectivity entry $C[e,i]$ simply provides the address in the global vector where the ith entry of the load vector for element e goes. We assume that the connectivity array is written in terms of degrees of freedom. In actual routines the connectivity array contains node numbers that are transformed to degrees of freedom for the currently assembled element.

14.1.3 Assembly Algorithm for Matrices

Assembly of element stiffness matrices is necessary for establishing the global stiffness matrix, which, together with the global load vector, sets up the finite element equation system. An example of the assembly of element stiffness matrix (element 1 of the mesh shown in Figure 14.1a) into the global stiffness matrix is depicted in Figure 14.4. In the figure, element connectivity numbers are placed near the top row and near the left column of the element stiffness matrix. A pair of connectivity numbers determines a place (row and column) in the global matrix where a coefficient of the element matrix should be added. For example, coefficient k_{34} is added to the intersection of the fifth row and the fourth column in the global stiffness matrix. Several entries from different element matrices can be accumulated at the same address in the global matrix.

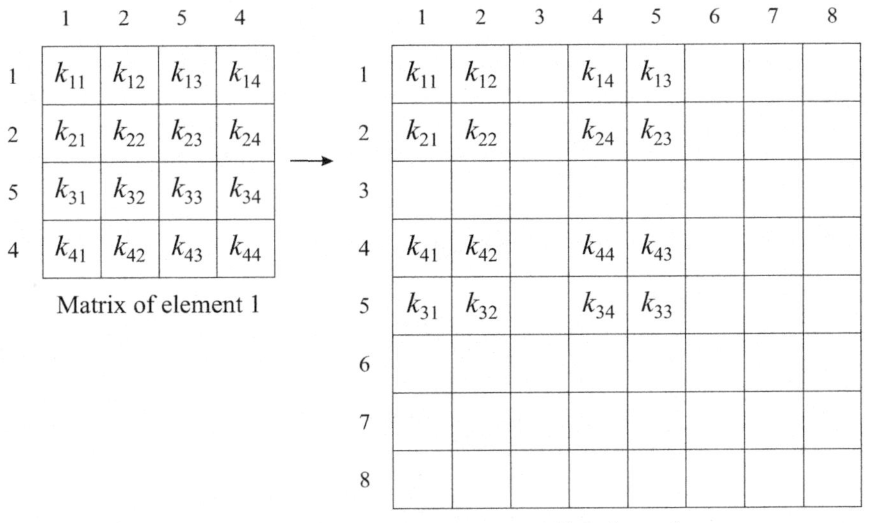

Matrix of element 1

Global matrix

Fig. 14.4 Assembly of element stiffness matrix into the global stiffness matrix

An algorithm of assembly of the global stiffness matrix K from contributions of element stiffness matrices k can be expressed by the following pseudocode:

Assembly of the global matrix

n = number of degrees of freedom per element
N = total number of degrees of freedom in the domain
E = number of elements
$C[E,n]$ = connectivity array
$k[n,n]$ = element stiffness matrix
$K[N,N]$ = global stiffness matrix

do $i = 1, N$
   do $j = 1, N$
      $K[i,j] = 0$
   end do
end do
do $e = 1, E$
   generate k
   do $i = 1, n$
      do $j = 1, n$
         $K[C[e,i],C[e,j]] = K[C[e,i],C[e,j]] + k[i,j]$
      end do
   end do
end do

Here, for simplicity, element matrices are assembled in the full square global matrix. Since element stiffness matrices are symmetric and each finite element is connected to just a few neighbors, the global stiffness matrix is symmetric and sparse. These properties can be used to economize space and time in finite element solvers.

14.2 Displacement Boundary Conditions

Displacement boundary conditions were not accounted for in the functional of the total potential energy. They can be applied to the global equation system after its assembly.

Let us consider application of the displacement boundary condition

$$Q_m = d \qquad (14.1)$$

to the global equation system. Two methods can be used for the specification of the displacement boundary condition.

14.2.1 Explicit Specification of Displacement Boundary Conditions

In the explicit method, we substitute the known value of the displacement $Q_m = d$ in the mth column and move this column to the right-hand side. Then, we put zeros into the mth column and mth row of the matrix except for the main diagonal element, which is replaced by units.

Explicit method:

$$\begin{aligned}
&F_i = F_i - K_{im}d, \quad i = 1...N, \ i \neq m, \\
&F_m = d, \\
&K_{mj} = 0, \quad j = 1...N, \\
&K_{im} = 0, \quad i = 1...N, \\
&K_{mm} = 1.
\end{aligned} \qquad (14.2)$$

The explicit method is easy to implement for the global stiffness matrix stored as a full two-dimensional array. However, in a real finite element program the global stiffness matrix is stored in some compact format. It is not always easy to access matrix rows and columns when a matrix is in compact format. Therefore, it is desirable to have a method that requires fewer modifications of the stiffness matrix for boundary-condition specifications.

14.2.2 Method of Large Number

The method of large number uses the fact that computer computations have limited precision. The results of double-precision computations contain about 15–16 digits. So, addition $1.0 + 1e^{-17}$ produces 1.0 as the result.

Method of large number ($M >> K_{ij}$):

$$K_{mm} = M,$$
$$F_m = Md. \tag{14.3}$$

After placing large numbers on the main diagonal and at the right-hand side, all other coefficients of the current equation are too small to affect computations. The method of large number is simpler than the explicit method of displacement boundary condition specification. The solution of the finite element problem is the same for both methods.

14.3 Solution of Finite Element Equations

Practical applications of the finite element method lead to large systems of simultaneous linear algebraic equations.

Fortunately, finite element equation systems possess some properties that allow reduction of storage and computing time. The finite element equation systems are symmetric, positive-definite, and sparse. Symmetry allows storing of only half of the matrix, including diagonal entries. Positive-definite matrices are characterized by large positive entries on the main diagonal. Solution can be carried out without pivoting. A sparse matrix contains more zero entries than nonzero entries. Sparsity can be used to economize storage and computations.

Solution methods for linear equation systems can be divided into two large categories: direct methods and iterative methods. Direct solution methods are usually used for problems of moderate size. For large problems, iterative methods require less computing time and hence they are preferable.

Matrix storage formats are closely related to solution methods. So, our plan of programming implementation of solvers consists in programming abstract class `Solver`, which knows nothing about system storage or solution method, and particular solver classes, which hide storage formats and solution procedures from outside.

Abstract class `Solver` is presented here. In the next chapters, we consider implementation of two solution methods that are widely used in finite element computer codes. The first method is the direct LDU solution with the profile global stiffness matrix. The second one is the preconditioned conjugate gradient method with sparse row-wise format of matrix storage.

14.4 Abstract Solver Class

Abstract class `Solver` is a parent Java^{TM} class for particular solvers that realize different methods for solution of finite element equation systems. A particular solver should extend class `Solver` and implement some methods necessary for assembly and solution of the finite element equation system.

```
1  package solver;
2
3  import elem.*;
4  import model.*;
5  import fea.FE;
6
7  // Solution of the global equation system
8  public abstract class Solver {
9
10     static FeModel fem;
11     // number of equations
12     static int neq;
13     // length of global stiffness matrix
14     public static int lengthOfGSM;
15     // elem connectivities - degrees of freedom
16     int[] indf;
17     // number of degrees of freedom for element
18     int nindf;
19     // Indicator of new global matrix
20     boolean newMatrix;
21
22     public static enum Solvers {
23         ldu {Solver create()
24                 {return new SolverLDU();} },
25         pcg {Solver create()
26                 {return new SolverPCG();} };
27         abstract Solver create();
28     }
29
30     public static Solvers solver = Solvers.ldu;
31
32     public static Solver newSolver(FeModel fem) {
33         Solver.fem = fem;
34         neq = fem.nEq;
35         return solver.create();
36     }
37
38     // Assemble global stiffnes matrix
39     public void assembleGSM() {
40         Element elm;
41         indf = new int[FE.maxNodesPerElem*fem.nDf];
42
43         for (int iel=0; iel<fem.nEl; iel++) {
44             for (int i=0; i<fem.elems[iel].ind.length; i++) {
45                 for (int k=0; k<fem.nDf; k++)
46                     indf[fem.nDf*i+k] =
```

```
47                                (fem.elems[iel].ind[i]-1)*fem.nDf+k+1;
48                  }
49              nindf = fem.elems[iel].ind.length*fem.nDf;
50              elm = fem.elems[iel];
51              elm.setElemXy();
52              elm.stiffnessMatrix();
53              assembleESM();
54          }
55          // Indicate that new global matrix appeared
56          newMatrix = true;
57      }
58
59      // Add element stiffness matrix to GSM
60      void assembleESM() {}
61
62      // Solve global equation system
63      // x - right-hand side/solution (in/out)
64      public int solve(double x[]) {
65          return 0;
66      }
67
68 }
```

In the beginning of class `Solver` there are declarations of the following items:

`fem` – `FeModel` object describing the finite element model;

`neq` – number of equations in the finite element equation system;

`lengthOfGSM` – length of the global stiffness matrix in words set by a particular solver;

`indf` – array of element connectivities expressed as degrees of freedom;

`nindf` – number of degrees of freedom for current element;

`newMatrix` – boolean variable, an indicator of the matrix state; for a newly created matrix it has `true` value.

It is supposed that other solver classes will be placed in package `solver` and all the above data will be available for them.

Java enum `Solvers` is used for storing symbolic names of solvers and for calling solver constructors (lines 22–28). It contains two solvers, which we are going to implement further:

`ldu` – direct equation solver based on LDU matrix decomposition;

`pcg` – preconditioned conjugate gradient solver.

Each solver record implements method `create` that returns a corresponding solver object. If we want to add another solver then it is necessary to create a class for this solver and to include its reference in enum `Solvers`.

Line 30 specifies that the default solver is the LDU direct solver. Static method `newSolver` (lines 32–36) returns a `Solver` object depending on the value of `solver`, which can be specified during input of data for the finite element model.

Method `assembleGSM` performs assembly of the global stiffness matrix. The loop over all finite elements starts in line 43. Lines 44–48 create element connectivities `indf` in terms of degrees of freedom. Both node numbers in element connectivities `ind` and degrees of freedom in array `indf` start from 1. For example, node 3 corresponds to three degrees of freedom 7, 8, 9.

`Element` object `elm` is set in line 48. Element methods `setElemXy` and `stiffnessMatrix` (lines 51–52) set element nodal coordinates and compute element stiffness matrix. Method `assembleESM` assembles element stiffness matrix into the global stiffness matrix. Line 56 sets variable `newMatrix` to a true value that indicates that a new global matrix has been asembled.

Lines 59–66 contain two empty methods, which should be implemented in a particular solver class. Method `assembleESM()` assembles an element stiffness matrix into a global stiffness matrix. It uses connectivities `indf[nindf]` for assembly. Method `solve` solves the global equation system using `x` as the input for the right-hand side and as the output for the solution vector. The method returns one integer parameter that can be used for returning the error state or other information.

14.5 Adding New Equation Solver

The design of abstract class `Solver` makes it relatively easy to implement an additional equation solver in our finite element system. In order to add a new equation solver it is necessary:

1. To include solver factory method in enum `Solvers`, similar to lines 19–20,

   ```
   solverName {Solver create()
           {return new SolverClass();} },
   ```

   where `solverName` is a solver name in a data input file and `SolverClass` is a solver class name. Specification of data item

   ```
   solver = solverName
   ```

   will result in using this solver for this problem.

2. To develop class `SolverClass` that implements methods `assembleESM` and `solve`, shown in lines 59–66.

   `assembleESM` – assembles the stiffness matrix for current finite element. Element connectivities as degrees of freedom are available from array `indf` in class `Solver`. The $i - j$ entry of the element stiffness matrix is accessed as `Element.kmat[i][j]`;

   `int solve(double x[])` – solves the global equation system. Parameter `x` represents the right-hand side of the equation system. It should contain the solution vector after return from method `solve`. The method returns the number of iterations if it implements an iterative method, and zero otherwise.

Since both element stiffness matrix assembly and solution of the global equation system are performed by a particular solver, other classes have no knowledge about the storage format of the global stiffness matrix and solution algorithm.

Problems

14.1. Perform disassembly of the global displacement vector $\{Q\} = \{Q_1\ Q_2\ \cdots\ Q_6\}$ into three element displacement vectors $\{q^{e1}\}$, $\{q^{e2}\}$, and $\{q^{e3}\}$ for the mesh shown below.

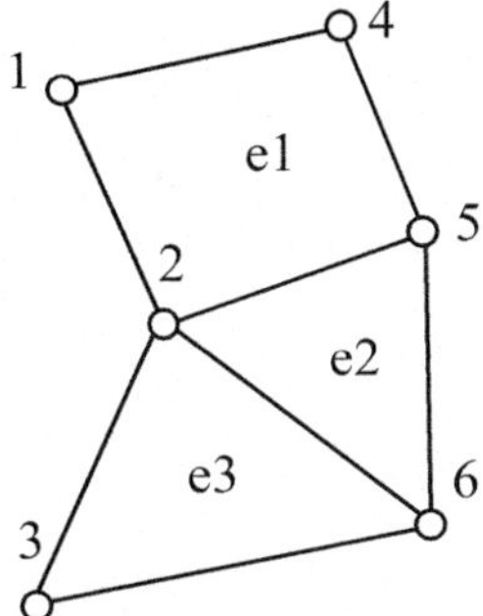

Assume one degree of freedom per node. When creating element connectivities, start from the lowest node in an element and list nodes in an anticlockwise direction.

14.2. For the mesh given in the previous problem, assemble the element force vectors $\{f^e\}$ into a global force vector $\{F\}$. All entries of the element force vectors $\{f^{e1}\}$, $\{f^{e2}\}$, and $\{f^{e3}\}$ have values 1, 2, and 3, respectively.

14.3. For the mesh of Problem 14.1, assemble element stiffness matrices $[k^e]$ into a global stiffness matrix $[K]$. The coefficients of element stiffness matrices $[k^{e1}]$, $[k^{e2}]$, and $[k^{e3}]$ have values 10, 20, and 30, respectively.

14.4. For the global stiffness matrix assembled in Problem 14.3 and for the global force vector obtained in Problem 14.2, apply displacement boundary conditions
$$Q_1 = 0, \quad Q_2 = 0.01, \quad Q_6 = 0.02$$
using a) the explicit method, and b) the method of large number. Assume that computations are performed using Java `double` type.

14.5. The finite element model has 100 nodes. Each node has two degrees of freedom. The global vector consists of the following values: $\{Q\} = \{1, 2, 3, ..., 200\}$. Using disassembly, select element vector $\{q\}$ from $\{Q\}$ for the element with nodal connectivities $\{25, 16, 17, 33\}$.

Chapter 15
Direct Equation Solver

Abstract A direct equation solver based on LDU decomposition is developed. According to the LDU method, a matrix of the equation system is decomposed into lower triangular, diagonal, and upper triangular matrices. In our implementation of the LDU solver, the equation system matrix is stored in a symmetric profile format. Algorithms of matrix assembly, decomposition, forward reduction, and back-substitution are presented. Java^{TM} class `SolverLDU` implements algorithms of the LDU solver. Tuning of the solver using a blocking technique increases solution speed by several times.

15.1 LDU Solution Method

Direct equation solvers are computationally efficient for finite element systems of moderate size. Direct solution methods are preferable for engineering purposes since they are reliable and the solution time can be easily predicted.

Let us consider the LDU algorithm for the solution of an equation system,

$$[A]\{x\} = \{b\}, \tag{15.1}$$

where $[A]$ is the equation matrix with coefficients A_{ij}, $\{b\}$ is the right-hand side, and $\{x\}$ is the unknown vector.

According to the LDU method, solution of the symmetric linear algebraic system consists of three stages [13]:

$$\text{Factorization}: \quad [A] = [U]^{\mathrm{T}}[D][U],$$

$$\text{Forward solution}: \quad \{y\} = [U]^{-\mathrm{T}}\{b\}, \tag{15.2}$$

$$\text{Back} - \text{substitution}: \quad \{x\} = [U]^{-1}[D]^{-1}\{y\}.$$

First, matrix $[A]$ is decomposed into lower triangular $[U]^T$, diagonal $[D]$, and upper triangular $[U]$ matrices. The decomposition phase takes most computing time and does not require the right-hand side. Forward solution treats the right-side vector. Back-substitution produces the solution vector.

A matrix of the finite element equation system has good features, which ease its solution. It is *symmetric*, *sparse*, and usually *positive-definite*. The symmetric matrix structure allows storage of the symmetric half of the matrix and to operate on coefficients of this part. Positive-definiteness means that larger coefficients are grouped near the main diagonal. This helps to simplify the solution procedure because pivoting is not necessary.

Sparseness of the matrix tremendously decreases storage requirements and operation count in comparison to a fully populated matrix. If the nodes of the finite element model are numbered in a proper way then nonzero coefficients are located near the main diagonal of the matrix. The simplest format to store a matrix with coefficients near the main diagonal is symmetric band storage. However, the only coefficient located far from the main diagonal makes the bandwidth unacceptable. A possible and relatively simple way to improve the storage scheme is to use a profile-storage format.

15.2 Assembly of Matrix in Symmetric Profile Format

Using a symmetric profile format, the global stiffness matrix $[A] = A_{ij}$ of order N is stored by columns. Each column starts from the first top nonzero element and ends at the diagonal element.

The matrix is represented by two arrays:

`pcol[N+1]` – integer pointer array for columns;

`A[pcol[N]]` – array of doubles containing matrix coefficients.

The ith element of `pcol` contains the address of the first column element. The length of the ith column is given by `pcol[i+1]-pcol[i]`. The length of the array A is equal to `pcol[N]`. We assume that the indices of array A begin from 1.

Filling the global stiffness matrix for the mesh of Figure 14.1 is depicted in Figure 15.1a. Storage of this matrix in the symmetric profile format is shown in Figure 15.1b. Array `pcol` has the following contents:

```
pcol = {0, 1, 3, 5, 9, 14, 19, 22, 26} .
```

The first entry of array `pcol` is always zero and the last entry contains the number of coefficients in array A (or pointer to a nonexistent column after the matrix end).

It should be noted that proper node ordering can decrease the matrix profile significantly. Usually, node-ordering algorithms are based on some heuristic methods because the full ordering problem seeking the global minimum is too time consuming.

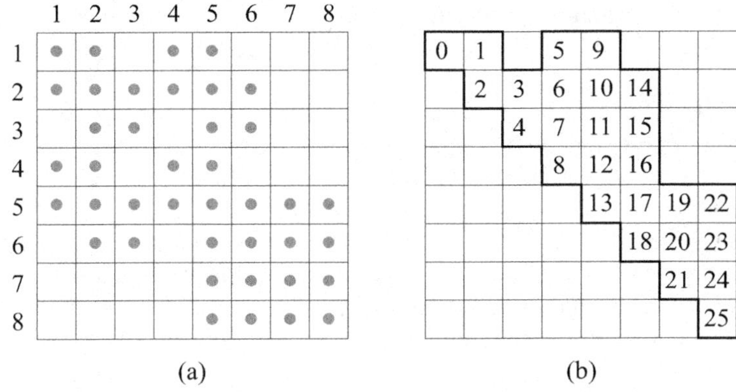

Fig. 15.1 Fill of the global stiffness matrix (a) and its symmetric profile storage by columns (b)

An algorithm for the creation of pointer array `pcol` using connectivities of finite elements can be explained with the following pseudocode.

Profile of symmetric part of the matrix
n = number of degrees of freedom per element
N = total number of degrees of freedom in the domain
E = number of elements
$C[E,n]$ = connectivity array
$p[N+1]$ = pointers to column tops
do $i = 1, N+1$
   $p[i] = 0$
end do
Compute column heights
do $e = 1, E$
   do $j = 1, n$
      do $k = 1, n$
         $p[C[e,j]+1] = max(p[C[e,j]+1], C[e,j] - C[e,k])$
      end do
   end do
end do
Transform column height to pointers
do $i = 2, N+1$
   $p[i] = p[i-1] + p[i]$
end do

As usual, we present an algorithm using indexes counted from one as in Figure 15.1a. However, pointer array p starts from zero as in Figure 15.1b. First, we put zeros at all entries of pointer array p. Then, in a loop over all elements, heights of columns from the main diagonal of the matrix are computed as the maximum differences between connectivity numbers. Column heights are stored in array p with a shift $(+1)$, i.e., the height for column i is placed in $p[i+1]$. Finally, column heights are transformed to column pointers by simple accumulation of previous column heights.

In the algorithm, it is assumed that we use connectivities for degrees of freedom. In the program implementation it is more efficient and not complicated to employ nodal connectivities for column-height calculation and to make conversion from nodes to degrees of freedom during transformation to column pointers.

We shall present solution algorithms in full matrix notation A_{ij}. Thus, it is necessary to have relations between two-index notation for the global stiffness matrix A_{ij} and one-dimensional array A used in a Java program. The location of the first nonzero element in the ith column of the matrix $[A]$ is given by the function

$$FN(i) = i - (pcol[i+1] - pcol[i]) + 1). \qquad (15.3)$$

The following correspondence relations can be easily obtained for a transition from two-index notation A_{ij} to Java notation for a one-dimensional array A:

$$A_{ij} \rightarrow A[i - j + pcol[j+1] - 1]. \qquad (15.4)$$

Here, i, j are row and column numbers, respectively.

Below, we present the beginning of class `SolverLDU`, which contains class constructor and methods for creation of the matrix profile and assembly of the global stiffness matrix. The class belongs to package `solver` and extends abstract class `Solver`.

```java
 1 package solver;
 2
 3 import fea.*;
 4 import model.*;
 5 import elem.*;
 6
 7 import java.util.ListIterator;
 8
 9 // Profile LDU symmetric solver.
10 // Upper symmetric part of the global stiffness matrix is
11 // stored by columns of variable height (profile storage).
12 public class SolverLDU extends Solver {
13
14     // Pointers to matrix columns
15     private int[] pcol;
16     // Global stiffness matrix
17     private double[] A;
18
19     // Constructor for LDU symmetric solver.
20     public SolverLDU() {
21         // Create profile of the global stiffness matrix
22         setProfile();
23         A = new double[pcol[neq]];
24     }
25
26     // Calculate profile of GSM: set column pointers pcol[]
```

```java
27      private void setProfile() {
28          pcol = new int[neq + 1];
29          for (int i = 0; i < neq+1; i++) pcol[i] = 0;
30
31          for (int iel = 0; iel < fem.nEl; iel++) {
32              for (int jind : fem.elems[iel].ind) {
33                  // Calculate profile: nodal hypercolumn length
34                  // is in the first entry for a hypercolumn
35                  if (jind > 0) {
36                      int ncol = (jind - 1) * fem.nDf + 1;
37                      int icolh = pcol[ncol];
38                      for (int kind : fem.elems[iel].ind) {
39                          if (kind > 0) icolh =
40                                  Math.max(icolh, jind - kind);
41                      }
42                      pcol[ncol] = icolh;
43                  }
44              }
45          }
46          //  Transform hypercolumn lengths to column addresses
47          int ic = 0;
48          for (int i = 0; i < fem.nNod; i++) {
49              int icolh = pcol[i*fem.nDf + 1]*fem.nDf;
50              for (int j = 0; j < fem.nDf; j++) {
51                  ic++;
52                  icolh++;
53                  pcol[ic] = pcol[ic - 1] + icolh;
54              }
55          }
56          lengthOfGSM = pcol[neq];
57      }
58
59      // Assemble element matrix to the global stiffness matrix
60      public void assembleESM() {
61          // Assemble all contributions to the upper part of GSM
62          for (int j = 0; j < nindf; j++) {
63              int jj = indf[j] - 1;
64              if (jj >= 0) {
65                  for (int i = 0; i < nindf; i++) {
66                      int ii = indf[i] - 1;
67                      if (ii >= 0) {
68                          // Profile format (columns top->bottom)
69                          if (ii <= jj) {
70                              int iad = pcol[jj+1] - jj + ii - 1;
71                              if (i <= j)
72                                  A[iad] += Element.kmat[i][j];
73                              else
74                                  A[iad] += Element.kmat[j][i];
75                          }
76                      }
77                  }
78              }
79          }
80      }
```

The array of pointers to matrix columns `pcol` and the symmetric part of the global stiffness matrix `A` are declared in lines 15 and 17. Constructor `SolverLDU` creates a profile of the equation system matrix (method `setProfile`) and allocates memory for the global stiffness matrix `A`.

Method `setProfile` calculates pointers to columns of matrix `A` in order to store the matrix in the symmetric profile format. Line 29 sets entries of column pointer array `pcol` to zero.

A loop over elements of the finite element model in lines 31–45 calculates column heights using nodal connectivities. Each node of the finite element model corresponds to a hypercolumn in matrix `A` (two columns in the two-dimensional case or three columns in the three-dimensional case). Nodal heights of hypercolumns are placed as the first entry of the next hypercolumn. Lines 47–55 transform nodal heights to actual column heights for all columns of the symmetric part of matrix `A`. Statement 56 sets the length of the global stiffness matrix that is later used for output.

Method `assembleESM` assembles the stiffness matrix of the current element into the global stiffness matrix. This method is necessary in any solver. The method uses element connectivities `indf`, which correspond to degrees of freedom. Assembly operation is performed for nonzero connectivity numbers (reminder: zero connectivity numbers can be used for absent midside nodes). An address in the global stiffness matrix (line 70) is determined according to Equation 15.4. All entries of the element stiffness matrix are considered in the assembly process. It is noted that element classes provide just the upper symmetric part of element stiffness matrices. If an entry from the lower part of the element stiffness matrix is necessary for computations then the symmetric entry is employed (compare statements in lines 72 and 74).

15.3 LDU Solution Algorithm

Development of the solver requires implementation of method `solve`, which performs solution of the finite element equation system.

```
82      // Solve equation system by direct LDU method.
83      // x - right-hand side/solution (in/out)
84      public int solve(double x[]) {
85
86          if (newMatrix) {
87              displacementBC();
88              if (FE.tunedSolver) lduDecompositionTuned();
89              else                lduDecomposition();
90              newMatrix = false;
91          }
92          lduFrwdBksb(x);
93          return 0;
94      }
95
```

```
96         // Apply displacement boundary conditions
97         private void displacementBC() {
98
99             ListIterator it = fem.defDs.listIterator(0);
100            Dof d;
101            while (it.hasNext()) {
102                d = (Dof) it.next();
103                A[pcol[d.dofNum] - 1] = FE.bigValue;
104            }
105        }
```

Method $\texttt{solve}$ (lines 84–94) obtains the right-hand side vector through its head, and the resulting solution is placed in the same array. If the matrix is newly assembled ($\texttt{newMatrix = true}$) then application of displacement boundary conditions (line 87) and LDU decomposition is performed. We will program two variants of a decomposition method – one normal and the other tuned. After matrix decomposition, variable $\texttt{newMatrix}$ is set to false. The solution vector x is provided by method $\texttt{lduFrwdBksb}$ that performs forward reduction and back-substitution for the right-hand side.

Method $\texttt{displacementBC}$ shown in lines 97–105 applies displacement boundary conditions with the use of the large number algorithm (14.3). $\texttt{ListIterator}$ object $\texttt{defDs}$ is used to obtain data on degrees of freedom and values of specified displacements.

Considering computer implementation of the *LDU* solution algorithm, we first present an algorithm using two-index notation A_{ij} for the system matrix $[A]$ stored in the symmetric profile format. The LDU algorithm of factorization of the system matrix $[A]$ is as follows:

LDU factorization

do $j = 2, N$
  $Cdivt(j)$
  do $i = j, N$
    $CCmod(j, i)$
  end do
end do
do $j = 2, N$
  $Cdiv(j)$
end do

$Cdivt(j) =$
  do $i = FN(j)), j - 1$
    $t_i = A_{ij}/A_{ii}$
  end do

$CCmod(j, i) =$
  do $k = max(FN(i), FN(j)), j - 1$
    $A_{ji} = A_{ji} - t_k A_{ki}$

end do

$$
Cdiv(j) = \\
\quad \textbf{do } i = FN(j)), j - 1 \\
\qquad A_{ij}/ = A_{ii} \\
\quad \textbf{end do}
$$

Here, N is the number of equations, *Cdivt*, *CCmod*, and *Cdiv* are procedures performing operations on columns of the matrix, $FN(i)$ is the function (15.3) that determines a row in the full matrix containing the first nonzero entry in column i. Most computing time during solution is spent for multiply-add operation in procedure *CCmod*, since this multiply-add operation appears inside a triple do loop.

Programming implementation of the *LDU* factorization is shown below.

```
107        // LDU decomposition for symmetric matrix
108        private int lduDecomposition() {
109            // Working array
110            double[] w = new double[neq];
111
112            // UtDU decomposition
113            for (int j = 1; j < neq; j++) {
114                int jfirst = j - (pcol[j+1] - pcol[j]) + 1;
115                int jj = pcol[j+1] - j - 1;
116                for (int i = jfirst; i < j; i++)
117                    w[i] = A[i+jj]/A[pcol[i+1] - 1];
118                for (int i = j; i < neq; i++) {
119                    int ifirst = i - (pcol[i+1] - pcol[i]) + 1;
120                    int ii = pcol[i+1] - i - 1;
121                    double s = 0.0;
122                    for (int m = Math.max(jfirst, ifirst);
123                            m < j; m++) s += A[m+ii]*w[m];
124                    A[j+ii] -= s;
125                }
126            }
127            for (int j = 0; j < neq; j++) {
128                int jfirst = j - (pcol[j+1] - pcol[j]) + 1;
129                int jj = pcol[j+1] - j - 1;
130                for (int i = jfirst; i < j; i++)
131                    A[i+jj] /= A[pcol[i+1] - 1];
132            }
133            return 0;
134        }
```

Implementation of procedures *Cdivt*, *CCmod*, and *Cdiv* is contained in lines 116–117, 119–124 and 128–131, respectively. Values of integer variables `ifirst`, `jfirst` correspond to values produced by $FN(i)$ and $FN(j)$. Equation 15.4 is employed for transforming two-index array entries A_{ij} into members of `A[]`.

Forward reduction and back-substitution for right-hand vector b are given by the pseudocode:

Forward reduction

do $j = 2, N$
   do $i = FN(j), j - 1$
      $b_j = b_j - A_{ij} * b_i$
   end do
end do

Back substitution

do $j = 1, N$
   $b_j = b_j / A_{jj}$
end do
do $j = N, 1, -1$
   do $i = FN(j), j - 1$
      $b_i = b_i - A_{ij} * b_j$
   end do
end do

The resulting solution is obtained in vector b replacing the specified right-hand side. The listing of method `lduFrwdBksb`, which implements the forward reduction and back-substitution algorithm, is shown below.

```
136        // Forward reduction and backsubstitution
137        private void lduFrwdBksb(double[] x) {
138
139            //   b = (U)-T*b
140            for (int j = 1; j < neq; j++) {
141                int jfirst = j - (pcol[j+1] - pcol[j]) + 1;
142                int jj = pcol[j+1] - j - 1;
143                for (int i = jfirst; i < j; i++)
144                    x[j] -= A[i+jj]*x[i];
145            }
146            //   b = (U)-1*(D)-1*b
147            for (int i = neq - 1; i >= 0; i--) {
148                x[i] /= A[pcol[i+1] - 1];
149                for (int j = i + 1; j < neq; j++) {
150                    int jfirst = j - (pcol[j+1] - pcol[j]) + 1;
151                    if (i >= jfirst) {
152                        int jj = pcol[j+1] - j - 1;
153                        x[i] -= A[i+jj]*x[j];
154                    }
155                }
156            }
157        }
```

Forward reduction of the right-hand side x is performed in lines 140–145. Back-substitution loop in lines 147–156 gives the solution in array x.

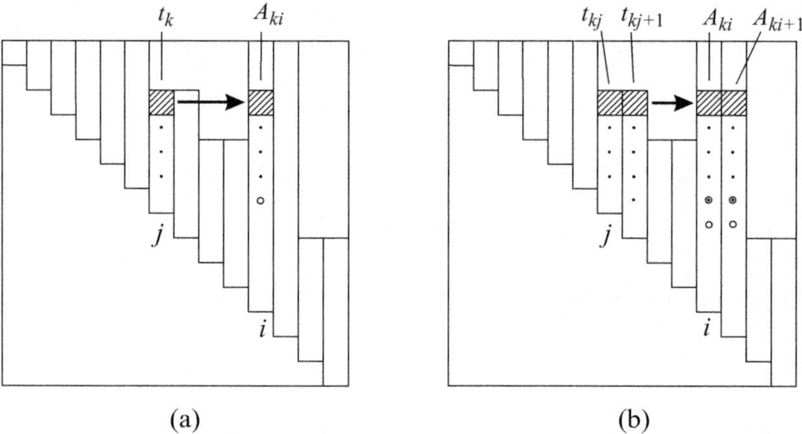

Fig. 15.2 In normal LDU decomposition two operands are involved in one floating-point operation inside the inner loop (a). Tuning with block size 2: four operands and four operations (b)

15.4 Tuning of the LDU Factorization

The triple do loop that takes most time of LDU decomposition is contained in the procedure *CCmod* presented in Section 15.3. One column of the matrix is used to modify another column inside the inner do loop as shown in Figure 15.2a. Operands t_k and A_{ki} are used in the algorithm *CCmod* of a column modification. Array t contains the modified matrix column j. Two operands should be loaded from memory in order to perform one floating-point multiply-add operation. Modern processors overlap data loads from memory and data stores to memory with arithmetic operations. Usually, load (store) time is equal to time of multiply-add operation. If the number of data loads and stores is larger than the number of arithmetic operations the processor idles for some time waiting for data transfer.

Data loads can be economized by tuning LDU factorization CCmod with the use of a blocking technique. With a blocking technique [23] one column block modifies the other column block. An algorithm of tuned LDU decomposition with block size $d = 2$ is given below.

Tuned LDU factorization

do $j = 1, N, d$
    $Bdivt(j, d)$
    do $i = j + d, N, d$
        $BBmod(j, i)$
    end do
end do
do $j = 2, N$
    $Cdiv(j)$

end do

$$Bdivt(k,d) =$$
    do $j = k, k+d-1$
        do $i = FN(k)), j-1$
            $t_{ij} = A_{ij}/A_{ii}$
        end do
        do $i = j, k+d-1$
            do $l = max(FN(j), FN(i)), j-1$
                $A_{ji} = A_{ji} - t_{lj}A_{li}$
            end do
        end do
    end do

$$BBmod(j,i,d=2) =$$
    do $k = max(FN(j), FN(i)), j-1$
        $A_{ji} = A_{ji} - t_{kj}A_{ki}$
        $A_{j+1i} = A_{j+1i} - t_{kj+1}A_{ki}$
        $A_{ji+1} = A_{ji+1} - t_{kj}A_{ki+1}$
        $A_{j+1i+1} = A_{j+1i+1} - t_{kj+1}A_{ki+1}$
    end do
    if $j >= FN(j)$**then**
        $A_{j+1i} = A_{j+1i} - t_{jj+1}A_{ji}$
        $A_{j+1i+1} = A_{j+1i+1} - t_{jj+1}A_{ji+1}$
    end if

Procedure $BBmod(j,i,d)$ performs modification of a column block, which starts from column i, by a column block, which starts from column j and contains d columns. Operands of the inner triple loop t_{kj}, t_{kj+1}, A_{ki} and A_{ki+1} are shown in Figure 15.2b. The pseudocode given above reveals that for block size $d = 2$ the four operands are used in four multiply-add operations. This provides a better balance between data loading and processing. Such a block size is suitable for solution of two-dimensional problems since the block size is equal to the number of degrees of freedom per node. In three-dimensional problems, where each node contains three degrees of freedom, block size $d = 3$ is used. In this case, nine arithmetic operations are done on six data items. The tuned algorithm assumes that columns in the block start at the same row of the full matrix A. This is fulfilled automatically if the column block contains columns, which are related to one node of the finite element model. Implementation of the tuned LDU decomposition for two- and three-dimensional finite element problems is as follows.

```
159        // Tuned LDU decomposition for symmetric matrix
160        // (block-block tuning, block size = 2 - 2D; 3 - 3D)
161        private int lduDecompositionTuned() {
162            double s0, s1, s2, t0, t1, t2, u0, u1, u2;
163            int ndf = fem.nDf;
164            double w[] = new double[ndf*neq];
```

```
165        double z[] = new double[neq];
166
167    int maxcol = 0;
168    for (int i = 0; i < neq; i++)
169        maxcol = Math.max(maxcol, pcol[i+1] - pcol[i]);
170
171    // UtDU decomposition
172    for (int i = 0; i < neq; i++)
173        z[i] = 1.0/A[pcol[i+1]-1];
174
175    for (int jc = 1; jc <= neq; jc += ndf) {
176        int jfirst = jc - (pcol[jc+1-1] - pcol[jc-1]) + 1;
177        int n1 = pcol[jc+ndf-1] - pcol[jc+ndf-2];
178        int jw = n1 - (jc + ndf - 1);
179        for (int j = jc; j < jc + ndf; j++) {
180            int jj = pcol[j] - j;
181            for (int i = jfirst; i < j; i++)
182                w[i+jw+n1*(j-jc)-1] = A[i+jj-1]*z[i-1];
183            for (int i = j; i < jc + ndf; i++) {
184                int ifirst = i - (pcol[i] - pcol[i-1]) + 1;
185                int ii = pcol[i] - i;
186                s0 = 0.0;
187                for (int m = Math.max(jfirst, ifirst);
188                         m < j; m++)
189                    s0 += A[m+ii-1]*w[m+jw+n1*(j-jc)-1];
190                A[j+ii-1] -= s0;
191            }
192            z[j-1] = 1.0/A[pcol[j]-1];
193            w[(j-jc+1)*n1-1] = z[j-1];
194        }
195        if (ndf == 2) {
196            for (int i = jc+ndf;
197                     i<Math.min(neq, jc+ndf+maxcol); i+=ndf) {
198                int ifirst = i - (pcol[i] - pcol[i-1]) + 1;
199                int ii = pcol[i]-i;
200                int ii1 = pcol[i+1]-(i+1);
201                s0 = s1 = t0 = t1 = 0;
202                for (int m = Math.max(jfirst, ifirst)-1;
203                                     m < jc-1; m++) {
204                    s0 += A[m+ii]*w[m+jw];
205                    s1 += A[m+ii]*w[m+jw+n1];
206                    t0 += A[m+ii1]*w[m+jw];
207                    t1 += A[m+ii1]*w[m+jw+n1];
208                }
209                if (jc >= ifirst) {
210                    A[jc+ii-1] -= s0;
211                    A[jc+ii1-1] -= t0;
212                    s1 += A[jc+ii-1]*w[jc+jw+n1-1];
213                    A[jc+1+ii-1] -= s1;
214                    t1 += A[jc+ii1-1]*w[jc+jw+n1-1];
215                    A[jc+1+ii1-1] -= t1;
216                }
217            }
218        }
```

```java
219                    else {     // tuned for nDf = 3
220                        for (int i = jc+ndf;
221                             i<Math.min(neq, jc+ndf+maxcol); i+=ndf) {
222                            int ifirst = i - (pcol[i] - pcol[i-1]) + 1;
223                            int ii = pcol[i]-i;
224                            int ii1 = pcol[i+1]-(i+1);
225                            s0 = s1 = t0 = t1 = 0;
226                            int ii2 = pcol[i+2]-(i+2);
227                            s2 = t2 = u0 = u1 = u2 = 0;
228                            for (int m = Math.max(jfirst, ifirst)-1;
229                                                     m < jc-1; m++) {
230                                s0 += A[m+ii]*w[m+jw];
231                                s1 += A[m+ii]*w[m+jw+n1];
232                                s2 += A[m+ii]*w[m+jw+2*n1];
233                                t0 += A[m+ii1]*w[m+jw];
234                                t1 += A[m+ii1]*w[m+jw+n1];
235                                t2 += A[m+ii1]*w[m+jw+2*n1];
236                                u0 += A[m+ii2]*w[m+jw];
237                                u1 += A[m+ii2]*w[m+jw+n1];
238                                u2 += A[m+ii2]*w[m+jw+2*n1];
239                            }
240                            if (jc >= ifirst) {
241                                A[jc+ii-1] -= s0;
242                                A[jc+ii1-1] -= t0;
243                                s1 += A[jc+ii-1]*w[jc+jw+n1-1];
244                                A[jc+1+ii-1] -= s1;
245                                t1 += A[jc+ii1-1]*w[jc+jw+n1-1];
246                                A[jc+1+ii1-1] -= t1;
247                                A[jc+ii2-1] -= u0;
248                                u1 += A[jc+ii2-1]*w[jc+jw+n1-1];
249                                A[jc+1+ii2-1] -= u1;
250                                s2 += A[jc+ii-1]*w[jc+jw+2*n1-1]
251                                    +A[jc+1+ii-1]*w[jc+jw+1+2*n1-1];
252                                A[jc+2+ii-1] -= s2;
253                                t2 += A[jc+ii1-1]*w[jc+jw+2*n1-1]
254                                    +A[jc+1+ii1-1]*w[jc+jw+1+2*n1-1];
255                                A[jc+2+ii1-1] -= t2;
256                                u2 += A[jc+ii2-1]*w[jc+jw+2*n1-1]
257                                    +A[jc+1+ii2-1]*w[jc+jw+1+2*n1-1];
258                                A[jc+2+ii2-1] -= u2;
259                            }
260                        }
261                    }
262            }
263        for (int j = 1; j <= neq; j++) {
264            int jfirst = j - (pcol[j+1-1] - pcol[j-1]) + 1;
265            int jj = pcol[j+1-1] - j;
266            for (int i = jfirst; i < j; i++)
267                A[i+jj-1] *= z[i-1];
268        }
269        return 0;
270    }
271
272 }
```

Line 164 declares working array w, which contains ndf vectors with length equal to the number of equations. Parameter ndf is the number of degrees of freedom per node. The method employs array w to store a column block for improving decomposition efficiency. Array z allocated in line 165 is used for storing an inverse of the diagonal matrix entries in order to avoid repeated divisions (division takes a much longer time compared to multiplication).

Statements in lines 167–169 calculate maximum column height $maxcol$ among all columns of the matrix profile. The $maxcol$ is used to restrict a limit of do loop for the block-block modification.

Block procedure $Bdivt$ is implemented in lines 179–194. Procedure $BBmod$ performing block-block column modification in the two-dimensional case corresponds to lines 195–218. Lines 219–262 contain statements of $BBmod$ for the three-dimensional case.

Block-block tuning of the LDU decomposition significantly affects the solution speed. For C codes the solution time can be decreased by about a factor of two. For Java code, a tuned solution can take as little as a quarter of the time of an untuned algorithm.

Problems

15.1. Global stiffness matrix $[K]$ has the following nonzero coefficients:

$$[K] = \begin{bmatrix} 2 & -1 & & & & & & & \\ -1 & 3 & & -2 & & 1 & & & \\ & & 4 & & 1 & -1 & & & \\ & -2 & & 4 & & & 1 & & \\ & & 1 & & 3 & & & & \\ & 1 & -1 & & & & 2 & -2 & -1 \\ & & & 1 & & & -2 & 4 & 1 \\ & & & & & & -1 & 1 & 2 \end{bmatrix}.$$

Write down this matrix in the symmetric profile format using pointer array $pcol$ and matrix coefficient array A. Start the pointer array with zero.

15.2. Calculate the location of the first nonzero coefficient in the last column of the global stiffness matrix given in Problem 15.1. Use Equation 15.3 and the pointer array $pcol$ from the solution of the previous problem.

15.3. Using Equation 15.4 calculate the index of matrix entry A_{45} in the one-dimensional array A where the matrix is stored in the symmetric profile format. Employ the data of Problem 15.1.

15.4. Create a main method in class $SolverLDU$ that tests methods for LDU decomposition $lduDecomposition$ and forward reduction and back-substitution $lduFrwdBksb$ on some simple equation system.

Chapter 16
Iterative Equation Solver

Abstract This chapter describes solution of finite element equation systems using an iterative preconditioned gradient method. Since an equation system matrix is not changed during solution it is useful to store it in in sparse-row format. Algorithms of matrix assembly and iterative solution using sparse-row format are presented. A preconditioned gradient solver is implemented as Java^{TM} class `SolverPCG`.

16.1 Preconditioned Conjugate Gradient Method

A simple and efficient iterative method widely used for the solution of sparse equation systems is the conjugate gradient (CG) method. In many cases the convergence of the CG method can be too slow for practical purposes. The convergence rate can be improved by preconditioning the equation system,

$$[M]^{-1}[A]x = [M]^{-1}\{b\}, \tag{16.1}$$

where $[M]^{-1}$ is the inverse of the preconditioning matrix, which in some sense approximates $[A]^{-1}$. The simplest preconditioning is diagonal preconditioning, in which $[M]$ contains only the diagonal entries of the matrix $[A]$. An algorithm of the preconditioned conjugate gradient (PCG) method can be presented as the following sequence of computations [13]:

PCG algorithm

Compute $[M]$
$\{x_0\} = 0$
$\{r_0\} = \{b\}$
do $i = 0, 1...$
    $\{w_i\} = [M]^{-1}\{r_i\}$
    $\gamma_i = \{r_i\}^{\text{T}}\{w_i\}$
    if $i = 0$ $\{p_i\} = \{w_i\}$

$$\textbf{else } \{p_i\} = \{w_i\} + (\gamma_i/\gamma_{i-1})\{p_{i-1}\}$$
$$\{w_i\} = [A]\{p_i\}$$
$$\beta_i = \{p_i\}^{\mathrm{T}}\{w_i\}$$
$$\{x_i\} = \{x_{i-1}\} + (\gamma_i/\beta_i)\{p_i\}$$
$$\{r_i\} = \{r_{i-1}\} - (\gamma_i/\beta_i)\{p_i\}$$
$$\textbf{if } \gamma_i/\gamma_0 < \varepsilon \textbf{ exit}$$
end do

Compared to the direct LDU solution method, the PCG algorithm requires less memory for storage of the equation system. Also, the computational efficiency of the PCG method is lower than that of the LDU method. Thus, the PCG method is better suited for the solution of large finite element equation systems.

16.2 Assembly of Matrix in Sparse-row Format

In the above PCG algorithm, matrix $[A]$ is not changed during computations. This means that no additional fill arises in the solution process. Theoretically, it is possible to store just the symmetric part of the global stiffness matrix in a compact form. However, the storage scheme should allow fast matrix-vector multiplication, since this is the most expensive operation inside the iteration loop of the PCG method. Recovering some coefficients of a row using symmetry may take considerable time.

Because of this, the sparse-row format for the entire matrix is an efficient storage scheme for the PCG iterative method. In this scheme, the values of nonzero entries of matrix $[A]$ are stored by rows along with their corresponding column indexes. An additional array points to the beginning of each row.

Thus, the matrix in the sparse-row format is determined by the following three arrays:

$prow[N+1]$ = pointers to the beginning of each row;
$coln[prow[N]]$ = column numbers for nonzero matrix entries;
$A[prow[N]]$ = array containing nonzero entries of the matrix.

The ith element of `prow` is the address of the first entry of a row. The number of entries in the ith row is given by `prow[i+1]-prow[i]`. The length of arrays `coln` and A is equal to `prow[N]`.

Fill of the global stiffness matrix for the mesh of Figure 14.1 is depicted in Figure 16.1a. Storage of this matrix in sparse-row format is shown in Figure 16.1b. Arrays `pcol` and `col` have the following contents:

```
prow = {0,  4,  10,  14,  18,  26,  32,  36,  40},
coln = {0,  1,  3,  4;  0,  1,  2,  3,  4,  5;
        1,  2,  4,  5;  0,  1,  3,  4;
        0,  1,  2,  3,  4,  5,  6,  7;  1,  2,  4,  5,  6,  7;
        4,  5,  6,  7;  4,  5,  6,  7 } .
```

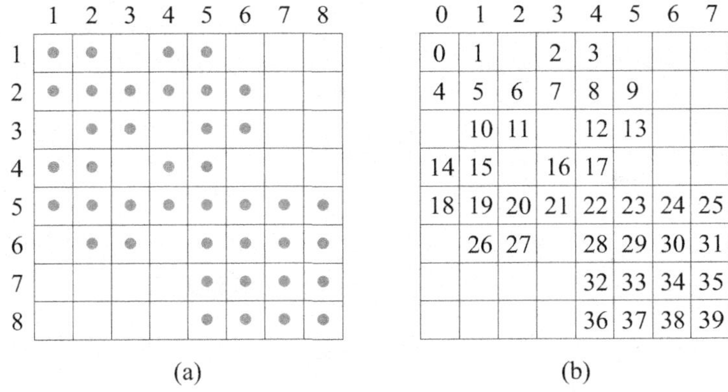

Fig. 16.1 Fill of the global stiffness matrix (a) and its storage in the sparse-row format (b)

The first entry of array `prow` is always zero and the last entry contains the number of coefficients in array A.

An algorithm for creation of pointer array `prow` and column array `coln` using connectivities of finite elements can be presented with the following pseudocode.

Data for the sparse-row format
n = number of degrees of freedom per element
N = total number of degrees of freedom in the domain
E = number of elements
$C[E,n]$ = connectivity array
$p[N+1]$ = pointers to rows
$c[N][..]$ = lists of column numbers for matrix rows
Determine lists of column numbers
do $e = 1, E$
    do $i = 1, n$
        do $j = 1, n$
            if $(C[e,j] - 1)$ is not in $c[C[e,i]][..]$
                store $(C[e,j] - 1)$ in $c[C[e,i]][..]$
            end if
        end do
    end do
end do
Compute pointers
$p[0] = 0$
do $i = 2, N+1$
    $p[i] = p[i-1] +$ **length of** $c[i-1]$
end do

In the algorithm, loop indexes are counted from one, as in Figure 16.1a. Pointers p and columns c start from zero as in Figure 16.1b. The resulting column numbers are placed in $c[N][..]$, an array of row lists. Using a list for each row is convenient since the number of nonzero entries in a row is not known in advance.

A triple loop as in assembly is used to form the column numbers. A combination of element connectivity numbers indicates the row and column numbers. If the column number is not present in the column list for this row then the column number is added to the list. The pointer array is created by accumulating lengths of row arrays for column numbers.

The partial source code of class `SolverPCG` containing the constructor and the method for creating the sparse-row structure of the global stiffness matrix is given below.

```java
 1 package solver;
 2
 3 import fea.*;
 4 import model.*;
 5 import elem.*;
 6 import util.UTIL;
 7
 8 import java.util.ListIterator;
 9
10 // Preconditioned conjugate gradient (PCG) solver.
11 // Matrix storage: sparse-row format
12 public class SolverPCG extends Solver {
13
14     // Pointers to matrix rows
15     private int[] prow;
16     // Column numbers for non-zero values in A
17     private int[] coln;
18     // Nonzero values of the global stiffness matrix by rows
19     private double[] A;
20     // Working arrays
21     private double b[], r[], w[], p[], md[];
22
23     // Constructor for PCG solver.
24     public SolverPCG() {
25
26         // Set structure of matrix A
27         setSparseRowStructure();
28
29         A = new double[prow[neq]];
30         b = new double[neq];
31         r = new double[neq];
32         w = new double[neq];
33         p = new double[neq];
34         md = new double[neq];
35     }
36
37     // Set sparse row structure for storage of nonzero
38     //   coefficients of the global stiffness matrix.
39     private void setSparseRowStructure() {
40         prow = new int[neq+1];
41         int lrow = fem.nDf *((fem.nDf == 2) ?
42                             FE.maxRow2D : FE.maxRow3D);
```

```
43            coln = new int[neq*lrow];
44            for (int i = 0; i <= fem.nNod; i++)
45                prow[i] = i*lrow*fem.nDf;
46
47            // Create nodal sparse-row matrix structure
48            for (int i = 0; i < prow[fem.nNod]; i++) coln[i] = -1;
49            // Diagonal entry - first in row
50            for (int i = 0; i < fem.nNod; i++)  coln[prow[i]] = i;
51            for (int iel = 0; iel < fem.nEl; iel++) {
52                for (int anInd : fem.elems[iel].ind) {
53                    if (anInd == 0) continue;
54                    int ii = anInd - 1;   // Hyperrow
55                    for (int anInd1 : fem.elems[iel].ind) {
56                        if (anInd1 == 0) continue;
57                        int jj = anInd1 - 1;   // Hypercolumn
58                        int k;
59                        for (k=prow[ii]; k<prow[ii+1]; k++) {
60                            // If column already exists
61                            if (coln[k] == jj) break;
62                            if (coln[k] == -1) {
63                                coln[k] = jj;
64                                break;
65                            }
66                        }
67                        if (k==prow[ii+1]) UTIL.errorMsg(
68                            "PCG sparse-row structure: "+
69                            "not enough space for node "+ii);
70                    }
71                }
72            }
73
74            // Compress
75            int p = 0;
76            for (int i = 0; i < fem.nNod; i++) {
77                int k = prow[i];
78                prow[i] = p;
79                for (int j = k; j < prow[i + 1]; j++) {
80                    if (coln[j] == -1) break;
81                    coln[p++] = coln[j];
82                }
83            }
84            prow[fem.nNod] = p;
85
86            // Transform to degrees of freedom
87            int pdof = p*fem.nDf *fem.nDf;
88            for (int i = fem.nNod - 1; i >= 0; i--) {
89                int deln = (prow[i+1] - prow[i]);
90                p -= deln;
91                for (int k = fem.nDf; k > 0; k--) {
92                    prow[i*fem.nDf+k] = pdof;
93                    pdof -= deln*fem.nDf;
94                    for (int j = prow[i+1]-prow[i]-1; j>=0; j--) {
95                        for (int m = fem.nDf-1; m >= 0; m--) {
96                            coln[pdof+j*fem.nDf+m] =
```

```
 97                                     coln[p+j]*fem.nDf + m;
 98                            }
 99                         }
100                     }
101                 }
102             lengthOfGSM = (int) (prow[neq]*1.5);
103         }
```

In the beginning of class `SolverPCG` arrays `prow`, `coln`, and `A` describing the global stiffness matrix in sparse-row format are declared. The constructor calls the method creating arrays `prow` and `coln` (line 27), and allocates all other arrays necessary for assembly and solution.

Method `setSparseRowStructure` creates the sparse-row structure for the global stiffness matrix according to the algorithm given above. In the algorithm we used the array of lists for storing column numbers for nonzero matrix entries. This makes the algorithm simpler but implies creation of a large number of objects. To avoid this we implement the code in a more efficient way using just one-dimensional arrays. Line 40 allocates array `prow`, the length of which is known.

The length of column array `coln` is not known in advance. Because of this we allocate the same memory for all rows of the matrix using the specified maximum average number of nodes per row, `FE.maxRow2D` and `FE.maxRow3D` for two- and three-dimensional problems, respectively. In our implementation the value of `FE.maxRow2D` is defined as 21. For a regular mesh of eight-node elements, a row corresponding to an inner corner node contains entries from twenty one nodes and a row for a midside node contains entries from thirteen nodes. In the three-dimensional case, the corresponding numbers of nodes per row are 117 for an inner vertex node and 57 for a midside node. The value `FE.maxRow3D` is set to 117. Selection of corner and vertex values as the average number of nodes per row creates extra space in array `coln` allowing some irregular nodes surrounded by a larger number of elements.

Column array `coln` is allocated in line 43. The average length for each row is calculated as the number of nodes per row multiplied by the number of degrees of freedom per node `nDf`.

Memory for the first `nNod` rows is used to create column information `coln` in terms of nodes. Variable `nNod` contains the number of nodes for the finite element model. In line 48, `nNod` rows of `coln` are filled with -1.

The loop in lines 51–72 forms lists of nodes connected with the current node through the surrounding elements. Lines 75–84 compress nodal pointer array `prow` and nodal column array `coln`. Transformation of nodal arrays to degrees of freedom is done in lines 87–101. Filling arrays corresponding to degrees of freedom in reverse order helps preserve nodal arrays located in the beginning of the same arrays.

Statement 102 sets the value of variable `lengthOfGSM` that is the length of the global stiffness matrix in double words. Coefficient 1.5 is used to account for column pointer `pcol` consisting of integers.

Method `assembleESM` performs assembly of an element stiffness matrix to the global stiffness matrix stored in the sparse-matrix format:

```
105        // Assemble element matrix to the global stiffness matrix
106        public void assembleESM() {
107            for (int j = 0; j < nindf; j++) {
108                int jj = indf[j] - 1;
109                if (jj >= 0) {
110                    for (int i = 0; i < nindf; i++) {
111                        int ii = indf[i] - 1;
112                        if (ii >= 0) {
113                            // Sparse row format (full matrix)
114                            int k;
115                            for (k=prow[jj]; k<prow[jj+1]; k++)
116                                if (coln[k] == ii) break;
117                            if (i <= j) A[k] += Element.kmat[i][j];
118                            else        A[k] += Element.kmat[j][i];
119                        }
120                    }
121                }
122            }
123        }
```

Connectivities in terms of degrees of freedom `indf` are employed for assembly.
A pair of connectivity numbers defines a row and a column in the global stiffness
matrix. Row number (jj) is used to organize looping over row entries. Column
number ii is sought in array entries `coln` for the current row. When found, the
coefficient of element stiffness matrix `kmat` is added to global stiffness matrix A
(lines 117–118). Two statements are necessary since the element methods provide
only the upper symmetric part of the element stiffness matrix.

16.3 PCG Solution

Solution of the equation system by the preconditioned gradient method is performed
by method `solve`, which should be implemented in each solver class. Listing of
this method and a method applying displacement boundary conditions is given below.

```
125        // Solve equation system by PCG method.
126        // x - right-hand side/solution (in/out)
127        public int solve(double x[]) {
128
129            if (newMatrix) {
130                displacementBC();
131                newMatrix = false;
132            }
133            return pcg(x);
134        }
135
136        // Apply displacement boundary conditions:
137        private void displacementBC() {
138
139            ListIterator it = fem.defDs.listIterator(0);
```

```
140            Dof d;
141            while (it.hasNext()) {
142                d = (Dof) it.next();
143                int j = d.dofNum-1;
144                for (int k = prow[j]; k < prow[j+1]; k++)
145                    if (coln[k] == j) A[k] = FE.bigValue;
146                    else              A[k] = 0.0;
147            }
148        }
```

Method `solve` obtains a right-hand side `x` as a parameter and replaces it by a solution vector. If the global stiffness matrix is newly formed then displacement boundary conditions are applied in line 130. Solution itself is done by method `pcg`, which is called in line 132.

Method `displacementBC` presented in lines 137–148 performs application of displacement boundary conditions using the method of a large value. According to this method a number that is much larger than the matrix coefficients is placed on the matrix main diagonal for a degree of freedom with a prescribed displacements.

The PCG method is implemented in method `pcg`. This method and two other methods (computing diagonal preconditioner and matrix-vector product) are presented below.

```
150        // PCG solution method.
151        // x - right-hand side/solution (in/out)
152        public int pcg(double x[]) {
153
154            diagonalPreconditioner();
155
156            // Save x[] in b[] and calculate initial x
157            for (int i = 0; i < neq; i++) {
158                b[i] = x[i];
159                x[i] = x[i]*md[i];
160            }
161
162            // r = b - A*x and initinal error
163            matrixVectorProduct(x, r);
164            for (int i = 0; i < neq; i++) r[i] = b[i] - r[i];
165            double gamma0 = 1;
166            double gammai = 1;
167            int iter;
168            for (iter = 0; iter < FE.maxIterPcg; iter++) {
169                //   w = (M-1)*r
170                for (int i = 0; i < neq; i++)
171                    w[i] = md[i]*r[i];
172                // gam = (r,w)
173                double gammai1 = gammai;
174                gammai = 0;
175                for (int i = 0; i < neq; i++)
176                    gammai += r[i]*w[i];
177                if (iter == 0) {
178                    gamma0 = gammai;
179                    System.arraycopy(w, 0, p, 0, neq);
180                }
```

```
181             else {
182                 double rg = gammai /gammai1;
183                 for (int i = 0; i < neq; i++)
184                     p[i] = w[i] + rg*p[i];
185             }
186             // w = A*p
187             matrixVectorProduct(p, w);
188             double beta = 0;
189             for (int i = 0;  i < neq; i++) beta += p[i]*w[i];
190             double alpha = gammai/beta;
191             // Update x and r, calculate error
192             for (int i = 0; i < neq; i++) {
193                 x[i] += alpha*p[i];
194                 r[i] -= alpha*w[i];
195             }
196             double err = Math.sqrt(gammai /gamma0);
197             if (err < FE.epsPCG) return (iter + 1);
198         }
199         return (iter);
200     }
201
202     // Diagonal preconditioner md = (Diag(A))-1.
203     private void diagonalPreconditioner() {
204         for (int j = 0; j < neq; j++) {
205             int i;
206             for (i = prow[j]; i < prow[j+1]; i++)
207                     if (coln[i] == j) break;
208             md[j] = 1.0/A[i];
209         }
210     }
211
212     //  Sparse matrix-vector product y = A*x.
213     private void matrixVectorProduct(double x[], double y[]) {
214         if (FE.tunedSolver) {
215             if (fem.nDf == 2) {  // tuned for nDf = 2
216                 for (int j = 0; j < neq; j++) {
217                     double s = 0;
218                     for (int i = prow[j]; i < prow[j+1]; i+=2)
219                         s += A[i   ]*x[coln[i   ]]
220                             + A[i+1]*x[coln[i+1]];
221                     y[j] = s;
222                 }
223             }
224             else {      // tuned for nDf = 3
225                 for (int j = 0; j < neq; j++) {
226                     double s = 0;
227                     for (int i = prow[j]; i < prow[j+1]; i+=3)
228                         s += A[i   ]*x[coln[i   ]]
229                             + A[i+1]*x[coln[i+1]]
230                             + A[i+2]*x[coln[i+2]];
231                     y[j] = s;
232                 }
233             }
234         }
```

```
235              else {              // not tuned
236                  for (int j = 0; j < neq; j++) {
237                      double s = 0;
238                      for (int i = prow[j]; i < prow[j+1]; i++)
239                          s += A[i]*x[coln[i]];
240                      y[j] = s;
241                  }
242              }
243          }
244
245  }
```

Computing a diagonal preconditioner is done in line 154. Lines 203–210 contain method `diagonalPreconditioner`, which composes the preconditioner vector of the inverse diagonal entries of the global stiffness matrix.

Solution of the equation system is sought inside the iteration loop (lines 168–198) according to the PCG algorithm described in Section 16.1. During the iteration process, the most expensive operation is the matrix-vector multiplication at line 187.

Multiplication of sparse matrix A stored in the sparse-row format by vector x is performed by method `matrixVectorProduct`. The result is placed in vector y. Lines 236–241 implement a conventional algorithm for the sparse matrix-vector product. Matrix entries `A[i]` are used sequentially, and entries of vector x are used more or less randomly.

It appears that the Java compiler is far from producing efficient code. A sparse matrix-vector product can be tuned in a very simple manner. Tuning is achieved by unrolling the inner loop and by explicit summation of two (two-dimensional case, lines 216–222) or three (three-dimensional case, lines 225–232) terms related to one node. Such tuning increases the speed of the PCG solver by about 1.5 times [24]. It is interesting to note that unrolling two loops (multiplication for a square block related to a node) decreases the tuning efficiency.

Problems

16.1. A global stiffness matrix $[K]$ has the following appearance:

$$[K] = \begin{bmatrix} 2 & -1 & & & & & \\ -1 & 3 & & -2 & & 1 & \\ & & 4 & & 1 & -1 & \\ & -2 & & 4 & & & \\ & & 1 & & 3 & & \\ & 1 & -1 & & & 2 \end{bmatrix}.$$

Represent this matrix in sparse-row format by creating row pointers `prow`, column numbers `coln`, and matrix coefficient array A. Index the row pointer array from zero.

16.2. The matrix of Problem 16.1 is stored in the sparse-row format. Find the index of matrix entry A_{35} in one-dimensional array A using row pointers `prow` and column numbers `coln`.

16.3. Consider a regular mesh of two-dimensional quadrilateral elements. Determine the maximum number of nonzero coefficients in a row of the global stiffness matrix provided that the mesh consists of a) linear four-node elements, b) quadratic eight-node elements.

16.4. A regular three-dimensional mesh is composed of twenty-node brick-type elements. Find the maximum number of nonzero coefficients in a row of the global stiffness matrix if this row is related to a) a vertex node, b) a midside node.

Chapter 17
Load Data and Load Vector Assembly

Abstract Data for loading cases in solid mechanics problems is described. The following external loading factors can be specified: concentrated nodal forces, distributed surface forces, and thermal loading. Java^{TM} class `FeLoadData` declares load data items. Class `FeLoad` contains methods for input and handling load data. It includes a method for global load vector assembly.

17.1 Data Describing the Load

External loading factors cause deformation of the finite element model. A global load vector is the right-hand side of the global finite element equation system. It is assembled from element and nodal contributions due to various external loading factors. In this program we consider the following external loading factors:

- concentrated forces applied at nodes;
- distributed forces applied at element faces;
- thermal loading specified as the nodal temperature field.

The data on the load is contained in class `FeLoadData` belonging to package `model`. The class listing is given below.

```java
1  package model;
2
3  import util.FeScanner;
4  import java.util.LinkedList;
5
6  // Load data
7  public class FeLoadData {
8
9      FeScanner RD;
10     public static String loadStepName;
11     // Load scale multiplier
12     double scaleLoad;
13     // Relative residual norm tolerance
```

```
14       static double residTolerance = 0.01;
15       // Maximum number of iterations (elastic-plastic problem)
16       static int maxIterNumber = 100;
17       // Degrees of freedom with node forces
18       LinkedList nodForces;
19       // Element face surface loads
20       LinkedList surForces;
21       // Temperature increment
22       public static double[] dtemp;
23
24       // Increment of force load
25       static double[] dpLoad;
26       // Total force load
27       static double[] spLoad;
28       // Increment of fictitious thermal loading
29       static double[] dhLoad;
30       // Displacement increment
31       static double[] dDispl;
32       // Total displacements
33       static double[] sDispl;
34       // Right-hand side of global equation system
35       public static double[] RHS;
36
37       // Working arrays
38       static int[] iw = new int[8];
39       static double[] dw = new double[8];
40       static double[][] box = new double[2][3];
41
42       enum vars {
43           loadstep, scaleload,
44           residtolerance, maxiternumber,
45           nodforce, surforce, boxsurforce, nodtemp,
46           includefile, end
47       }
48
49  }
```

Data describing the load is declared in lines 10–22:

loadStepName – name for current load step;
scaleLoad – load scale multiplier used for scaling previous load;
residTolerance – relative residual norm tolerance used to stop elastic–plastic iterations;
maxIterNumber – maximum number of iterations in elastic–plastic problems;
nodForces – linked list for storing nodal forces;
surForces – linked list for storing element face forces;
dtemp – increment of nodal temperature field.

Parameter loadStepName is used for load-step identification. Output files have names, which include load step names as extensions. In linear elastic problems, there is not much sense in using scaling of the previous load for obtaining the current load since the results can also be obtained by scaling. However, in elastic–plastic problems, using parameter scaleLoad is often useful because it is desirable to divide the total load into several increments even for proportional loading. Parameters `residTolerance` and `maxIterNumber` are also related to elastic–plastic problems. The relative residual norm is determined as the norm of the residual vector divided by the norm of the force load increment. Elastic–plastic iterations stop when the relative residual norm becomes smaller than the value of `residTolerance`. The number of iterations can also be restricted by `maxIterNumber`.

Objects of concentrated nodal forces `nodForces` and distributed element face forces `surForces` are stored as linked lists since their numbers are not known in advance. Thermal loading is specified by defining temperature increment values at the nodes.

Lines 42–47 declare load data names of `enum` type:

`loadstep` – load step name;
`scaleload` – scaling factor for previous load;
`residtolerance` – relative residual norm tolerance;
`maxiternumber` – maximum allowable number of iterations;
`nodforce` – concentrated nodal forces;
`surforce` – distributed forces applied to element faces;
`boxsurforce` – distributed surface forces applied inside specified box;
`nodtemp` – temperature field specified at nodes;
`includefile` – include file containing load data;
`end` – end of data for current load step.

17.2 Load Data Input

Class `FeLoad` extends class `FeLoadData` and contains methods for input and handling data describing the load. The constructor of class `Feload` is presented below.

```
1  package model;
2
3  import fea.*;
4  import elem.*;
5  import util.*;
6
7  import java.util.ListIterator;
8  import java.util.LinkedList;
9
10 // Load increment for the finite element model
11 public class FeLoad extends FeLoadData  {
12
13     // Finite element model
14     private static FeModel fem;
15     ListIterator itnf, itsf;
16
17     // Construct finite element load.
18     // fem - finite element model
```

```
19      public FeLoad(FeModel fem) {
20          FeLoad.fem = fem;
21          RD = FeModel.RD;
22
23          spLoad  = new double[fem.nEq];
24          dpLoad  = new double[fem.nEq];
25          dhLoad  = new double[fem.nEq];
26          sDispl  = new double[fem.nEq];
27          dDispl  = new double[fem.nEq];
28          RHS     = new double[fem.nEq];
29
30          if (fem.thermalLoading) {
31              dtemp = new double[fem.nNod];
32          }
33      }
```

The constructor stores instances of the finite element model `fem` and the finite element data scanner RD (lines 20–21). Lines 23–28 allocate six arrays with dimension equal to the number of equations in the global equation system. Array `spLoad` contains entries of the accumulated force load, i.e., the total load minus the fictitious thermal load. Arrays `dpLoad` and `dhLoad` are increments of the global force load and the fictitious thermal load. Arrays `sDispl` and `dDispl` are the total displacement vector and its increment at the current load step. Array RHS is used as a working array for the right-hand side of the finite element equation system. An array of nodal temperatures `dtemp` is allocated if thermal loading is requested when data for the finite element model is specified.

Load data input is performed by calling method `readData`. The source code of this method and related methods follow.

```
35      // Read data describing load increment.
36      // returns  true if load data has been read
37      public boolean readData( ) {
38
39          return readDataFile(RD, true);
40      }
41
42      // Read data fragment for load increment.
43      // newLoad = true - beginning of new load,
44      //         = false - continuation of load.
45      // returns  true if load data has been read
46      private boolean readDataFile(FeScanner es,
47                                   boolean newLoad) {
48          if (newLoad) {
49              scaleLoad = 0;
50              nodForces = new LinkedList();
51              itnf = nodForces.listIterator(0);
52              surForces = new LinkedList();
53              itsf = surForces.listIterator(0);
54              if (fem.thermalLoading) {
55                  for (int i = 0; i < dtemp.length; i++)
56                      dtemp[i] = 0.0;
57              }
58              for (int i=0; i<dDispl.length; i++) dDispl[i] = 0;
```

```java
59              }

61          if (!es.hasNext()) return false;  // No load data

63          vars name = null;
64          String s;

66          while (es.hasNext()) {
67              String varName = es.next();
68              String varNameLower = varName.toLowerCase();
69              if (varName.equals("#")) {
70                  es.nextLine(); continue; }
71              try {
72                  name = vars.valueOf(varNameLower);
73              } catch (Exception e) {
74                  UTIL.errorMsg(
75                      "Variable name is not found: " + varName);
76              }

78          switch (name) {

80          case loadstep:
81              loadStepName = es.next();
82              break;

84          case scaleload:
85              scaleLoad = es.readDouble();
86              break;

88          case residtolerance:
89              residTolerance = es.readDouble();
90              break;

92          case maxiternumber:
93              maxIterNumber = es.readInt();
94              break;

96          case nodforce:
97              readNodalForces(es);
98              break;

100         case surforce:
101             readSurForces(es);
102             break;

104         case boxsurforce:
105             createBoxSurForces(es);
106             break;

108         case nodtemp:
109             dtemp = new double[fem.nNod];
110             for (int i = 0; i < fem.nNod; i++)
111                 dtemp[i] = es.readDouble();
112             break;
```

```
113
114              case includefile:
115                  s = es.next().toLowerCase();
116                  FeScanner R = new FeScanner(s);
117                  readDataFile(R, false);
118                  break;
119
120              case end:
121                  return true;
122              }
123          }
124          return true;
125      }
126
127
128      // Read data for specified nodal forces
129      private void readNodalForces(FeScanner es) {
130
131          String s = es.next().toLowerCase();
132          int idf = UTIL.direction(s);
133          if (idf == -1) UTIL.errorMsg("nodForce" +
134              " direction should be x/y/z. Specified:"+s);
135
136          if (!es.hasNextDouble()) UTIL.errorMsg(
137              "nodForce value is not a double: " + es.next());
138          double vd = es.nextDouble();
139
140          itnf = es.readNumberList(itnf, idf, fem.nDim, vd);
141      }
142
143      // Read data for surface forces (element face loading):
144      // direction, iel, nFaceNodes, faceNodes, forcesAtNodes.
145      private void readSurForces(FeScanner es) {
146
147          String s = es.next().toLowerCase();
148          int dir = UTIL.direction(s);
149          if (dir == -1) UTIL.errorMsg("surForce" +
150              " direction should be x/y/z/n. Specified:"+s);
151          int iel = es.readInt();
152          int nFaceNodes = es.readInt();
153          for (int i=0; i<nFaceNodes; i++)
154              iw[i] = es.readInt();
155          for (int i=0; i<nFaceNodes; i++)
156              dw[i] = es.readDouble();
157          itsf.add(new ElemFaceLoad(iel-1,nFaceNodes,dir,iw,dw));
158      }
159
160      // Create data for distributed surface load
161      // specified inside a box
162      private void createBoxSurForces(FeScanner es) {
163          int[][] faces;
164          String s = es.next().toLowerCase();
165          int dir = UTIL.direction(s);
166          if (dir == -1) UTIL.errorMsg("boxSurForce" +
```

```
167                    " direction should be x/y/z/n. Specified:" + s);
168
169            if (!es.hasNextDouble()) UTIL.errorMsg(
170                "boxSurForce value is not a double: " + es.next());
171            double force = es.nextDouble();
172
173            for (int i = 0; i < 2; i++)
174                for (int j = 0; j < fem.nDim; j++)
175                    box[i][j] = es.readDouble();
176
177            for (int iel=0; iel<fem.nEl; iel++) {
178                Element el = fem.elems[iel];
179                faces = el.getElemFaces();
180                FACE:
181                for (int[] face : faces) {
182                    int nNodes = face.length;
183                    for (int inod = 0; inod < nNodes; inod++)
184                        iw[inod] = 0;
185                    for (int inod = 0; inod < nNodes; inod++) {
186                        int iGl = el.ind[face[inod]];
187                        if (iGl > 0) {
188                            for (int j = 0; j < fem.nDim; j++) {
189                                double x =
190                                        fem.getNodeCoord(iGl-1,j);
191                                if (x < box[0][j] || x > box[1][j])
192                                    continue FACE;
193                            }
194                            iw[inod] = iGl;
195                        }
196                    }
197                    itsf.add(
198                        new ElemFaceLoad(iel,nNodes,dir,iw,force));
199                }
200            }
201    }
```

Method `readData` just calls method `readDataFile`, passing to it data scanner RD and indicating that input of new load data is requested. Actual load data input is performed in method `readDataFile`. The first parameter of the method is data scanner `es`. If the second parameter `newLoad` is asserted then statements in lines 49–58 are executed, thus initializing parameters for the new load increment. The method returns `false` if no data is available. If some data is read the `true` value is returned.

The data input loop includes lines 66–123. This loop continues while input items are available for scanner `es`. The text data item is read to string `varName` (line 67). Then it is transformed to lower case `varNameLower`. The next line checks if the item is the comment symbol #. In the case of a comment we proceed to the next line of the scanner and try to read the next data item. If the data item is not the comment symbol then we try to find the `name` that corresponds to the string `varNameLower` among the enumerated `vars`. If no correspondence is found then an error message is communicated using static method `errorMsg` of class `UTIL`.

If `varNameLower` is discovered among predetermined values in `vars` then a `switch` statement is executed in line 78 and a particular case statement performs input of the second part of an input statement, which may contain one or more (sometimes many) input items.

The statements in lines 81, 85, 89, and 93 simply input scalar data: `loadStepName` (name for the current load step), `scaleLoad` (parameter for scaling the previous load), `residTolerance` (tolerance for the relative residual norm), and `maxIterNumber` (maximum number of iterations).

Data on nodal forces (line 97) is processed by method `readNodalForces` (lines 129–141). A direction coded as `x/y/z` and a double value common for subsequent node numbers are read in lines 132 and 138. In line 140 method `readNumberList` reads the list of node numbers, generates `Dof` objects, and adds them to the linked list `nodForces`. Each degree of freedom object `Dof` contains a degree of freedom number and a nodal force value.

The statement in line 101 reads surface forces for element faces with the help of method `readSurForces`. This method shown in lines 145–158 reads the direction of the surface load (line 148), the element number (line 151), the number of nodes on this face (line 152), and the face node numbers and corresponding load values (lines 153–156). The force direction can have values `x/y/z` or `n`. The latter means that a positive force acts along the external normal to the element face. The statement in line 157 creates `ElemFaceLoad` object and adds it to the linked list `surForces`. Class `ElemFaceLoad` designed for storing element face loads will be presented later in this chapter.

Another possibility for input of the distributed face load is implemented in line 105 that calls method `createBoxSurForces`. In this case the user specifies a diagonal of a box. Element faces fully located inside this box are considered loaded by the distributed force with given direction and value. The source code of method `createBoxSurForces` is presented in lines 162–201. The method reads a direction and a value of the distributed load in lines 165 and 171. Four or six coordinate values for the box diagonal is input in lines 173–175. Loop `iel` iterates over all elements in the model and test all element faces against the box. If all nodes of a face are inside the box then they are stored in array `iw`. This array is used to create element face load object `ElemFaceLoad`, which is added to the linked list `surForces`.

The temperature field is set by reading the array on nodal temperatures `dtemp` in lines 109–111.

Inserting a file containing partial of full load data is implemented in lines 115–117 (data statement `includeFile fileName`). A recursive call of method `readDataFile` with new data scanner `R` helps to process load data from the separate file.

Data statement `end` (line 120) causes termination of data input for the current load step.

17.3 Load Vector Assembly

The global load vector, the right-hand side of the global equation system, is assembled by method `assembleRHS` that is presented next.

```
203    // Assemble right-hand side of the global equation system
204    public void assembleRHS() {
205
206        if (scaleLoad != 0.0) {
207            for (int i = 0; i < fem.nEq; i++) {
208                dpLoad[i] *= scaleLoad;
209                dhLoad[i] *= scaleLoad;
210                RHS   [i] = dpLoad[i] + dhLoad[i];
211            }
212            return;
213        }
214        for (int i = 0; i < fem.nEq; i++) {
215            dpLoad[i] = 0.0;
216            dhLoad[i] = 0.0;
217        }
218
219        // Nodal forces specified directly
220        itnf = nodForces.listIterator(0);
221        Dof d;
222        while (itnf.hasNext()) {
223            d = (Dof) itnf.next();
224            dpLoad[d.dofNum-1] = d.value;
225        }
226
227        // Surface load at element faces
228        itsf = surForces.listIterator(0);
229        ElemFaceLoad efl;
230        Element elm;
231        while (itsf.hasNext()) {
232            efl =(ElemFaceLoad) itsf.next();
233            elm = fem.elems[efl.iel];
234            elm.setElemXy();
235            if (elm.equivFaceLoad(efl)==-1)
236                UTIL.errorMsg("surForce" +
237                    " does not match any face of element: "
238                    + efl.iel);
239            elm.assembleElemVector(Element.evec,dpLoad);
240        }
241
242        // Temperature field
243        if (fem.thermalLoading) {
244            for (int iel = 0; iel < fem.nEl; iel++) {
245                elm = fem.elems[iel];
246                elm.setElemXyT();
247                elm.thermalVector();
248                elm.assembleElemVector(Element.evec,dhLoad);
249            }
250        }
```

```
251
252            // Right-hand side = actual load + fictitious load
253            for (int i = 0; i < fem.nEq; i++)
254                RHS[i] = dpLoad[i] + dhLoad[i];
255
256            // Displacement boundary conditions for right-hand side
257            ListIterator itdbc = fem.defDs.listIterator(0);
258            while (itnf.hasNext()) {
259                d = (Dof) itdbc.next();
260                RHS[d.dofNum-1] = FE.bigValue * d.value;
261            }
262        }
263
264 }
```

A load vector for the current load step can be obtained by scaling the previous load vector or by assembling a new load vector. If parameter `scaleLoad` is nonzero then previous vectors of force load `dpLoad` and thermal load `dhLoad` are scaled and the right-hand side of the equation system RHS is obtained as their sum in lines 207–211. Otherwise, we start with resetting to zero both `dpLoad` and `dhLoad` vectors.

Assembly of concentrated nodal forces is done in lines 220–225. The list iterator `itnf` is used to access all nodal forces stored in degrees of freedom objects `Dof`. Values of nodal forces are directly added to the force load vector `dpLoad` according to the degrees of freedom associated with them.

Lines 228–240 perform assembly of the surface load specified at element faces. In this case we look through linked list `surForces` and rely on element methods for main operations with surface forces. For each `ElemFaceLoad` object, nodal coordinates for the specified element are set (method `setElemXy`) and method `equivFaceLoad` is called to compute the nodal equivalent of the surface load. Method `assembleElemVector` in line 239 assembles element vector `evec` to global vector `dpLoad`.

Deformation of the finite element model due to the temperature field is modeled by computing the fictitious element nodal equivalents of temperature expansion and assembling them to the global thermal vector in lines 243–250. For each finite element, nodal coordinates and nodal temperatures are set by method `setElemXyT`. Method `thermalVector` computes the element thermal vector and method `assembleElemVector` assembles it to the global thermal vector `dhLoad` (lines 247–248).

The sum of global force `dpLoad` and thermal `dhLoad` vectors gives the right-hand side vector of the global equation system RHS.

Finally, the right-hand side vector is modified in lines 257–261 to take into account displacement boundary conditions. Values of specified displacements are taken from linked list `defDs`, and, after multiplication by the large number, are used for replacement of coefficients in the right-hand side.

17.4 Element Face Load

Class `ElemFaceLoad` is designed for storing data on a distributed load at an element face. The following variables and arrays describe the element face load:

`iel` – element number;
`direction` – force direction, x, y, z – along coordinate axes, n – along the external normal to an element face;
`faceNodes` – global numbers of face nodes in arbitrary order;
`forceAtNodes` – values of distributed force intensity in the same order as the face nodes.

```
 1 package model;
 2
 3 // Element face load
 4 public class ElemFaceLoad {
 5     // Element number (start with 0)
 6     public int iel;
 7     // Direction: 1-x, 2-y, 3-z, 0-normal
 8     public int direction;
 9     public int[] faceNodes;
10     public double[] forceAtNodes;
11
12     ElemFaceLoad(int iel, int nFaceNodes, int direction,
13                  int[] faceNodes, double[] forceAtNodes) {
14         this.iel = iel;
15         this.direction = direction;
16         this.faceNodes = new int[nFaceNodes];
17         this.forceAtNodes = new double[nFaceNodes];
18
19         for (int i=0; i<nFaceNodes; i++) {
20             this.faceNodes[i] = faceNodes[i];
21             this.forceAtNodes[i] = forceAtNodes[i];
22         }
23     }
24
25     ElemFaceLoad(int iel, int nFaceNodes, int direction,
26                  int[] faceNodes, double force) {
27         this.iel = iel;
28         this.direction = direction;
29         this.faceNodes = new int[nFaceNodes];
30         this.forceAtNodes = new double[nFaceNodes];
31
32         for (int i=0; i<nFaceNodes; i++) {
33             this.faceNodes[i] = faceNodes[i];
34             this.forceAtNodes[i] = force;
35         }
36     }
37
38     // Rearrange surface load (faceNodes[] and ForcesAtNodes[])
39     // according to order in element faces.
40     // faces - local numbers (from zero) of element faces,
```

```
41          // ind - element connectivities.
42          // returns  loaded face number or -1 if no match
43          //  between ind[] and load data.
44       public int rearrange(int[][] faces, int[] ind) {
45
46              int perm[] = new int[8];
47              double fw[] = new double[8];
48              int loadedFace = -1;
49
50              FACE: for (int iface=0; iface<faces.length; iface++) {
51                  int nNodes = faces[iface].length;
52                  for (int inod = 0; inod < nNodes; inod++)
53                      perm[inod] = -1;
54                  for (int inod = 0; inod < nNodes; inod++) {
55                      int iGlob = ind[faces[iface][inod]];
56                      if (iGlob > 0) {
57                          boolean EQ = false;
58                          int i;
59                          for (i = 0; i < nNodes; i++)
60                              if (faceNodes[i] == iGlob) {
61                                  EQ = true;
62                                  break;
63                              }
64                          if (!EQ) continue FACE;
65                          perm[inod] = i;
66                      }
67                  }
68                  loadedFace = iface;
69                  for (int inod = 0; inod < nNodes; inod++) {
70                      faceNodes[inod] = ind[faces[iface][inod]];
71                      fw[inod] = forceAtNodes[inod];
72                  }
73                  for (int inod = 0; inod < nNodes; inod++)
74                      forceAtNodes[inod] =
75                          (perm[inod] == -1) ? 0.0 : fw[perm[inod]];
76              }
77          return loadedFace;
78       }
79
80 }
```

Two constructors of the class simply store element face data. In the first constructor, forces at nodes can have different values. The second constructor gets a constant force intensity and stores it for each face node.

The class includes method `rearrange`, which rearranges face nodes and force values according to the order used in element classes. This method is called by element methods responsible for computing nodal equivalents of distributed load. The method gets local node numbers for all element faces `faces` and element connectivities `ind`. This data allows global node numbers to be obtained for faces of this element. Variable `iGlob` in line 55 represents the global node number for local node number `inod` at face `iface`. Comparing `iGlob` with node numbers in array `faceNodes` helps to create the permutation array `perm`. Changing the order of force values at face nodes `forceAtNodes` is performed in lines 73–75 using the permutation array.

Problems

17.1. Class `FeLoad` implements three load types: concentrated nodal forces, distributed forces applied to element faces, and thermal loading. Propose other load types that can be useful in finite element programs.

17.2. Suppose a new scalar parameter `s0` is needed in class `FeLoad`. The parameter should be read from the input file. Suggest code modifications necessary for introduction of new data item.

17.3. Write down load data for a three-dimensional mesh consisting of two twenty-node hexahedral elements.

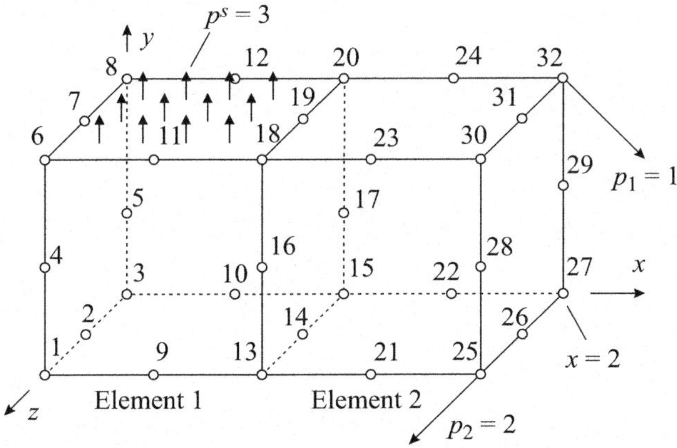

A distributed load with intensity $p^s = 3$ normal to the surface is applied to the upper face of element 1. Concentrated nodal force $p_1 = 1$ acts in plane xy at angle $45°$ to the x-axis, and concentrated nodal force $p_2 = 2$ is directed along z.

17.4. Prepare load data for the distributed load $p^s = 3$ of the previous problem as equivalent nodal forces.

Chapter 18
Stress Increment, Residual Vector and Results

Abstract This chapter presents classes for computing stress increment, residual vector and results. When a stress increment is estimated, the physical law used depends on the material model and can represent elastic or elastic–plastic material. Element methods for equivalent stress vector are employed during computation of the residual vector. These classes complete an elastic portion of the finite element processor. A simple elastic problem solution is demonstrated.

18.1 Computing Stress Increment

Java^{TM} class `FeStress` contains methods for computing the stress increment, for estimating equilibrium of the finite element model, and for writing results files.

Below is a source code fragment containing method `computeIncrement`, which performs computation of the stress increment due to the displacement increment.

```
 1  package model;
 2
 3  import elem.*;
 4  import material.*;
 5  import util.*;
 6
 7  import java.io.*;
 8  import java.util.ListIterator;
 9
10  // Stress increment due to displacement increment
11  public class FeStress {
12
13      public static double relResidNorm;
14      private FeModel fem;
15
16      // Constructor for stress increment.
17      // fem - finite element model
18      public FeStress(FeModel fem) {
```

```
19              this.fem = fem;
20          }
21
22          // Compute stress increment for the finite element model
23          public void computeIncrement() {
24
25              // Accumulate solution vector in displacement increment
26              for (int i=0; i<fem.nEq; i++)
27                  FeLoad.dDispl[i] += FeLoad.RHS[i];
28
29              // Compute stresses at reduced integration points
30              for (int iel = 0; iel < fem.nEl; iel++) {
31                  Element elm = fem.elems[iel];
32                  elm.setElemXyT();
33                  elm.disAssembleElemVector(FeLoad.dDispl);
34
35                  for (int ip = 0; ip < elm.str.length; ip++) {
36                      Material mat =
37                          (Material) fem.materials.get(elm.matName);
38                      mat.strainToStress(elm, ip);
39                  }
40              }
41          }
```

Constructor FeStress just stores a reference to the finite element model fem.
Method computeIncrement first accumulates the solution vector in the displacement increment vector dDispl. In elastic problems, this operation means
just copying the solution vector since the number of iterations is equal to one. The
elastic–plastic iteration procedure involves multiple iterations. In this case, the loop
of lines 26–27 allows the displacement increment from the beginning of the current
load step to be accumulated. Computing the stress increment for the displacement
increment since the last equilibrium state is important because estimation of the
stress increment for each iteration with accumulation of the stress increment can
lead to false unloading and consequently to incorrect results.

Stress increments at element reduced-integration points are calculated inside
the loop over finite elements in lines 30–40. Methods of classes Element and
Material are used to perform operations. Different problems are treated in this
way depending on the particular element and material methods. Line 32 sets element nodal coordinates and temperatures. The next statement selects the displacement increment for the current element from the global displacement increment
vector. The loop, which starts in line 35, iterates over reduced integration points
of the current element. Material method strainToStress computes the stress
increment according to the constitutive equations of the material model. A variant
of the method for the elastic material gets the strain increment by calling element
method getStrainsAtIntPoint. The stress increment is evaluated according
to Hooke's law using the elastic fraction of strains. The stress increment vector is
stored at an element integration point ip.

18.2 Residual Vector

The residual vector $\{\psi\}$ characterizes the imbalance of the equilibrium equation expressed through stresses (3.22):

$$\{\psi\} = \{p\} - \int_V [B]^\mathrm{T}\{\sigma\}dV. \tag{18.1}$$

Here, $\{p\}$ is the current level of the force load, $\{\sigma\}$ is the stress vector, $[B]$ is the displacement differentiation matrix. The volume integral is estimated for each finite element and assembled into a global vector. In nonlinear problems a relative residual norm is used for checking equilibrium:

$$\frac{\|\psi\|}{\|\Delta p\|} \leq \varepsilon. \tag{18.2}$$

Iterations are terminated when the residual norm $\|\psi\|$ becomes small in comparison to the norm of the force vector increment $\|\Delta p\|$ at this load step.

Method equilibrium below computes the residual vector and checks equilibrium for the finite element model.

```
43      // Check equilibrium and assemble residual vector.
44      // iter - number of iterations performed
45      public boolean equilibrium(int iter) {
46
47          if (fem.physLaw == FeModel.PhysLaws.elastic ||
48                  iter == FeLoad.maxIterNumber) return true;
49          // Assemble residual vector to right-hand side
50          for (int i=0; i<fem.nEq; i++)
51              FeLoad.RHS[i] = FeLoad.spLoad[i]+FeLoad.dpLoad[i];
52          Element elm;
53          for (int iel = 0; iel < fem.nEl; iel++) {
54              elm = fem.elems[iel];
55              elm.setElemXy();
56              elm.equivStressVector();
57              elm.assembleElemVector(Element.evec,FeLoad.RHS);
58          }
59          // Displacement boundary conditions
60          ListIterator it = fem.defDs.listIterator(0);
61          while (it.hasNext()) {
62              Dof d = (Dof) it.next();
63              FeLoad.RHS[d.dofNum-1] = 0;
64          }
65          // Relative residual norm
66          double dpLoadNorm = vectorNorm(FeLoad.dpLoad);
67          if (dpLoadNorm < 1e-30)
68              dpLoadNorm = vectorNorm(FeLoad.dhLoad);
69          relResidNorm = vectorNorm(FeLoad.RHS)/dpLoadNorm;
70          return relResidNorm < FeLoad.residTolerance;
71      }
72
73      // Returns norm of a vector v
```

```
74        double vectorNorm(double[] v) {
75
76            double norm = 0;
77            for (double aV : v) norm += aV * aV;
78            return Math.sqrt(norm);
79        }
80
81        // Accumulate loads, temperature and stresses
82        public void accumulate() {
83
84            for (int i=0; i<fem.nEq; i++) {
85                FeLoad.spLoad[i] += FeLoad.dpLoad[i];
86                FeLoad.sDispl[i] += FeLoad.dDispl[i];
87            }
88            for (int iel = 0; iel < fem.nEl; iel++)
89                fem.elems[iel].accumulateStress();
90        }
```

If the problem is elastic then we assume that stress equilibrium is fulfilled automatically. In the elastic case and in the case when the number of iterations has reached the specified maximum, the method returns `true`, which means that the current load step is finished (lines 47–48).

For the elastic–plastic case with an allowed iteration number, we put the full force load (`spLoad[i]+dpLoad`) into the right-hand side RHS in lines 50–51 and start estimating a relative residual norm. The loop over elements in lines 53–58 contains computation of the element equivalent stress vector (with negative sign) and its assembly to the right-hand side vector. Since we previously put the full force vector in the right-hand side, then now we have there the global residual vector, which is the difference between the force vector and the equivalent stress vector. Displacement boundary conditions (constraints) are applied to the residual vector in lines 60–64.

Lines 66–68 estimate the norm that is used as a divider for determining the relative residual norm. First, we try to use the norm of the force vector increment (line 66). If a problem has pure thermal loading than this norm is zero. In this case, the norm of the global thermal vector is computed in line 68. The relative residual norm is determined in line 69. The method returns `true` if this norm is sufficiently small (less than the user-specified tolerance `residTolerance`).

Method `vectorNorm` (lines 74–79) returns the norm of the vector, which is a square root of a scalar product of the vector by itself.

Method `accumulate` adds increments of the force load vector and of the displacement vector to their accumulated vectors. The element loop in lines 88–89 accumulates stress increments and possibly (in elastic–plastic problems) the equivalent plastic strain increments. The method is called at the end of the load step.

18.3 Results

The output of results is performed by method `writeResults` and the input of results is done by method `readResults`.

```
92      // Write results to a file.
93      public void writeResults() {
94
95          String fileResult = fea.Jfem.fileOut + "."
96                          + FeLoad.loadStepName;
97          PrintWriter PR =
98                  new FePrintWriter().getPrinter(fileResult);
99
100         PR.printf("Displacements\n\n");
101         if (fem.nDim == 2)
102             PR.printf(" Node                 ux                uy");
103         else
104             PR.printf(" Node                 ux                uy"
105                                         + "                uz");
106         for (int i = 0; i < fem.nNod; i++) {
107             PR.printf("\n%5d", i + 1);
108             for (int j = 0; j < fem.nDim; j++)
109                 PR.printf("%15.6e", FeLoad.sDispl[fem.nDim*i+j]);
110         }
111
112         PR.printf("\n\nStresses\n");
113         for (int iel = 0; iel < fem.nEl; iel++) {
114             if (fem.nDim == 2)
115                 PR.printf("\nEl %4d       sxx              syy"
116                     +"              sxy             szz"
117                     +"              epi", iel+1);
118             else
119                 PR.printf("\nEl %4d       sxx              syy"
120                     +"              szz             sxy"
121                     +"              syz             szx"
122                     +"              epi", iel+1);
123             for (StressContainer aStr : fem.elems[iel].str) {
124                 PR.printf("\n");
125                 for (int i = 0; i < 2 * fem.nDim; i++)
126                     PR.printf("%15.8f", aStr.sStress[i]);
127                 PR.printf("%15.8f", aStr.sEpi);
128             }
129         }
130         PR.close();
131     }
132
133     // Read results from a file.
134     // displ - displacements for the finite element model (out)
135     public void readResults(String resultFile,double[] displ) {
136
137         if (resultFile==null) return;
138
139         FeScanner RD = new FeScanner(resultFile);
```

```
140            // Read displacements
141            RD.moveAfterLineWithWord("node");
142            for (int i = 0; i < fem.nNod; i++) {
143                RD.readInt();
144                for (int j = 0; j < fem.nDim; j++)
145                    displ[fem.nDim*i + j] = RD.readDouble();
146            }
147            // Read stresses
148            for (int iel = 0; iel < fem.nEl; iel++) {
149                RD.moveAfterLineWithWord("el");
150                for (StressContainer aStr : fem.elems[iel].str) {
151                    for (int i = 0; i < 2 * fem.nDim; i++)
152                        aStr.sStress[i] = RD.readDouble();
153                    aStr.sEpi = RD.readDouble();
154                }
155            }
156            RD.close();
157        }
158
159 }
```

Method `writeResults` writes the results of the finite element solution for the
current load step into a text file. Lines 95–96 create a results file name using an
output file name for the problem and a name for the current load step. Print writer
PR is constructed in lines 97–98.

Lines 100–110 print the total nodal displacements. Each line contains a node
number and two or three displacement components along the coordinate axes.
Stresses are printed in lines 112–129 inside a loop over the elements. Each line
contains an element number, normal and shear stress components, and equivalent
plastic strain (always zero in elastic problems) at one reduced integration point.
Thus, four lines of stress results are printed for each two-dimensional quadratic element and eight lines for each three-dimensional quadratic element. Line 130 closes
the print writer.

It is quite clear that too few possibilities are provided for results output. However, our finite element program serves for educational purposes and it is difficult to
create here a fully functional rich user interface because we want to keep the amount
of source code reasonable for reading and understanding. The lack of output possibilities is partly ameliorated by visualization code `Jvis`, which is considered later.

Method `readResults` is employed by `Jvis` code to read results for their visualization. The results file name `resultFile` is passed as a method parameter. For
data input, our finite element scanner `FeScanner` is used. To read displacements
the word `node` is found in the results file (line 141). Then, all nodal displacements
are read in a loop over the nodes.

When reading stresses, the word `el` is sought for each element, then stresses
for element integration points are copied to their containers in finite elements (lines
148–155). Line 156 calls method `close` for the finite element data scanner.

18.4 Solution of a Simple Test Problem

We already considered classes that allow production of working finite element code
for solution of elastic problems. Let us create a directory `Jfea` with two subdirectories – `src` for Java source files and `classes` for compiled Java code. Directory
`src` in turn contains directories named as packages of our finite element code. If
we put Java files in directories with names corresponding to package names, then it
is possible to compile Java code `Jfem` using the following command

```
javac -sourcepath src -d classes src/fea/Jfem.java
```

Here, `javac` is a call to the Java compiler, option `-sourcepath` specifies that
source files are in directory `src`, option `-d` sets directory `classes` as output for
the compiled code, and finally `src/fea/Jfem.java` specifies the main class of
the finite element processor. It is possible to request messages about the compilation
process with an option `-verbose`. As a result of compilation, class files appear in
directory `classes`. These class files are placed in subdirectories with names of
respective packages.

Let us test our finite element code on a simple example, shown in Figure 5.2. A
two-dimensional rectangular bar with width 2 and height 1 is subject to a distributed
load with intensity 1 along the x-axis and to a temperature field of magnitude 10.
We do not specify any units for data. The finite element code does not use any units.
The user is responsible for using consistent units for data.

Data describing this problem is given in Section 5.2.4. Let us take this data and
put it in file `f.fem`. In order to solve the problem it is possible to place file `f.fem`
in directory `fea` and to execute the finite element code as

```
java -cp classes fea.Jfem f.fem f.lst
```

Finite element processor `fea.Jfem` receives parameter `f.fem` (input data) and
parameter `f.lst` (output file).

The results of the solution are in files `f.lst` and `f.lst.1`. The first file contains just brief information about the solution. Actual results output is done for each
load step. The names of results files are created by adding loadstep names to the
name of the output file specified as the second parameter for the code. In our case,
displacements and stresses for the finite element model are in file `f.lst.1` since
the loadstep name is "1". The contents of the file are shown below.

```
    Displacements

        Node              ux                  uy
           1    1.666667e-65      8.433758e-80
           2    6.666667e-65      3.500000e-01
           3    1.666667e-65      7.000000e-01
           4    1.000000e+00     -7.168695e-80
           5    1.000000e+00      7.000000e-01
           6    2.000000e+00     -6.747007e-80
           7    2.000000e+00      3.500000e-01
           8    2.000000e+00      7.000000e-01
```

```
 9    3.000000e+00     7.801226e-80
10    3.000000e+00     7.000000e-01
11    4.000000e+00    -9.962377e-80
12    4.000000e+00     3.500000e-01
13    4.000000e+00     7.000000e-01
```

```
Stresses
```

```
El    1    sxx              syy              sxy              szz
         1.00000000      0.00000000       0.00000000       0.00000000
         1.00000000      0.00000000       0.00000000       0.00000000
         1.00000000     -0.00000000      -0.00000000       0.00000000
         1.00000000     -0.00000000       0.00000000       0.00000000
El    2    sxx              syy              sxy              szz
         1.00000000     -0.00000000      -0.00000000       0.00000000
         1.00000000     -0.00000000      -0.00000000       0.00000000
         1.00000000      0.00000000       0.00000000       0.00000000
         1.00000000      0.00000000      -0.00000000       0.00000000
```

It is easy to check that the results are correct. Normal stress $\sigma_x = 1$, all other stresses are zero. Nodal displacements are due to deformation because of the applied distributed force and uniform temperature expansion. Some nodal displacements should be zero because of boundary conditions (nodes 1–3 along x, nodes 1, 4, 6, 9, and 11 along y). In the output file we can see small numbers instead of pure zeros. Such "approximate" zeros are caused by the method of large numbers used for implementation of displacement boundary conditions in our finite element code (see Section 14.2.2). Such small numbers instead of absolute zeros do not affect computation of results such as strains and stresses since the difference of orders is larger than the number of significant digits in the computer representation of double-precision numbers.

Problems

18.1. Assume a two-dimensional finite element is a square of unit size. After application of external loading, the element deforms with an increase of size in the x-direction by 0.13 and a decrease of size in the y-direction by the same amount. The deformed element shape is indicated by a dashed line in the following figure.

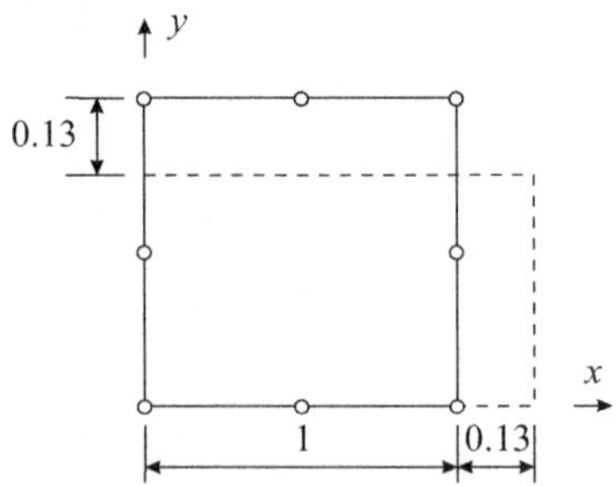

Find the stresses in the finite element for plane stress conditions. The material properties are: elasticity modulus $E = 0.1$ and Poisson's ratio $v = 0.3$.

18.2. Read and understand method `computeIncrement` of class `FeStress`. Pay attention to use of method `setElemXyT` in line 32. Explain why nodal temperatures are necessary in addition to nodal coordinates for evaluation of stress increments.

18.3. Review method `equilibrium`. Displacement boundary conditions are applied to the right-hand side (lines 60–64) before computing the residual norm. Why is application of displacement boundary conditions necessary?

Chapter 19
Elastic–Plastic Problems

Abstract Solution of elastic–plastic problems is considered. Constitutive relations for a von Mises elastic–plastic material (flow theory) are presented. Algorithms of computing finite stress increments based on subincrementation and on midpoint integration are explained. The algorithms are implemented in Java^{TM} class `ElasticPlasticMaterial`. Nonlinear solution procedures using the Newton–Raphson method and initial stress method are reviewed.

19.1 Constitutive Relations for Elastic–Plastic Material

Consider an elastic–plastic material with behavior described by the flow theory with isotropic hardening and von Mises yield surface [33]. For an elastic–plastic body it is postulated that:

the strain increment can be decomposed into elastic $\{\varepsilon^{\mathrm{e}}\}$, plastic $\{\varepsilon^{\mathrm{p}}\}$, and thermal $\{\varepsilon^{\mathrm{t}}\}$ fractions:

$$\{d\varepsilon\} = \{d\varepsilon^{\mathrm{e}}\} + \{d\varepsilon^{\mathrm{p}}\} + \{d\varepsilon^{\mathrm{t}}\}, \tag{19.1}$$

the stress increment $\{d\sigma\}$ can be calculated through the elastic strain increment $\{d\varepsilon^{\mathrm{e}}\}$ and the elasticity matrix $[E]$ according to Hooke's law:

$$\{d\sigma\} = [E]\{d\varepsilon^{\mathrm{e}}\} = [E](\{d\varepsilon\} - \{d\varepsilon^{\mathrm{p}}\} - \{d\varepsilon^{\mathrm{t}}\}), \tag{19.2}$$

the thermal strain increment is specified through the thermal-expansion coefficient α and the temperature increment dT:

$$\{d\varepsilon^{\mathrm{t}}\} = \alpha dT\{1\ 1\ 1\ 0\ 0\ 0\}, \tag{19.3}$$

and the plastic deformation does not change the body volume:

$$\sum \varepsilon_{ii}^{\mathrm{p}} = 0. \tag{19.4}$$

Let us assume that the elastic–plastic material behavior is governed by the von Mises yield function (von Mises yield surface):

$$f = \sigma_i - Y(\kappa) = 0. \tag{19.5}$$

Here, σ_i is the equivalent stress,

$$\begin{aligned}\sigma_i &= \sqrt{\frac{3}{2} s_{ij} s_{ij}} \\ &= \frac{1}{\sqrt{2}} \sqrt{(\sigma_x - \sigma_y)^2 + (\sigma_x - \sigma_y)^2 + (\sigma_x - \sigma_y)^2 + 6(\tau_{xy}^2 + \tau_{yz}^2 + \tau_{zx}^2)},\end{aligned} \tag{19.6}$$

and Y is the instantaneous yield stress depending on the hardening parameter κ,

$$\kappa = \int d\varepsilon_i^{\mathrm{p}}. \tag{19.7}$$

In the above equations s_{ij} are the deviatoric stress components,

$$s_{ij} = \sigma_{ij} - \frac{1}{3}\sigma_{ii}, \tag{19.8}$$

and $d\varepsilon_i^{\mathrm{p}}$ is the increment of the equivalent plastic strain,

$$\begin{aligned}d\varepsilon_i^{\mathrm{p}} &= \sqrt{\frac{2}{3} d\varepsilon_{ij}^{\mathrm{p}} d\varepsilon_{ij}^{\mathrm{p}}} \\ &= \frac{\sqrt{2}}{3} \sqrt{(d\varepsilon_x^{\mathrm{p}} - d\varepsilon_y^{\mathrm{p}})^2 + (\ldots)^2 + (\ldots)^2 + \frac{3}{2}[(d\gamma_{xy}^{\mathrm{p}})^2 + (\ldots)^2 + (\ldots)^2]}.\end{aligned} \tag{19.9}$$

Let us introduce the vector of yield function derivatives $\{a\}$ as

$$\{a\} = \left\{\frac{\partial f}{\partial \sigma}\right\} = \frac{3}{2\sigma_i}\{s_x\ s_y\ s_z\ 2\tau_{xy}\ 2\tau_{yz}\ 2\tau_{zx}\}. \tag{19.10}$$

During plastic flow the stresses must remain on the yield surface. With the above-introduced vector $\{a\}$ the condition for the stress–strain state on the yield surface can be written as

$$df = \frac{\partial f}{\partial \sigma_{ij}} d\sigma_{ij} - \frac{\partial Y}{\partial \kappa} d\kappa = \{a\}^{\mathrm{T}}\{d\sigma\} - H d\varepsilon_i^{\mathrm{p}} = 0, \tag{19.11}$$

where $H = \partial Y/\partial \kappa$ is the slope of the material yield curve with respect to the hardening parameter κ. According to the Prandtl–Reuss flow rule (normality of plastic deformation to the flow surface), the increment of plastic strains is proportional to the derivatives of the yield function,

$$d\varepsilon_{ij}^{\mathrm{p}} = \lambda \frac{\partial f}{\partial \sigma_{ij}} = \lambda \{a\}, \tag{19.12}$$

where λ is the plastic strain-rate multiplier. Substitution of the above equation into (19.9) gives the following simple expression for the plastic multiplier,

$$d\varepsilon_i^{\mathrm{p}} = \lambda. \tag{19.13}$$

Now Equation 19.11 becomes

$$df = \{a\}^{\mathrm{T}}\{d\sigma\} - H\lambda = 0. \tag{19.14}$$

Substitution of Equation 19.2 into Equation 19.14 provides the relation

$$\lambda = \frac{\{a\}^{\mathrm{T}}[E]\{d\varepsilon^{\mathrm{ep}}\}}{\{a\}^{\mathrm{T}}[E]\{a\} + H}, \tag{19.15}$$

where $\{d\varepsilon^{\mathrm{ep}}\} = \{d\varepsilon^e\} + \{d\varepsilon^p\}$. Using this expression for the plastic multiplier λ and Equation 19.12, Hooke's law (19.2) can be transformed to the following constitutive equation relating the stress increment and increment of elastic–plastic strains:

$$\{d\sigma\} = \left([E] - \frac{[E]\{a\}\{a\}^{\mathrm{T}}[E]}{\{a\}^{\mathrm{T}}[E]\{a\} + H} \right) \{d\varepsilon^{\mathrm{ep}}\}. \tag{19.16}$$

It is possible to take into account that for the von Mises yield function the following equality is preserved

$$\{a\}^{\mathrm{T}}[E]\{a\} = 3G, \tag{19.17}$$

where G is the shear modulus. Finally, the constitutive relation for increments of stresses and elastic–plastic strains is

$$\{d\sigma\} = [E^{\mathrm{ep}}]\{d\varepsilon^{\mathrm{ep}}\},$$

$$[E^{\mathrm{ep}}] = [E] - \frac{[E]\{a\}\{a\}^{\mathrm{T}}[E]}{3G + H}. \tag{19.18}$$

19.2 Computing Finite Stress Increments

The stress–strain state of the finite element model is controlled at reduced integration points inside finite elements since displacement derivatives have better precision there. For example, stresses and accumulated plastic strains are kept at $2 \times 2 \times 2$ Gauss integration points in the twenty-node hexahedral element.

Equation 19.18 is precise for infinitesimally small strain increments. In practice, displacement increments during iterative solution of nonlinear problems can be quite large. Using (19.18) for finite strain increments can lead to large solution errors. To overcome this difficulty, two approaches can be employed. One is subincrementation, based on dividing the strain increments into sufficiently small subincrements. The other approach is the integration of constitutive relations with the help of a midpoint integration algorithm.

Another problem in computing finite stress increments is related to increments, which are started in an elastic state and are finished in an elastic–plastic state. Elastic Hooke's law should be used for the elastic fraction of such increments and the elastic–plastic constitutive law should be used after entering the plastic state.

19.2.1 Determining Elastic Fraction of Stress Increment

Let $\{\sigma_0\}$ be the stress vector at the beginning of the increment and let $\{\Delta\sigma^e\}$ be the elastic stress increment due to the strain increment $\{\Delta\varepsilon\}$. We consider the case when for the original stress vector the state is elastic,

$$f_0 = f(\{\sigma_0\}) < 0, \tag{19.19}$$

and for the final stress vector

$$f_e = f(\{\sigma_0\} + \{\Delta\sigma^e\}) > 0. \tag{19.20}$$

We need to determine the scalar parameter r that yields

$$f_e = f(\{\sigma_0\} + r\{\Delta\sigma^e\}) = 0. \tag{19.21}$$

It is possible to use the linear initial estimate for the parameter r

$$r_0 = -\frac{f_0}{f_e - f_0}, \tag{19.22}$$

and then to improve its value in an iterative manner using a truncated Taylor series:

$$\Delta r_i = -\frac{f(r_{i-1})}{\{a(r_{i-1})\}\{\Delta\sigma^e\}}, \tag{19.23}$$

$$r_i = r_{i-1} + \Delta r_i.$$

19.2.2 Subincrementation for Computing Stress Increment

The subincrementation technique for computing the finite stress increment is based on dividing a finite strain increment $\{\Delta\varepsilon\}$ into m subincrements $\{d\varepsilon\}$

$$\{d\varepsilon\} = \frac{\{\Delta\varepsilon\}}{m}. \tag{19.24}$$

The number of subincrements m can be estimated as

$$m = \frac{f(\{\sigma_0\} + \{\Delta\sigma^e\})}{\beta\sigma_Y} + 1, \tag{19.25}$$

where $\{\sigma_0\}$ is the stress vector at the beginning of the stress increment (on the yield surface), $\{\Delta\sigma^e\}$ is the elastic stress increment, σ_Y is the material yield stress, and β is a tolerance (a typical value of β might be 0.05).

The algorithm of subincrementation for computing the finite elastic–plastic stress increment is given by the following pseudocode:

$$\{d\varepsilon^{ep}\} = \{\Delta\varepsilon^{ep}\}/m$$

do $i = 1, m$

$$\{d\sigma_i\} = [E^{ep}(\{\sigma_{i-1}\})]\{d\varepsilon^{ep}\} \tag{19.26}$$

$$\{\sigma_i\} = \{\sigma_{i-1}\} + \{d\sigma_i\}$$

end do

Subincrements are smaller than increments. However, strain subincrements are still finite. Because of this the stress at the end of each subincrement departs slightly from the yield surface. Using the truncated Taylor series it is possible to improve the stress at the end of a subincrement in the following way:

$$\{\sigma_i^1\} = \{\sigma_i\} - \frac{f(\{\sigma_i\})\{a\}}{\{a\}^T\{a\}}. \tag{19.27}$$

Here, $\{\sigma_i^1\}$ is the corrected stress at the increment end.

19.3 Material Deformation Curve

The material deformation curve for an elastic–plastic material consists of two parts – elastic and elastic–plastic. The material properties in the elastic range are described by elasticity modulus E and Poisson's ratio ν.

Simple description of the material curve in the elastic–plastic range adopted here follows a power function

$$\sigma = \sigma_Y + k(\varepsilon_p)^m, \tag{19.28}$$

where σ is the uniaxial stress, σ_Y is the yield stress where plastic strains appear, ε_p is the magnitude of plastic strain, k is the material hardening coefficient, and m is the material hardening power.

It is assumed that the material deformation curve determined from a uniaxial specimen test can be used in constitutive plasticity relations after replacement of σ and ε_p with equivalent stress σ_i and accumulated equivalent plastic strain ε_i^p. The yield function according to the von Mises yield criterion is

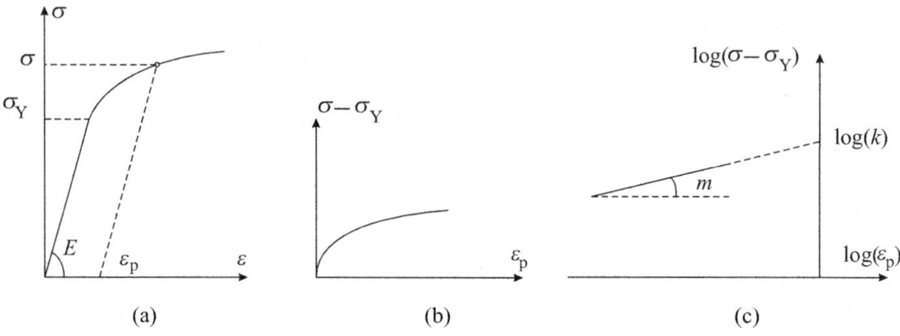

Fig. 19.1 Determining elastic–plastic parameters k and m: deformation curve $\varepsilon - \sigma$ (a), curve $\varepsilon_p - (\sigma - \sigma_Y)$ (b), and approximation of $\log \varepsilon_p - \log(\sigma - \sigma_Y)$ by a straight line (c)

$$f = \sigma_i - Y(\varepsilon_i^p),$$
$$Y(\varepsilon_i^p) = \sigma_Y + k(\varepsilon_i^p)^m. \tag{19.29}$$

The yield function f is negative in the elastic range. During elastic–plastic material deformation, it is always zero: $f = 0$. The slope of the material yield curve is determined by

$$H = \frac{\partial f}{\partial \varepsilon_p} = km(\varepsilon_p)^{m-1}. \tag{19.30}$$

Let us show how to determine parameters k and m using the material deformation strain–stress curve shown in Figure 19.1a. Moving σ_Y to the left side of (19.28) and taking logs of both sides obtains

$$\log(\sigma - \sigma_Y) = \log(k) + m\log(\varepsilon_p). \tag{19.31}$$

To find the deformation curve parameters, the curve $\varepsilon - \sigma$ is transformed into a curve $\varepsilon_p - (\sigma - \sigma_Y)$ by subtraction of the plastic strain, as depicted in Figure 19.1b. Then, the latter curve is presented in a $\log - \log$ plot and is approximated by a straight line. Figure 19.1c illustrates the determination of the hardening coefficient logarithm $\log(k)$ as the intersection of the straight line with vertical axis and the hardening power m as the slope of the straight line.

19.4 Implementation of Elastic–Plastic Material Relations

Constitutive relations for an elastic–plastic material described by the flow theory are implemented in class `ElasticPlasticMaterial`, which inherits from class `ElasticMaterial`. The source code presented below implements calculation of the elastic–plastic stress increment for a finite strain increment using a tangent material matrix and subincrementation.

```java
 1 package material;
 2
 3 import elem.Element;
 4 import fea.FE;
 5 import util.UTIL;
 6
 7 // Constitutive relations for elastic-plastic material
 8 public class ElasticPlasticMaterial extends ElasticMaterial {
 9
10     // Stress at the beginning of increment
11     private static double[] sig0 = new double[6];
12     // Stress at the end of increment
13     private static double[] sig = new double[6];
14     // Derivatives of yield function
15     private static double[] a = new double[6];
16     // Elasticity matrix
17     private static double[][] Emat = new double[6][6];
18     // Ea = Emat*a
19     private static double[] Ea = new double[6];
20     // Plastic strain increment
21     private static double[] depsp = new double[6];
22     // Equivalent plastic strain
23     private static double epi;
24     // Shear modulus
25     private static double G;
26     private static final double beta = 0.1;
27     private Midpoint midpoint;
28
29     public ElasticPlasticMaterial(String stressState) {
30
31         super(stressState);
32         if (stressState.equals("plstress"))
33             UTIL.errorMsg("Elastic-plastic material is not "
34             + "implemented for plane stress");
35         if (!FE.epIntegrationTANGENT)
36             midpoint = new Midpoint();
37     }
38
39     // Elastic-plastic stress increment.
40     // elm - element,
41     // ip - integration point within element
42     public void strainToStress(Element elm, int ip) {
43
44         elasticityMatrix(Emat);
45         G = getMu();
46
47         // Elastic stress increment dsig due to deps
48         super.strainToStress(elm, ip);
49
50         for (int i = 0; i < lv; i++) {
51             sig0[i] = elm.str[ip].sStress[i];
52             sig[i] = sig0[i] + dsig[i];
53         }
54         // Equivalent plastic strain
```

```
55            epi = elm.str[ip].sEpi;
56            double epi0 = epi;
57            double f1 = yieldFunction(sig, epi);
58
59            // Elastic point
60            if (f1 < 0.0) return;
61
62            // Elastic-plastic point
63            double f0 = yieldFunction(sig0, epi);
64            double r;
65            if (f0 < 0.0) {
66                r = -f0/(f1 - f0);
67                for (int i = 0; i < lv; i++)
68                    sig[i] = sig0[i] + dsig[i]*r;
69                double f = yieldFunction(sig, epi);
70                derivYieldFunc(sig, a);
71                double c1 = 0.0;
72                for (int i = 0; i < lv; i++) c1 += a[i]*dsig[i];
73                r = r - f/c1;
74            }
75            else r = 0.0;
76
77            // Number of subincrements ( = 1 for midpoint method)
78            int nsub = (FE.epIntegrationTANGENT) ?
79                    (int) (f1/(beta*sY))+1 : 1;
80
81            for (int i = 0; i < lv; i++) {
82                sig[i] = sig0[i] + dsig[i]*r;
83                dsig[i] = (1.0 - r)*dsig[i]/nsub;
84                deps[i] = (1.0 - r)*deps[i]/nsub;
85            }
86            // Subincrement loop: tangent or midpoint method
87            for (int isub = 0; isub < nsub; isub++) {
88                if (FE.epIntegrationTANGENT)
89                    tangentStressIncrement();
90                else midpoint.stressIncrement();
91            }
92            for (int i = 0; i < lv; i++)
93                elm.str[ip].dStress[i] = sig[i] - sig0[i];
94            elm.str[ip].dEpi = epi - epi0;
95    }
```

At the beginning of the class, static arrays and scalars are declared, which are necessary during stress-increment computing. The class constructor calls the constructor of superclass `ElasticMaterial`. Because of certain complications we do not consider implementation of the elastic–plastic material under plane stress conditions. An error message is generated when the user attempts to solve an elastic–plastic problem under plane stress conditions. If integration of elastic–plastic constitutive equations is performed with a midpoint method (see next section) then the constructor of inner class `Midpoint` is called.

The main logic of the final stress increment computation is contained in method `strainToStress`. The method obtains `Element` object `elm` and integration point number `ip`. The method should estimate the increments of stress and equivalent plastic strain at this integration point using a strain increment.

Lines 44–45 set the elasticity matrix `emat` and shear modulus G. The elastic stress increment is computed using a call to method `strainToStress` of parent class `ElasticMaterial` (line 48). After this call, the increment of strain `deps` and elastic increment of stress `dsig` are available. In lines 50–53 the stress level at the increment beginning is stored in array `sig0` and elastic stress prediction at the increment end is placed in array `sig`. The accumulated equivalent plastic strain is contained in variables `epi0` and `epi` (line 56).

If the yield function value `f1` computed in line 57 for stress at the increment end is negative then the state of the integration point is elastic and the method returns with the stress increment already computed by superclass `ElasticMaterial`.

Otherwise, the state of the integration point is elastic–plastic. In general, at the increment beginning, the integration point is in the elastic state and we need to determine the scalar parameter `r` of Equation 19.21, which indicates the elastic fraction of the increment. Determination of parameter `r` is done in lines 63–75 according to Equations 19.22 and 19.23. Improvement of the `r` value is restricted to one iteration. The array a contains derivatives of the yield function defined in (19.10).

Lines 78–79 determine the number of subincrements `nsub` according to (19.25). Note that the number of subincrements is equal to one for the midpoint integration of constitutive equations since this method does not require subincrementation. Stress update to the beginning of plastic flow and determination of elastic stress subincrement `dsig` and strain subincrement `deps` is performed in lines 81–85.

The subincrement loop in lines 87–91 contains conditional calls to the method that uses the tangent elastic–plastic matrix (line 89) or to the method that employs a midpoint integration procedure (line 90). The resulting stress increment and equivalent plastic strain increment is placed in element `dStress` (line 93) and `dEpi` (line 94), respectively.

The elastic–plastic stress increment with the use of the tangent material matrix (19.18) is computed in method `tangentStressIncrement` presented below.

```java
97       // Compute elastic-plastic increment by tangent method.
98       // Update stresses sig and equivalent plastic strain epi
99       private void tangentStressIncrement() {
100
101          double H = slopeH(epi);
102          derivYieldFunc(sig, a);
103          double dlambda = 0.0;
104          for (int i = 0; i < lv; i++) {
105              double s = 0.0;
106              for (int j = 0; j < lv; j++) s += Emat[i][j]*a[j];
107              Ea[i] = s;
108              dlambda += a[i]*dsig[i];
109          }
110          dlambda /= (H + 3*G);
111          if (dlambda < 0.0) dlambda = 0.0;
112          for (int i = 0; i < lv; i++) {
113              sig[i] += dsig[i] - dlambda*Ea[i];
114              depsp[i] = dlambda*a[i];
```

```
115            }
116            epi += eqPlastStrain(depsp);
117
118            // Stress correction
119            double f = yieldFunction(sig, epi);
120            double c1 = 0.0;
121            for (int i = 0; i < lv; i++) c1 += a[i]*a[i];
122            for (int i = 0; i < lv; i++) sig[i] -= a[i]*f/c1;
123        }
```

Lines 101–102 determine the slope H of the material deformation curve and derivatives of the yield function a. Plastic multiplier λ, denoted in the code as dlambda, is estimated in lines 103–111 according to Equation 19.15. Using the value of dlambda, stress sig is updated and the plastic strain increment depsp is calculated. These calculations are based on Equations 19.18 and 19.12. Equivalent plastic strain epi is incremented in line 116. Stress correction is done in lines 119–122 using (19.27).

The next code fragment contains methods that compute quantities and vectors used for performing an elastic–plastic increment.

```
125        // Yield function.
126        // s - stresses,
127        // epi - equivalent plastic strain,
128        // returns  yield function value
129        private double yieldFunction(double[] s, double epi) {
130            double sm, seq;
131            if (stressState.equals("threed")) {
132                sm = (s[0] + s[1] + s[2])/3;
133                seq = Math.sqrt(3.0*(0.5*((s[0] - sm)*(s[0] - sm)
134                        + (s[1] - sm)*(s[1] - sm)
135                        + (s[2] - sm)*(s[2] - sm))
136                        + s[3]*s[3] + s[4]*s[4] + s[5]*s[5])));
137            }
138            else {
139                sm = (s[0] + s[1] + s[3])/3;
140                seq = Math.sqrt(3.0*(0.5*((s[0] - sm)*(s[0] - sm)
141                        + (s[1] - sm)*(s[1] - sm)
142                        + (s[3] - sm)*(s[3] - sm)) + s[2]*s[2])));
143            }
144            return seq - yieldRadius(epi);
145        }
146
147        // Radius of yield surface Y = sY + k*ep^m
148        private double yieldRadius(double ep) {
149            if (ep <= 0.0) return sY;
150            else return (km*Math.pow(ep, mm) + sY);
151        }
152
153        // Derivatives of yield function.
154        // s - stresses (in),
155        // a - derivatives of yield function (out)
156        private void derivYieldFunc(double[] s, double[] a) {
157            double sm, seq;
```

```
158
159        if (stressState.equals("threed")) {
160            sm = (s[0] + s[1] + s[2])/3;
161            a[0] = s[0] - sm;
162            a[1] = s[1] - sm;
163            a[2] = s[2] - sm;
164            a[3] = 2*s[3];
165            a[4] = 2*s[4];
166            a[5] = 2*s[5];
167            seq = Math.sqrt(
168                    0.5*(a[0]*a[0] + a[1]*a[1] + a[2]*a[2])
169                    + s[3]*s[3] + s[4]*s[4] + s[5]*s[5]);
170        }
171        else {
172            sm = (s[0] + s[1] + s[3])/3;
173            a[0] = s[0] - sm;
174            a[1] = s[1] - sm;
175            a[2] = 2*s[2];
176            a[3] = s[3] - sm;
177            seq = Math.sqrt(0.5*(a[0]*a[0] + a[1]*a[1]
178                    + a[3]*a[3]) + s[2]*s[2]);
179        }
180
181        for (int i = 0; i < lv; i++)
182            a[i] = 0.5*Math.sqrt(3.0)/seq*a[i];
183    }
184
185    // Returns slope of deformation curve
186    // epi - equivalent plastic strain
187    private double slopeH(double epi) {
188        if (km == 0.0) return 0.0;
189        else if (mm == 1.0) return km;
190        else return (epi == 0.0) ?
191                0.0 : km*mm*Math.pow(epi, mm - 1.0);
192    }
193
194    // Returns equivalent plastic strain
195    // dp - pastic strains (in)
196    private double eqPlastStrain(double[] dp) {
197        if (stressState.equals("threed"))
198            return Math.sqrt(
199                (2*(dp[0]*dp[0] + dp[1]*dp[1] + dp[2]*dp[2])
200                + dp[3]*dp[3] + dp[4]*dp[4] + dp[5]*dp[5])/3.0);
201        else {
202            if (stressState.equals("plstress"))
203                dp[3] = -(dp[0] + dp[1]);
204            return Math.sqrt((2*(dp[0]*dp[0] + dp[1]*dp[1]
205                    + dp[3]*dp[3]) + dp[2]*dp[2])/3.0);
206        }
207    }
```

The yield function is estimated by method `yieldFunction` in lines 129–145 according to (19.5). The method obtains stresses `s` and equivalent plastic strain `epi` and uses the material-deformation curve in order to return the Mises yield function value (Equation 19.5). The radius of the yield surface is evaluated by method `yieldRadius` in lines 148–151. Method `derivYieldFunc` computes the array a containing derivatives of yield function (19.10). Method `slopeH` returns the slope of the material deformation curve for the specified value of equivalent plastic strain `epi`. Equation 19.9 for the increment of the equivalent plastic strain is realized in method `eqPlastStrain`.

19.5 Midpoint Integration of Constitutive Relations

The generalized midpoint integration rule is based on computing the increment of plastic strains $\{\Delta \varepsilon^{\mathrm{p}}\}$ using the vectors of the yield function derivatives $\{a\}$ at the beginning and end of the increment and on enforcing the zero-value constraint of the yield function f at the end of the increment. The generalized midpoint integration rule can be presented in the following form:

$$\{\sigma_1\} = [E](\{\varepsilon_1\} - \{\varepsilon_1^{\mathrm{p}}\} - \{\varepsilon_1^{\mathrm{t}}\}),$$

$$\{\Delta \varepsilon^{\mathrm{p}}\} = \lambda((1 - \alpha)\{a_0\} + \alpha\{a_1\}),$$

$$\Delta \varepsilon_i^{\mathrm{p}} = \sqrt{\frac{2}{3} \Delta \varepsilon_{ij}^{\mathrm{p}} \Delta \varepsilon_{ij}^{\mathrm{p}}},$$

$$f(\{\sigma_1\}, \varepsilon_{i0}^{\mathrm{p}} + \Delta \varepsilon_i^{\mathrm{p}}) = 0.$$

(19.32)

Here, the subscript zero denotes values at the beginning of the increment and the subscript one denotes values at the end of the increment. The integration parameter α may vary from 0 to 1. For $\alpha > 0$, the integration rule is implicit. The value $\alpha = 1$ leads to the so-called radial return algorithm with first-order accuracy. For $\alpha = 1/2$, the algorithm has second-order accuracy.

The integration algorithm in the form of (19.32) is a system of nonlinear algebraic equations, since derivatives of yield function $\{a_1\}$ and the plastic multiplier λ are unknown. It is shown in [22] that the above system of nonlinear equations can be solved after its reduction to a scalar nonlinear equation with respect to the parameter λ:

$$\varphi(\lambda) = \sqrt{\frac{3}{2}} \|s_\alpha\| - 3G\alpha\lambda - Y(\varepsilon_{i0}^{\mathrm{p}} + \Delta \varepsilon_i^{\mathrm{p}}) = 0,$$

$$\Delta \varepsilon_i^{\mathrm{p}} = \lambda \sqrt{\frac{2}{3}} \left\| (1 - \alpha)\{a_0\} + \alpha \sqrt{3/2}\{s_\alpha\} / \|s_\alpha\| \right\|,$$

(19.33)

where the deviatoric stresses $\{s_\alpha\}$ are equal to

$$\{s_\alpha\} = \{s_0\} + 2G(\{\Delta e\} - \lambda^{(i-1)}(1 - \alpha)\{a_0\})$$

(19.34)

and their norm is computed as

$$\|s\| = \left[s_{11}^2 + s_{22}^2 + s_{33}^2 + 2(s_{12}^2 + s_{23}^2 + s_{31}^2) \right]^{1/2}. \tag{19.35}$$

Solution of the nonlinear equation (19.33) is obtained numerically using the Newton–Raphson iterative procedure. Using the initial value $\lambda^{(0)}$ the successive approximation for parameter λ is

$$\lambda^{(i)} = \lambda^{(i-1)} - \frac{\varphi(\lambda^{(i-1)})}{\varphi'(\lambda^{(i-1)})}. \tag{19.36}$$

Iterations are stopped when an equation residual becomes small compared to the yield stress σ_Y.

The generalized midpoint algorithm for integration of constitutive relations for the von Mises hardening material starts with known stresses $\{\sigma_0\}$, equivalent plastic strain ε_i^p, and strain increment $\{\Delta\varepsilon\}$. It can be expressed by the following pseudo-code [22].

$$\{\Delta e\} = \{\Delta\varepsilon\} - \{\Delta\varepsilon_m\}$$
$$\{s_0\} = \{\sigma_0\} - \{\sigma_{m0}\}$$
$$\{a_0\} = 3\{s_0\}/(2\sigma_{i0})$$
$$\lambda^{(0)} = \sqrt{2/3}\,\|\Delta e\|$$

do

$$\{s_\alpha\} = \{s_0\} + 2G(\{\Delta e\} - \lambda^{(i-1)}(1-\alpha)\{a_0\})$$
$$\{\bar{s}_\alpha\} = \{s_\alpha\}/\|s_\alpha\|$$
$$\{b\} = (1-\alpha)\{a_0\} + \alpha\sqrt{3/2}\{\bar{s}_\alpha\}$$
$$\Delta\varepsilon_i^p = \lambda^{(i-1)}\sqrt{2/3}\,\|b\|$$
$$\varphi(\lambda^{(i-1)}) = \sqrt{3/2}\,\|s_\alpha\| - 3G\alpha\lambda^{(i-1)} - Y(\varepsilon_{i0}^p + \Delta\varepsilon_i^p) \tag{19.37}$$
$$\varphi'(\lambda^{(i-1)}) = -2\sqrt{3/2}G(1-\alpha)\{a_0\} : \{\bar{s}_\alpha\}$$
$$\qquad\qquad - 3G\alpha - \sqrt{2/3}(dY/d\varepsilon_i^p)\|b\|$$
$$\lambda^{(i)} = \lambda^{(i-1)} - \varphi(\lambda^{(i-1)})/\varphi'(\lambda^{(i-1)})$$

until $\left|\varphi(\lambda^{(i)})\right| < \varepsilon\sigma_Y$

$$\varepsilon_{i1}^p = \varepsilon_{i0}^p + \Delta\varepsilon_i^p$$
$$\sigma_{i1} = Y(\varepsilon_{i1}^p)$$
$$\{s_1\} = \{s_\alpha\}/(1 + 3G\alpha\lambda/\sigma_{i1})$$
$$\Delta\sigma_m = E/(1-2v)\Delta\varepsilon_m$$
$$\{\sigma_1\} = \{s_1\} + \{\sigma_{m0}\} + \{\Delta\sigma_m\}$$

Here, $\{\varepsilon_m\}$, $\{\sigma_m\}$ are the mean strain and mean stress, respectively, and an operation of the dyadic product is defined as

$$\{a\} : \{s\} = a_{11}s_{11} + a_{22}s_{22} + a_{33}s_{33} + 2(a_{12}s_{12} + a_{23}s_{23} + a_{31}s_{31}). \tag{19.38}$$

Implementation of the generalized midpoint algorithm for computing the stress increment is done in inner class `Midpoint`. The head of the class and its method for evaluating a stress increment follow.

```
209     // Midpoint method for integration of constitutive
210     // relations for the von Mises hardening material
211     class Midpoint {
212         // Strain deviator
213         double[] ed = new double[6];
214         // Stress deviator
215         double[] sd = new double[6];
216         // Trial stress deviator
217         double[] sdtr = new double[6];
218         // Yield function derivatives
219         double[] a0 = new double[6];
220
221         double[] sal = new double[6];
222         double[] salbar = new double[6];
223         double[] b = new double[6];
224         double alpha = 0.5;
225
226         // Elastic-plastic stress increment by midpoint method.
227         // Update stresses sig and
228         //     equivalent plastic strain epi
229         void stressIncrement() {
230             double SQ32 = Math.sqrt(1.5);
231             double SQ23 = Math.sqrt(2./3.);
232             double tolerance = 1.e-5;
233             // Transform strains to tensor components
234             if (lv == 4) deps[2] *= 0.5;
235             else for (int i = 3; i < 6; i++) deps[i] *= 0.5;
236             double depsm = deviator(deps, ed);
237             double sigm = deviator(sig, sd);
238             double sigeq0 = SQ32*norm(sd);
239             for (int i = 0; i < lv; i++) {
240                 sdtr[i] = sd[i] + 2*G*ed[i];
241                 a0[i] = 1.5*sd[i]/sigeq0;
242             }
243             double lambda = SQ23*norm(ed);
244             // Find lambda by Newton-Raphson iteration
245             double epi1, sigeq;
246             for (; ;) {
247                 for (int i = 0; i < lv; i++)
248                     sal[i] = sdtr[i] -
249                             2*G*lambda*(1 - alpha)*a0[i];
250                 double salmod = norm(sal);
251                 for (int i = 0; i < lv; i++) {
252                     salbar[i] = sal[i]/salmod;
253                     b[i] = (1 - alpha)*a0[i] +
254                             alpha*SQ32*salbar[i];
255                 }
256                 double bmod = norm(b);
257                 epi1 = epi + lambda*SQ23*bmod;
258                 sigeq = yieldRadius(epi1);
```

```
259              double phi = SQ32*salmod -
260                      3*G*alpha*lambda - sigeq;
261              if (Math.abs(phi) < tolerance*sY) break;
262              double phiPrime = -2.*SQ32*G
263                      *(1 - alpha)*dyadicProduct(a0, salbar)
264                      - 3.*G*alpha - slopeH(epi1)*SQ23*bmod;
265              double lambda1 = lambda - phi/phiPrime;
266              lambda = (lambda1 <= 0.) ? 0.5*lambda:lambda1;
267          }
268          epi = epi1;
269          for (int i = 0; i < lv; i++)
270              sd[i] = sal[i]/(1 + 3*G*alpha*lambda/sigeq);
271          double dsigm = depsm*e/(1. - 2.*nu);
272          stressFromDeviator(sd, sigm + dsigm, sig);
273      }
```

Method `stressIncrement` computes the increments of stresses and the equivalent plastic strain using the midpoint algorithm. The source code of the method closely follows the algorithm (19.37). In lines 234–235, the shear strain components are halved, thus transforming them into tensor components. This is because using the strain tensor entries is more convenient in the midpoint algorithm.

Strain and stress deviators and the equivalent stress are evaluated in lines 236–238. The value of the plastic parameter λ is sought in the loop of lines 246–267 using the Newton–Raphson iterative procedure. Line 268 stores an incremented value of the equivalent plastic strain `epi`. Incremented stresses `sig` are determined from the stress deviator and the mean stress at the increment end.

Several additional methods are designed for the midpoint algorithm.

```
275          // Compute deviator.
276          // s - stress,
277          // d - deviator (out),
278          // returns   mean value
279          double deviator(double[] s, double[] d) {
280              double sm;
281              if (lv == 4) {
282                  sm = (s[0] + s[1] + s[3])/3;
283                  d[0] = s[0] - sm;
284                  d[1] = s[1] - sm;
285                  d[3] = s[3] - sm;
286                  d[2] = s[2];
287              }
288              else {
289                  sm = (s[0] + s[1] + s[2])/3;
290                  d[0] = s[0] - sm;
291                  d[1] = s[1] - sm;
292                  d[2] = s[2] - sm;
293                  d[3] = s[3];
294                  d[4] = s[4];
295                  d[5] = s[5];
296              }
297              return (sm);
298          }
299
```

```java
300                 // Compute stress s = d + sm.
301                 // d - deviator,
302                 // sm - mean stress,
303                 // s - stress (out)
304                 void stressFromDeviator(double[] d, double sm,
305                                         double[] s) {
306                     if (lv == 4) {
307                         s[0] = d[0] + sm;
308                         s[1] = d[1] + sm;
309                         s[3] = d[3] + sm;
310                         s[2] = d[2];
311                     }
312                     else {
313                         s[0] = d[0] + sm;
314                         s[1] = d[1] + sm;
315                         s[2] = d[2] + sm;
316                         s[3] = d[3];
317                         s[4] = d[4];
318                         s[5] = d[5];
319                     }
320                 }
321
322                 // Returns norm = sqrt(Sij*Sij)
323                 double norm(double[] s) {
324                     if (lv == 4)
325                         return (Math.sqrt(s[0]*s[0] + s[1]*s[1] +
326                                 s[3]*s[3] + 2*s[2]*s[2]));
327                     else
328                         return (Math.sqrt(s[0]*s[0] + s[1]*s[1] +
329                                 s[2]*s[2] + 2*(s[3]*s[3] + s[4]*s[4] +
330                                 s[5]*s[5])));
331                 }
332
333                 // Returns dyadic product = aij*bij
334                 double dyadicProduct(double[] a, double[] b) {
335                     if (lv == 4)
336                         return (a[0]*b[0] + a[1]*b[1] + a[3]*b[3] +
337                                 2*a[2]*b[2]);
338                     else
339                         return (a[0]*b[0] + a[1]*b[1] + a[2]*b[2] +
340                                 2*(a[3]*b[3] + a[4]*b[4] + a[5]*b[5]));
341                 }
342
343         }
344
345 }
```

The method `deviator` presented in lines 279–298 computes a deviator from given tensor components and returns a mean value of diagonal entries of the tensor. Method `stressFromDeviator` (lines 304–320) restores the stress components from given devatoric components and mean stress. Methods `norm` and `dyadicProduct` return the norm of a tensor and the dyadic product of the components of two tensors.

19.6 Nonlinear Solution Procedure

Constitutive relation (19.18) contains the elastic–plastic constitutive matrix $[E^{\text{ep}}]$, which depends on the current stress-strain state at a given material point. Using this constitutive matrix it is possible to compute an elastic–plastic element stiffness matrix $[k^{\text{ep}}]$:

$$[k^{\text{ep}}] = \int_V [B]^{\text{T}}[E^{\text{ep}}][B]dV, \tag{19.39}$$

and an increment of the thermal vector $\{dh^{\text{ep}}\}$:

$$\{dh^{\text{ep}}\} = \int_V [B]^{\text{T}}[E^{\text{ep}}]\{d\varepsilon^{\text{t}}\}dV. \tag{19.40}$$

Then, the incremental finite element equation expressed through displacement increment $\{dq\}$ has the following appearance:

$$
\begin{aligned}
[k^{\text{ep}}]\{dq\} &= \{df\}, \\
\{df\} &= \{dp\} + \{dh^{\text{ep}}\}.
\end{aligned}
\tag{19.41}
$$

where $\{dp\}$ is an increment of the actual force load, which we consider independent of the strain–stress level.

The incremental finite element equation is valid for infinitesimally small increments of force or displacements. Usually, in elastic–plastic problems the applied load is divided into a number of increments. In some cases this division is necessary in order to reproduce load history. However, we cannot afford very small load increments because the solution of the global equation system is an expensive operation. Hence, we need a nonlinear solution procedure that works for finite load increments.

In general, nonlinear problems are formulated in terms of some unknown parameters. The finite element elastic–plastic problem can be stated as follows. Starting from an equilibrium state, for a given load increment it is necessary to find the displacement field that satisfies the stress equilibrium equation

$$\{p\} - \int_V [B]^{\text{T}}\{\sigma\}dV = 0. \tag{19.42}$$

This can be done using iterative techniques, which decrease a residual vector $\{\psi\}$ due to the imbalance of external and internal forces:

$$\{\psi\} = \{p\} - \int_V [B]^{\text{T}}\{\sigma\}dV. \tag{19.43}$$

Several iterative techniques are used for the solution of finite element nonlinear problems [7]. Among them the most famous techniques are the Newton–Raphson method and the initial stress method.

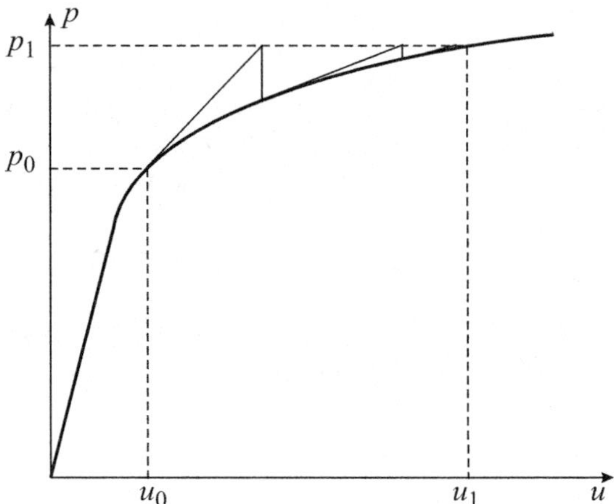

Fig. 19.2 Illustration of the Newton–Raphson method

19.6.1 Newton–Raphson Method

The Newton–Raphson method is considered the most rapidly convergent process for solution of nonlinear problems. The iterative procedure of the Newton–Raphson method for one load step is given by the following pseudocode:

Set $\{q_0\}$, $\{\sigma_0\}$ from the previous load step

$\{\psi_0\} = \{\Delta p\} + \{\Delta h^{\mathrm{ep}}\}$

do

$$
\begin{aligned}
[k^{\mathrm{ep}}_{i-1}] &= [k^{\mathrm{ep}}(\{\sigma_{i-1}\})] \\
\{\Delta q_i\} &= [k^{\mathrm{ep}}_{i-1}]^{-1}\{\psi_{i-1}\} \\
\{q_i\} &= \{q_{i-1}\} + \{\Delta q_i\} \\
\{\Delta\sigma_i\} &= \{\Delta\sigma^{\mathrm{ep}}(\{\Delta q_i\})\} \\
\{\sigma_i\} &= \{\sigma_{i-1}\} + \{\Delta\sigma_i\} \\
\{\psi_i\} &= \{p\} - \int_V [B]^{\mathrm{T}}\{\sigma_i\}dV
\end{aligned}
$$

\qquad\qquad(19.44)

until convergence

The Newton–Raphson method for the one-dimensional case of a nonlinear problem is illustrated in Figure 19.2, where u_0, p_0 are displacement and load, which represent an equilibrium state from the previous load step. The purpose of the solution is to determine displacement u_1 corresponding to the load value p_1. Iterations of the Newton–Raphson method start from the equilibrium stress–strain state, which is known at the end of the previous load step. A high convergence rate is provided by computing new tangent stiffness matrix $[k_{i-1}^{ep}]$, which corresponds to the strain–stress state of the previous iteration. A displacement increment is used for determining the stress increment according to the constitutive relation (19.18) with subincrementation or another method that provides sufficient accuracy. Residual $\{\psi\}$ due to the imbalance of external and internal forces is calculated using the accumulated stress. If the residual is small enough then convergence is reached and the iteration cycle is finished; otherwise the residual $\{\psi_i\}$ is employed as a load for the next iteration.

The modified Newton–Raphson method uses the same algorithm as the Newton–Raphson iterative procedure, but tries to economize computations by computing $[k_{i-1}^{ep}]$ only at the first iteration and by keeping it constant during the following iterations.

19.6.2 Initial Stress Method

The initial stress method uses an iterative procedure similar to that of the Newton–Raphson method. The main difference is that the initial stress method employs the elastic stiffness matrix for computing the displacement increment. The following pseudocode presents the iterative procedure of the initial stress method:

Set $\{q_0\}$, $\{\sigma_0\}$ from the previous load step

$$\{\psi_0\} = \{\Delta p\} + \{\Delta h^e\}$$

do

$$\{\Delta q_i\} = [k^e]^{-1}\{\psi_{i-1}\}$$
$$\{q_i\} = \{q_{i-1}\} + \{\Delta q_i\}$$
$$\{\Delta \sigma_i\} = \{\Delta \sigma^{ep}(\{\Delta q_i\})\}$$
$$\{\sigma_i\} = \{\sigma_{i-1}\} + \{\Delta \sigma_i\}$$
$$\{\psi_i\} = \{p\} - \int_V [B]^T\{\sigma_i\}dV$$

(19.45)

until convergence

Illustration of the initial stress method in a one-dimensional nonlinear problem is presented in Figure 19.3. It is evident that the number of iterations in the initial stress method can be considerable. It is worth noting that the iteration cost in the initial stress method is low because only a resolution procedure for the equation system is performed. Nevertheless, for the developed plasticity the convergence of the initial stress method can be too slow. While the method is very simple, it is recommended to use it with certain care. Probably, the initial stress method can be used for nonlinear problems with stress concentrators and moderate development of the plastic zone.

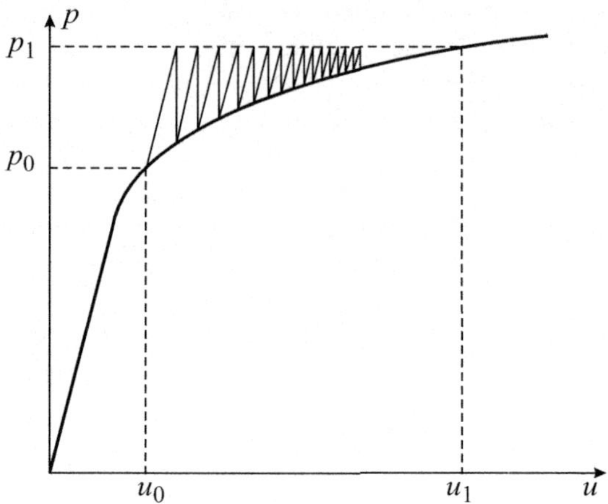

Fig. 19.3 Illustration of the initial stress method

19.6.3 Convergence Criteria

A natural convergence criterion for the iterative procedure of a nonlinear solution is to compute the norm of the residual vector $\{\psi\}$ and to compare it to the norm of the applied load $\{p\}$:

$$\frac{\|\psi\|}{\|p\|} < \varepsilon_\psi, \tag{19.46}$$

where ε_ψ is the error tolerance and the vector norm is determined as

$$\|\psi\| = \sqrt{\{\psi\}^{\mathrm{T}}\{\psi\}}. \tag{19.47}$$

The other possibility is to control the displacement increment during the current iteration. The convergence criterion based on the norm of the displacement increment is:

$$\frac{\|\Delta q\|}{\|q\|} < \varepsilon_q. \tag{19.48}$$

Here, $\{\Delta q\}$ is the displacement increment at the current iteration, and $\{q\}$ is the displacement vector accumulated during this load step.

Values of error tolerance for convergence can be selected as follows: $\varepsilon_\psi \approx 10^{-2}$ and $\varepsilon_q \approx 10^{-2} - 10^{-3}$.

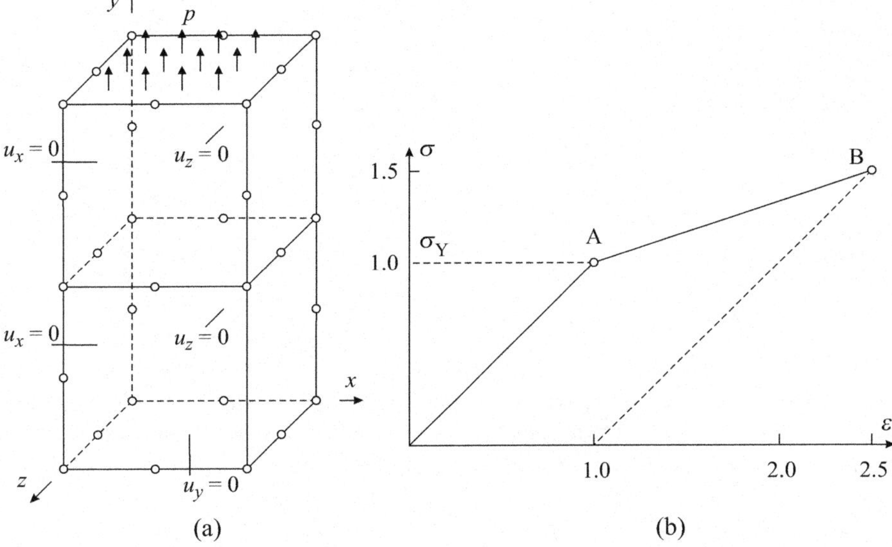

Fig. 19.4 Elastic–plastic problem of simple tension test: schematic of a problem (a) and strain–stress curve (b)

19.7 Example: Solution of an Elastic–Plastic Problem

Methods of elastic–plastic analysis are demonstrated on the solution of a simple problem – uniform tension of a rectangular prism with a square-cross section shown in Figure 19.4a. Let us select the following elastic–plastic material with linear hardening:

Elasticity modulus $E = 1.0$;
Poisson's ratio $v = 0.3$;
Yield stress $σ_Y = 1.0$;
Hardening coefficient $k = 0.5$;
Hardening power $m = 1.0$.

In the considered problem, the strain–stress state at any point of the specimen follows the material-deformation curve presented in Figure 19.4b.

First, a mesh of two twenty-node finite elements is prepared and placed in file f.mesh. For cube elements with edge size 1, the file contains

```
nNod = 32
nEl  =  2
nDim =  3


nodCoord
0.0 0.0 0.0    0.0 0.0 0.5    0.0 0.0 1.0    0.5 0.0 0.0
0.5 0.0 1.0    1.0 0.0 0.0    1.0 0.0 0.5    1.0 0.0 1.0
0.0 0.5 0.0    0.0 0.5 1.0    1.0 0.5 0.0    1.0 0.5 1.0
```

```
0.0 1.0 0.0      0.0 1.0 0.5      0.0 1.0 1.0      0.5 1.0 0.0
0.5 1.0 1.0      1.0 1.0 0.0      1.0 1.0 0.5      1.0 1.0 1.0
0.0 1.5 0.0      0.0 1.5 1.0      1.0 1.5 0.0      1.0 1.5 1.0
0.0 2.0 0.0      0.0 2.0 0.5      0.0 2.0 1.0      0.5 2.0 0.0
0.5 2.0 1.0      1.0 2.0 0.0      1.0 2.0 0.5      1.0 2.0 1.0

elCon
hex20   mat
 1   4   6 11 18 16 13  9  2  7 19 14  3  5  8 12 20 17 15 10
hex20   mat
13 16 18 23 30 28 25 21 14 19 31 26 15 17 20 24 32 29 27 22

end
```

The first three scalars are: `nNod` – number of nodes, `nEl` – number of elements
and `nDim` – number of space dimensions (three-dimensional problem). Array
`nodCoord` includes coordinates of thirty two nodes. Element connectivities that
follow name `elCon` contain element type `hex20`, material name `mat`, and element connectivity numbers. The word `end` signals the end of the file.

A second file with name `f.fem` contains other information on the problem.

```
# Elastic-plastic problem, simple tension, 3D
physLaw = elPlastic
includeFile f.mesh

#       Mat name  E  nu alpha  SY   k    m
material = mat    1 0.3   0.0  1.0 0.5 1.0

# Displacement boundary conditions
boxConstrDispl = x   0.0   -0.1 -0.1 -0.1   0.1 2.1 1.1
boxConstrDispl = y   0.0   -0.1 -0.1 -0.1   1.1 0.1 1.1
boxConstrDispl = z   0.0   -0.1 -0.1 -0.1   1.1 2.1 0.1
end

loadStep = A
boxSurForce = y 1.0   -0.1 1.9 -0.1   1.1 2.1 1.1
end

loadStep = B
scaleLoad = 0.5
residTolerance = 0.001
end
```

In the beginning, the elastic–plastic physical law is specified and the content of file
`s.mesh` is included in data. Statement `material` defines the elastic and elastic-
plastic properties of material `mat`. Displacement boundary conditions $u_x = 0$, $u_y = 0$
and $u_z = 0$ are applied to planes $x = 0$, $y = 0$ and $z = 0$, respectively, using instructions `boxConstrDispl`.

First, load step A defines the surface load $p = 1$ with direction along the z-axis.
As can be seen from Figure 19.4b this load corresponds to the yield stress on the
material-deformation curve (point A). Second load step B is in the elastic–plastic
range. The load increment is created by scaling the previous load with factor 0.5.

The total load at point B has a magnitude of 1.5. The residual norm tolerance is set to 0.001.

Execution of the Jfem code is performed with the command

```
java  -cp classes fea.Jfem f.fem f.lst
```

It is supposed that the current directory is Jfea and the data file f.fem is located in this directory. After program run three results files appear: f.lst – brief information about solution and f.lst.A and f.lst.B – results for load steps A and B.

Displacements u, v, and w for nodes 32 with coordinates $x = 1$, $y = 2$, $z = 1$, normal stresses σ_x, σ_y, and σ_z and the equivalent plastic strain ε_i^p are presented in Table 19.1. For each load step, finite element results (FEM) are compared to the exact solution.

Table 19.1 Results of elastic–plastic solution

Load	u_{32}	v_{32}	w_{32}	σ_x	σ_y	σ_z	ε_i^p
A-FEM	−0.30000	2.00000	−0.30000	0.00000	1.00000	0.00000	0.00000
A-exact	−0.3	2.0	−0.3	0.0	1.0	0.0	0.0
B-FEM	−0.94946	4.99786	−0.94946	0.00012	1.49975	0.00012	0.99925
B-exact	−0.95	5.0	−0.95	0.0	1.5	0.0	1.0

The load step A is elastic. The finite element method produced an exact solution. For elastic–plastic step B finite element results have small differences from the exact solution. The program performed 20 iterations to achieve a relative residual norm 0.001. The small differences between finite element and exact results are due to the limited number of iterations.

Problems

19.1. The material-deformation curve in the elastic–plastic region is represented by the relation

$$\sigma = 2 + \frac{5}{11}\varepsilon,$$

where σ is the stress and ε is the total strain. Find parameters k and m in the description of deformation curve (19.28), provided that the yield stress and the elasticity modulus are $\sigma_Y = 2$ and $E = 5$.

19.2. Solve the elastic–plastic problem presented in Section 19.7 in three-dimensional "plane strain" conditions when the deformation in the z-direction is constrained by specifying $u_z = 0$ at plane $z = 1$. Compare the results obtained with and without such a constraint. Explain the differences.

19.3. Create a finite element mesh and perform elastic–plastic analysis of Problem 19.2 using a two-dimensional approach under plane strain conditions. Confirm that the three- and two-dimensional results are identical.

19.4. Using Hooke's law for the elastic strain fraction and the condition of zero volume change for plastic deformation, determine strains ε_x and ε_z at the end of load step B in the elastic–plastic problem of Section 19.7.

Part III
Mesh Generation

Chapter 20
Mesh Generator

Abstract This chapter opens the book part about generation of finite element meshes. The general approach to mesh generation is based on the block decomposition method. The user divides a solution domain into multiple blocks, each of which is suitable for the local meshing process. The main class of the mesh generator, $Jmgen$, calls modules that perform tasks of mesh generation. New modules can be added without changing the existing Java^{TM} code.

20.1 Block Decomposition Method

Preparation of a finite element model is an important step in finite element analysis. Since the number of elements and nodes in finite element meshes used for solution of practical problems is usually large it is impossible to prepare the finite element mesh manually. In general, mesh generation is a complicated problem. Commercial mesh-generation programs usually include geometric modeling and numerous techniques for mesh generation including meshing for arbitrary domains. It is difficult to use such approaches here since the code will be very large. So, we selected the block decomposition method [12, 27] as a basis for mesh generation in our program $Jfea$.

In the block decomposition method, the user divides a solution domain into multiple blocks such that each block is suitable for the local meshing process. Mesh generation within blocks is performed by various mesh-generation modules.

Some blocks can be meshed using both two- and three-dimensional approaches. First, a two-dimensional mesh is created. Then, this mesh is swept in space to produce a three-dimensional mesh block. Besides mesh generation, program modules can perform other operations on mesh blocks such as transformations, copying, etc.

The generated mesh blocks are pasted together in order to create a total mesh. Instead of pasting all mesh blocks at once, it is easier to connect each time two blocks. A newly created block can be used as a normal mesh block in a block pair for pasting.

Mesh generation using the block decomposition method can be performed as the following steps:

1. Decompose the computational domain into simple blocks.

2. Define block interfaces (common edges or faces) and their subdivisions to ensure continuity of mesh across these shared boundaries.

3. Generate a mesh inside each block separately.

4. Create the final mesh by pasting (connecting) the mesh blocks.

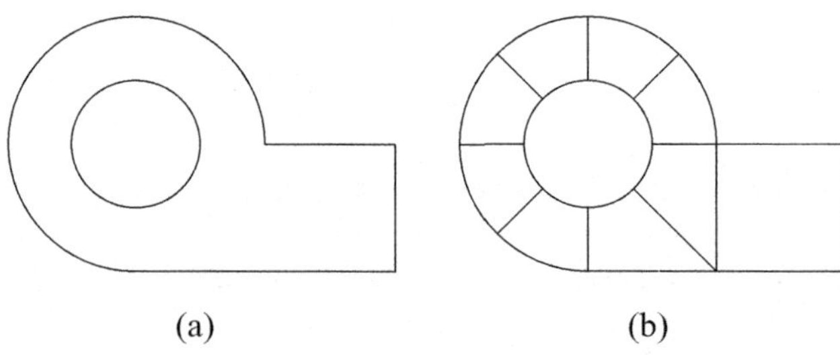

(a)   (b)

Fig. 20.1 Two-dimensional computational domain (a) and its subdivision into multiple blocks (b)

An example of a two-dimensional computational domain is depicted in Figure 20.1a. The domain can be subdivided into quadrilateral blocks for separate meshing as shown in Figure 20.1b. If quadrilateral blocks can have curved edges with quadratic interpolation, representation of circular boundaries can be quite satisfactory.

20.2 Class Structure

A class diagram of the finite element preprocessor is shown in Figure 20.2. Class Jmgen contains the main method and activates all other classes necessary for mesh creation. Each class performs some action on one or more finite element models. FeModel objects are stored in hash table blocks. For example, any mesh generator can create a mesh block as FeModel and can put it in blocks hash table under a name specified by the user. Since all modules are independent, the main class invokes them by the name of its class.

The source code of class Jmgen is given below.

```
1 package fea;
2
3 import util.*;
4
5 import java.io.*;
6 import java.util.HashMap;
7
```

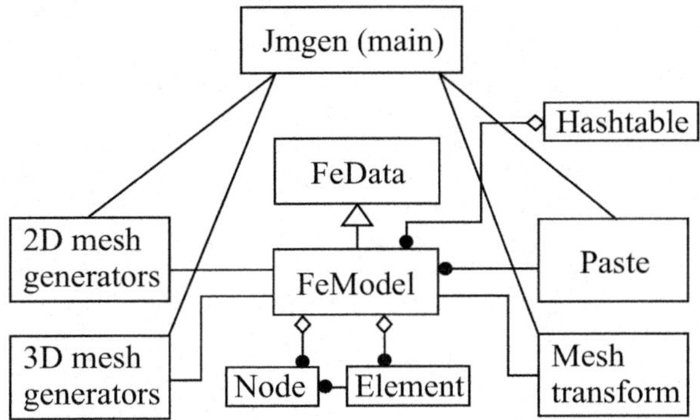

Fig. 20.2 Class diagram of the finite element preprocessor

```
 8  // Main class of the mesh generator
 9  public class Jmgen {
10
11      public static FeScanner RD;
12      public static PrintWriter PR;
13      public static HashMap blocks;
14
15      public static void main(String[] args) {
16
17          if (args.length == 0) {
18              System.out.println(
19                  "Usage: java fea.Jmgen FileIn [FileOut]\n");
20              return;
21          }
22          FE.main = FE.JMGEN;
23
24          RD = new FeScanner(args[0]);
25
26          String fileOut = (args.length == 1) ?
27                  args[0]+".lst" : args[1];
28          PR = new FePrintWriter().getPrinter(fileOut);
29
30          PR.println("fea.Jmgen: Mesh generator. Data file: "
31                  + args[0]);
32          System.out.println("fea.Jmgen: Mesh generator. "
33                  + "Data file: " + args[0]);
34
35          new Jmgen();
36      }
37
38      Jmgen() {
39
40          UTIL.printDate(PR);
41
42          // Hash table for storing mesh blocks
43          blocks = new HashMap();
44
```

```
45            while (RD.hasNext()) {
46
47                String name = RD.next().toLowerCase();
48                if (name.equals("#")) { RD.nextLine(); continue; }
49                PR.println("-------------------------------------");
50
51                try {
52                    Class.forName("gener." + name).newInstance();
53                } catch (Exception e) {
54                    UTIL.errorMsg("Class name not found: "+name);
55                }
56            }
57        PR.close();
58        }
59
60 }
```

If no parameters are specified by the user, a message is printed that the code Jmgem should be run with one or two parameters (lines 17–21). Line 22 specifies that the mesh generator is currently running. This information is used by the constructor of the finite element model (arrays related to stresses and strains are not employed during mesh generation). A scanner RD for reading input data from an ASCII file is constructed in line 24, and a printer PR for printing information into the ASCII file is created in lines 26–27. The main object Jmgem that performs calls to block mesh generators is created in line 35.

In the constructor Jmgem static method printDate prints the current date and time using print writer PR. Line 42 creates hash table blocks, which is used for storage of finite element model for mesh blocks.

The loop in lines 45–56 reads data from an input file. A read token name is transformed to lower case. If the token is symbol # then this line is considered a comment and a token from the next line is input. If string name is not the comment character it is supposed to be a class name. A new object name from package gener is created using methods forName and newInstance.

20.3 Mesh-generation Modules

Mesh-generation modules may be divided into several groups according to their functions. Mesh generators create meshes inside two- or three-dimensional blocks. Read/write modules read meshes from file, write meshes to a file, or print meshes. Transformation modules perform geometric transformations on meshes like translate, scale, or rotate. A module pasting two meshes into one plays an important role in the mesh-generation process since it allows production of complicated meshes composed of relatively simple fragments.

The following mesh-generation modules are present in mesh generator Jmgen:

rectangle – generate rectangular mesh inside rectangular region;

`genquad8` – generate topologically regular mesh inside curvilinear quadrilateral region;

`sweep` – generate three-dimensional mesh by sweeping two-dimensional mesh in space;

`readmesh` – read mesh from file;

`writemesh` – write mesh to file;

`copy` – copy mesh block;

`transform` – make transformations (translate, scale, rotate) for the mesh block;

`connect` – produce new mesh block by connecting two mesh blocks.

Each module is identified by its class name in a data file. The name of the module is read by the main method `Jmgen`. The module inputs all necessary data and performs operations of mesh generation, transformations, and others.

The data file for the mesh generation has in general the following appearance:

```
classA
data for classA

classB
data for classB

classC
data for classC

. . .
```

20.4 Adding New Module

Since mesh-generation modules are called by their names it is possible to add new modules without changing the existing code. The module developer should follow the following rules.

Finite element scanner `Jmgen.RD` is used for data input. Data printer `Jmgen.PR` is employed for printing data into output file. Finite element models are stored in the `Jmgen.blocks` hash table.

The finite element model with name `modelName` can be extracted from the hash table by executing the statement

```
FeModel m = (FeModel) Jmgen.blocks.get(modelName);
```

To put the finite element model `m` into the hash table under name `modelName`, one can use the statement

```
Jmgen.blocks.put(modelName,m);.
```

A simple mesh-generation module generates a mesh for a finite element model and puts the resulting model into hash table `Jmgen.blocks`. Later, other modules can use this model in their operations.

A program structure of a typical mesh-generation module is shown below.

```
public class meshGenModule {

    private FeModel m;

    public meshGenModule() {

        String modelName = Jmgen.RD.next();
        readData();
        printData();
        m = new FeModel(Jmgen.RD, Jmgen.PR);
        generateMesh();
        Jmgen.blocks.put(modelName, m);
    }

    private void readData() {
        ...
    }

    private void printData() {
        ...
    }

    private void generateMesh() {
        ...
    }
}
```

Module constructor reads `modelName` that is used for identification of this mesh block. Method `readData` inputs all the rest of the data used for description of the mesh block. The input data is printed by method `printData`. A generated mesh is placed in an `FeModel` object, so we construct such an object m. The mesh is generated in m by method `generateMesh` and placed in hash table `blocks` under name `modelName`.

Problems

20.1. Subdivide the two-dimensional square domain into quadrilateral blocks in such a way that it has two blocks on the left boundary and four blocks on the right boundary.

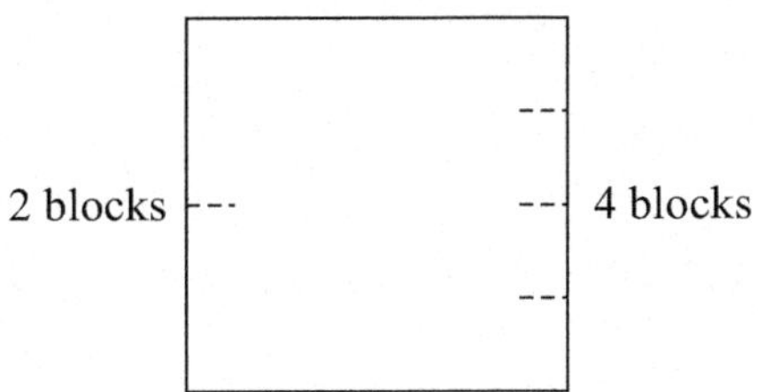

20.2. Represent a two-dimensional domains shaped as a regular triangle (a), a pentagon (b), and a hexagon (c) with convex quadrilateral blocks.

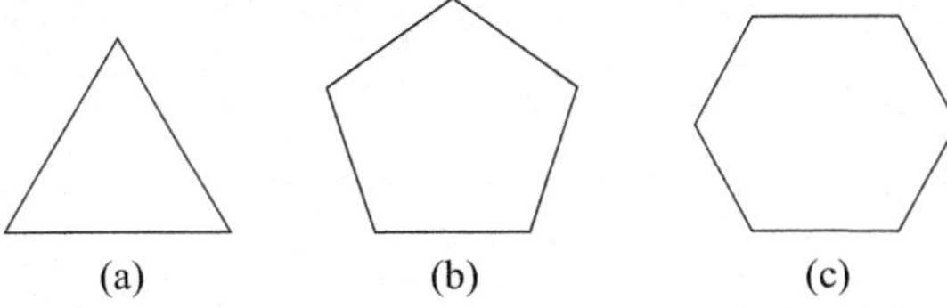

20.3. Propose a procedure for mesh generation inside a quadrilateral block with curved edges.

Chapter 21
Two-dimensional Mesh Generators

Abstract Two-dimensional mesh generators are described. A simple mesh generator `rectangle` creates a rectangular mesh of quadratic eight-node elements inside a rectangular block. Java^{TM} class `genquad8` is designed for mesh generation inside a quadrilateral area with curved boundaries. The area has the shape of the quadratic finite element with eight nodes. Element boundary placement and mesh refinement are achieved by double-quadratic transformation.

21.1 Rectangular Block

A simple mesh generator `rectangle` creates a rectangular mesh of quadratic eight-node elements inside a rectangular block. The locations of element boundaries are oriented vertically and horizontally, as shown in Figure 21.1a.

Input data consists of the following:

`rectangle blockName` – first data statement: mesh-generator name and mesh block name;

`nx, ny` – number of elements along `x` and `y`;

`xs[nx+1], ys[ny+1]` – locations of element boundaries on `x` and `y`;

`mat` – material name (default `mat=1`).

Geometrical data is necessary since this data has no default values. The contents of arrays `xs` and `ys` are shown as $x0...x3$ and $y0...y2$ in Figure 21.1a. Material name `mat` has default value 1. If not defined in the input stream, generated elements will have a default material name.

A constructor of class `rectangle` and methods for data input and data print are given below.

```
1 package gener;
2
3 import model.*;
4 import fea.*;
```

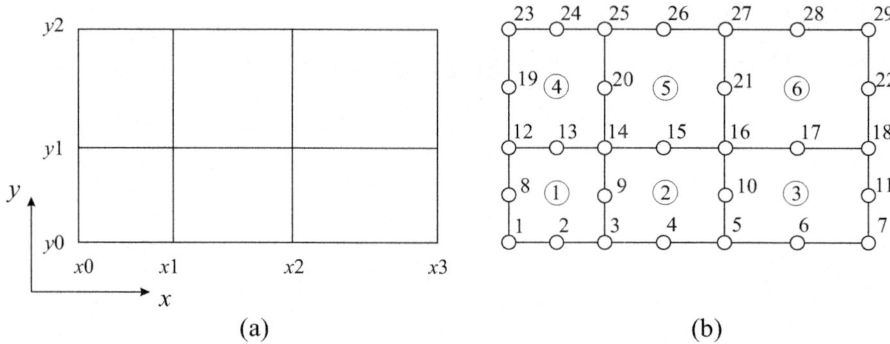

Fig. 21.1 Generation of a mesh inside a rectangular block: specification of mesh lines (a), generated mesh with node and element numbering (b)

```
 5  import util.*;
 6  import elem.*;
 7
 8  // Generate mesh of quadratic elements inside a rectangle.
 9  // Input: nx, ny - number of elements along x and y;
10  //   xs, ys - locations of element boundaries on x and y;
11  //   [mat] - material name.
12  public class rectangle {
13
14      private FeModel m;
15      enum vars {
16          nx, ny, xs, ys, mat, end
17      }
18
19      private vars name;
20
21      private int nx, ny;
22      String mat="1";
23      private double xs[], ys[];
24
25      public rectangle() {
26          String modelName = Jmgen.RD.next();
27          Jmgen.PR.printf("Rectangle: %s\n", modelName);
28          readData();
29          printData();
30          m = new FeModel(Jmgen.RD, Jmgen.PR);
31          generateMesh();
32          Jmgen.blocks.put(modelName,m);
33          Jmgen.PR.printf("Mesh " + modelName +
34                  ": nEl = %d  nNod = %d\n", m.nEl, m.nNod);
35      }
36
37      private void readData() {
38          String varName, varname;
39
40          while (Jmgen.RD.hasNext()) {
```

```
41
42                    varName = Jmgen.RD.next();
43                    varname = varName.toLowerCase();
44                    if (varName.equals("#")) {
45                        Jmgen.RD.nextLine(); continue;
46                    }
47                    try {
48                        name = vars.valueOf(varname);
49                    } catch (Exception e) {
50                        UTIL.errorMsg("Variable name is not found: "
51                                    + varName);
52                    }
53                    switch (name) {
54                        case nx:     nx = Jmgen.RD.readInt();
55                            break;
56                        case ny:     ny = Jmgen.RD.readInt();
57                            break;
58                        case xs:
59                            xs = new double[nx+1];
60                            for (int i = 0; i <= nx; i++)
61                                xs[i] = Jmgen.RD.readDouble();
62                            break;
63                        case ys:
64                            ys = new double[ny+1];
65                            for (int i = 0; i <= ny; i++)
66                                ys[i] = Jmgen.RD.readDouble();
67                            break;
68                        case mat:     mat = Jmgen.RD.next();
69                            break;
70                        case end:
71                            return;
72                    }
73                }
74            }
75
76    private void printData() {
77        Jmgen.PR.printf(" nx =%5d\n", nx);
78        Jmgen.PR.printf(" ny =%5d\n", ny);
79        Jmgen.PR.printf(" xs:   ");
80        for (int i = 0; i <= nx; i++)
81            Jmgen.PR.printf("%7.3f", xs[i]);
82        Jmgen.PR.printf("\n ys:   ");
83        for (int i = 0; i <= ny; i++)
84            Jmgen.PR.printf("%7.3f", ys[i]);
85        Jmgen.PR.printf("\n");
86    }
```

Input data names are placed in enumerated `vars` (lines 15–17). Item `end` is also included. It is necessary to mark the end of input data for the module. Line 26 reads the model name `modelName`. Line 30 creates a new finite element model `m`. Method `generateMesh` generates the mesh and puts the mesh data into model `m`. In line 32, the generated finite element model is placed in hash table `Jmgen.blocks`. Methods `readData` and `printData` input and print data.

Method `generateMesh` creates element connectivities and nodal coordinates.

```
88      private void generateMesh() {
89          int ind[] = new int[8];
90          m.nDim = 2;
91
92          // Connectivity array
93          m.nEl = nx*ny;
94          m.elems = new Element[m.nEl];
95
96          int el = 0;
97          for (int iy=0; iy<ny; iy++) {
98              for (int ix=0; ix<nx; ix++) {
99                  m.elems[el] = Element.newElement("quad8");
100                 int in0 = iy*(3*nx+2) + 2*ix;
101                 ind[0] = in0 + 1;
102                 ind[1] = in0 + 2;
103                 ind[2] = in0 + 3;
104                 int in1 = iy*(3*nx+2) + 2*nx + 1 + ix + 1;
105                 ind[3] = in1 + 1;
106                 ind[7] = in1;
107                 int in2 = (iy+1)*(3*nx+2) + 2*ix;
108                 ind[4] = in2 + 3;
109                 ind[5] = in2 + 2;
110                 ind[6] = in2 + 1;
111                 m.elems[el].setElemConnectivities(ind);
112                 m.elems[el].setElemMaterial(mat);
113                 el++;
114             }
115         }
116
117         // Node coordinate array
118         m.nNod = (3*nx+2)*ny + 2*nx + 1;
119         m.newCoordArray();
120         int n = 0;
121         for (int iy=0; iy<2*ny+1; iy++) {
122             int py = (iy+1)/2;
123             for (int ix=0; ix<2*nx+1; ix++) {
124                 int px = (ix+1)/2;
125                 if (ix%2==0 && iy%2==0) {
126                     m.setNodeCoord(n, 0, xs[px]);
127                     m.setNodeCoord(n, 1, ys[py]);
128                     n++;
129                 }
130                 else if (ix%2==1 && iy%2==0) {
131                     m.setNodeCoord(n,0,0.5*(xs[px-1]+xs[px]));
132                     m.setNodeCoord(n, 1, ys[py]);
133                     n++;
134                 }
135                 else if (ix%2==0 && iy%2==1) {
136                     m.setNodeCoord(n, 0, xs[px]);
137                     m.setNodeCoord(n,1,0.5*(ys[py-1]+ys[py]));
138                     n++;
139                 }
```

```
140                      }
141                }
142        }
143
144  }
```

Nodes and elements are numbered in a mesh by rows along the *x*-direction, as shown in Figure 21.1b. An array of elements is allocated in line 94. Element connectivities are generated inside the double loop in lines 97–115. The outer loop iterates on element rows; the element number in a row is changed in the inner loop. A current element of type `quad8` with eight nodes is constructed in line 99. The connectivity number of the left lower node is calculated in lines 100–101. In the statement of line 98, `(3*nx+2)` is the number of nodes in a row on `nx` eight-node elements without upper nodes. Lines 111–112 set the element connectivities and element material. The element number is incremented in line 113.

Line 118 sets the number of nodes in the mesh. Line 119 allocates a nodal coordinates array according to problem dimension `nDim` and the number of nodes `nNod`. Nodal coordinates are set in lines 121–141. Double loop `iy – ix` is organized along rows and columns of nodes, including rows and columns where midside nodes are located. Lines 125–129 set coordinates of element corner nodes. Lines 130–134 estimate coordinates of midside nodes on horizontal element sides and lines 135–139 on vertical element sides. Thus, combinations of indices `iy` and `ix` corresponding to element centers are not used, and coordinates are generated just for actually existing nodes.

In conclusion, let us show input data, which is necessary to generate the mesh shown in Figure 21.1b.

```
rectangle block1
   nx = 3   ny = 2
   xs = 0 1 2.5 4.5
   ys = 0 1 2.5
   mat = STEEL
end
```

The first line specifies that the mesh will be created by the `rectangle` mesh generator and will have the name `block1`. Line 2 sets the numbers of elements in the *x*- and *y*-directions. The locations of element boundaries are defined in lines 3 and 4. Line 5 assigns `STEEL` as the material name for all elements. The final statement `end` marks the end of data.

21.2 Mesh Inside Eight-node Macroelement

21.2.1 Algorithm of Double-quadratic Transformation

A block mesh generator for a quadrilateral area with curved boundaries can be created on the basis of the quadratic finite element with eight nodes. Interpolation in

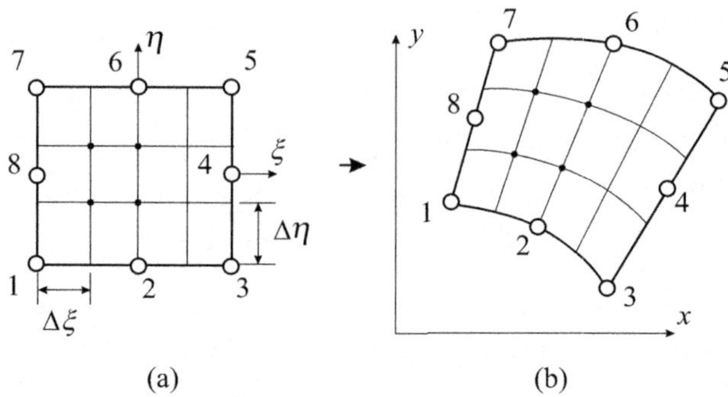

Fig. 21.2 Generation of a mesh inside a quadrilateral block with curved sides: regular subdivision of the parent square element (a), mesh after transformation to the global coordinates x, y (b)

this element is performed using quadratic shape functions (10.7). Such interpolation $z(\xi, \eta) = Q(\xi, \eta)$ can be written in the following form:

$$
\begin{aligned}
z(\xi, \eta) = &- 0.25(1 - \xi)(1 - \eta)(1 + \xi + \eta)z_1 \\
&- 0.25(1 + \xi)(1 - \eta)(1 - \xi + \eta)z_3 \\
&- 0.25(1 + \xi)(1 + \eta)(1 - \xi - \eta)z_5 \\
&- 0.25(1 - \xi)(1 + \eta)(1 + \xi - \eta)z_7 \\
&+ 0.5(1 - \xi)(1 + \xi)(1 - \eta)z_2 \\
&+ 0.5(1 - \eta)(1 + \eta)(1 + \xi)z_4 \\
&+ 0.5(1 - \xi)(1 + \xi)(1 + \eta)z_6 \\
&+ 0.5(1 - \eta)(1 + \eta)(1 - \xi)z_8,
\end{aligned}
\tag{21.1}
$$

where coordinates ξ, η have the range $[-1, 1]$; $z_1 ... z_8$ are nodal values of an interpolated quantity, and $z(\xi, \eta)$ is the value at point ξ, η obtained by the quadratic interpolation. Nodes are numbered in the anticlockwise order starting with any corner node, as shown in Figure 21.2. Mesh refinement can be achieved by shifting macroelement midside nodes from their central positions. However, specification of midside node locations is difficult since we do not know what element sizes will be produced after quadratic transformation. It is better to specify the size of the smallest element at an element edge and then to find locations of midside nodes.

In order to develop an algorithm for computing midside node location on the basis of the smallest element size, let us consider a one-dimensional quadratic transformation for arbitrary location of the midside node, as shown in Figure 21.3. An edge of the quadratic macroelement has three nodes with local coordinates $\xi_1 = -1$, $\xi_2 = 0$ and $\xi_3 = 1$. Subdivision of the side into n elements uses global coordinate x. Nodes 1 and 3 have global coordinates $x_1 = 0$ and $x_3 = l$. It is necessary to determine position c of the midside node, which yields the smallest element with size $e_{\min}$. Interpolation of x is done using the one-dimensional shape functions $N_i(\xi)$:

$$
x = \sum_{i=1}^{3} N_i(\xi)x_i = (1 - \xi^2)c + \frac{1}{2}\xi(1 + \xi)l.
\tag{21.2}
$$

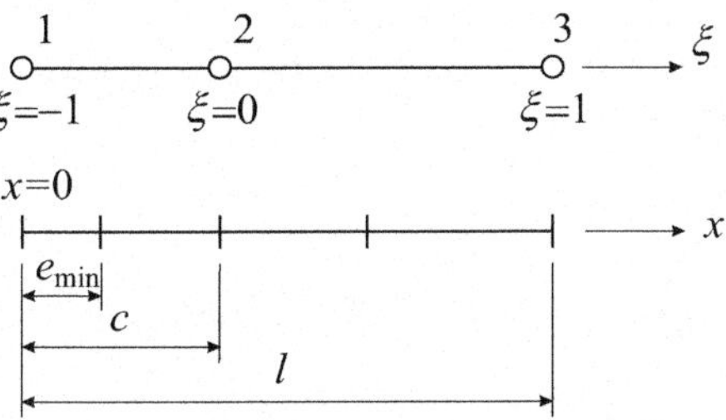

Fig. 21.3 Shift of macroelement midside node helps to achieve mesh refinement with minimum element size $e_{\min}$

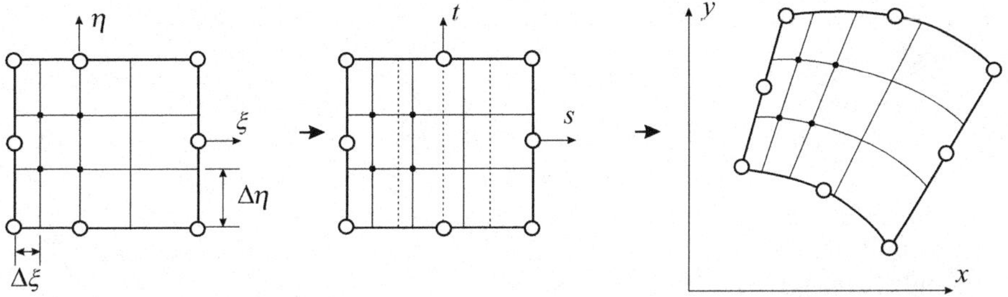

Fig. 21.4 Double-quadratic transformation. The first transformation is performed from the element with shifted midside nodes to the element with regular locations of midside nodes. The second transformation to global coordinates produces a mesh of elements within a curved quadrilateral

The value of c can be determined from the condition that x should be equal to the smallest element size at point $\xi = -1 + 2/n$:

$$x|_{\xi=-1+2/n} = e_{\min}.$$

Solving the above equation gives the location of the midside node c:

$$\frac{c}{l} = \frac{(e_{\min}/l)n^2 + n - 2}{4(n - 1)}, \tag{21.3}$$

where n is the number of elements on the side. This relation shows that as the number of element subdivisions tends to infinity, the maximum midside node shift (corresponding to zero size of the smallest element) is a quarter of the side length.

The shift of a midside node along a curved macroelement side is a complicated procedure, which in general can not be performed exactly. To overcome this difficulty, let us perform a double-quadratic transformation shown in Figure 21.4.

The first quadratic transformation is performed from element $-1 \leq \xi, \eta \leq 1$ with shifted midside nodes to element $-1 \leq s, t \leq 1$ with regular locations of midside nodes. This transformation provides mesh refinement corresponding to specified sizes of the smallest elements. Equation 21.1 is used to perform quadratic transformation. Midside nodes are shifted according to Equation 21.3.

The second quadratic transformation from element $-1 \leq s,t \leq 1$ to a curved element in global coordinates x,y produces a mesh of elements inside the specified region with specified refinements.

21.2.2 Implementation of Mesh Generation

Class `genquad8` implements double-quadratic transformation for generating a mesh of quadratic elements inside a macroelement with eight nodes.

The following input data should be specified for mesh generation:

`genquad8 blockName` – first data statement: mesh-generator name and mesh block name;

`nh, nv` – number of elements along "horizontal" and "vertical" directions. The "horizontal" direction is along macroelement side 1–2–3;

`x1,y1,x2,y2 ...x8,y8` – locations of eight nodes for macroelement definition in anticlockwise order starting from any corner node. If both coordinates of a midside node are specified as zeros, then they are interpolated linearly from neighboring corner nodes;

`res[4]` – relative sizes of smallest elements on macroelement sides. If the smallest element is located at the side end (anticlockwise order) then it is specified as (1-size). Default values for relative element sizes `res = 0, 0, 0, 0` - no refinement;

`mat` – material name (default `mat=1`).

Array `res` and material name `mat` may be omitted from the input data since they have default values.

A constructor of class `genquad8` and methods for data input and data print are as follows.

```
1 package gener;
2
3 import fea.*;
4 import util.*;
5 import model.*;
6 import elem.*;
7
8 // Generate mesh of quadratic elements inside
9 //      a macroelement with 8 nodes.
10 // Input: modelName - name of the finite element model;
11 //   nh, nv - number of elements in local coordinate directions;
12 //   xyp - coordinates of 8 macroelement nodes (x1,y1,x2,y2 ..);
13 //   [res] - relative sizes of smallest elements on
14 //           macroelemet edges;
15 //   [mat] - material name.
16 public class genquad8 {
17
18       private FeModel m;
```

```java
19      enum vars {
20          nh, nv, res, xyp, mat, end
21      }
22
23      private vars name;
24      private int nh, nv;
25      String mat = "1";
26      private double res[] = new double[4],
27                     xp[] = new double[8], yp[] = new double[8];
28      //  Nodes of parent element (coordinates xi, eta)
29      double xip[] = {-1, 0, 1, 1, 1, 0,-1,-1},
30             etp[] = {-1,-1,-1, 0, 1, 1, 1, 0};
31
32      public genquad8() {
33
34          String modelName = Jmgen.RD.next();
35          Jmgen.PR.printf("GenQuad8: %s\n", modelName);
36          readData();
37          printData();
38          m = new FeModel(Jmgen.RD, Jmgen.PR);
39          generateMesh();
40          Jmgen.blocks.put(modelName,m);
41          Jmgen.PR.printf("Mesh " + modelName +
42                  ": nEl = %d  nNod = %d\n", m.nEl, m.nNod);
43      }
44
45      private void readData() {
46          String varName, varname;
47
48          while (Jmgen.RD.hasNext()) {
49              varName = Jmgen.RD.next();
50              varname = varName.toLowerCase();
51              if (varName.equals("#")) {
52                  Jmgen.RD.nextLine(); continue;
53              }
54              try {
55                  name = vars.valueOf(varname);
56              } catch (Exception e) {
57                  UTIL.errorMsg("Variable name is not found: "
58                          + varName);
59              }
60              switch (name) {
61              case nh:    nh = Jmgen.RD.readInt();
62                  break;
63              case nv:    nv = Jmgen.RD.readInt();
64                  break;
65              case res:
66                  for (int i = 0; i < 4; i++)
67                      res[i] = Jmgen.RD.readDouble();
68                  break;
69              case xyp:
70                  for (int i = 0; i < 8; i++) {
71                      xp[i] = Jmgen.RD.readDouble();
72                      yp[i] = Jmgen.RD.readDouble();
```

```
 73                          }
 74                          // Interpolation of macroelement midside
 75                          // nodes if both coordinates are zeroes
 76                          for (int i = 1; i < 8; i += 2)
 77                              if (xp[i] == 0.0 && yp[i] == 0.0) {
 78                                  xp[i] = 0.5*(xp[i-1]+xp[(i+1)%8]);
 79                                  yp[i] = 0.5*(yp[i-1]+yp[(i+1)%8]);
 80                              }
 81                          break;
 82                      case mat:    mat = Jmgen.RD.next();
 83                          break;
 84                      case end:
 85                          return;
 86                      }
 87              }
 88      }
 89
 90      private void printData() {
 91          Jmgen.PR.printf(" nh =%5d\n", nh);
 92          Jmgen.PR.printf(" nv =%5d\n", nv);
 93          Jmgen.PR.printf(" res:   ");
 94          for (int i = 0; i < 4; i++)
 95              Jmgen.PR.printf("%7.3f", res[i]);
 96          Jmgen.PR.printf("\n xyp:   ");
 97          for (int i = 0; i < 8; i++) {
 98              Jmgen.PR.printf("%7.3f%7.3f", xp[i],yp[i]);
 99              if (i == 3) Jmgen.PR.printf("\n          ");
100          }
101          Jmgen.PR.printf("\n");
102      }
```

The class constructor reads a name of the mesh block that is being generated (line
34). Then it reads and prints data (lines 36 and 37), generates a mesh block by
calling method `generateMesh` (in line 39), and puts the mesh block in the hash
table `blocks` (line 40). Input data parameters are read in method `readData` (lines
45–88). Input of macroelement nodal coordinates in lines 70–73 is accompanied by
interpolation of midside nodes, which are specified by double zeros (lines 76–80).

Two methods – `midsideNodeShift` and `quadraticTransform` – implement shift of midside nodes for mesh refinement and quadratic coordinate transformation.

```
104      // Shift of midside nodes to have specified element size
105      private void midsideNodeShift() {
106          for (int edge = 0; edge < 4; edge++) {
107              // minElem = relative size of the smallest element
108              double minElem = res[edge];
109              int sign = 1;
110              if (minElem > 0.5) {
111                  minElem = 1.0 - minElem;
112                  sign = -1;
113              }
114              if (edge > 1) sign = -sign;
115              if (minElem > 0.0) {
```

```
116                        // ne = number of elements on the elem edge
117                        double ne = (edge%2 == 0) ?
118                                (double) nh : (double) nv;
119                        double ratio = 0.25*(minElem*ne*ne + (ne-2))
120                                                /(ne-1);
121                        double c = (-1.0 + ratio*2)*sign;
122                        if (edge%2 == 0) xip[2*edge + 1] = c;
123                        else             etp[2*edge + 1] = c;
124                    }
125                }
126            }
127
128        // Quadratic 2D mapping.
129        // z [8] - values at nodes,
130        // returns interpolated z at point xi, et.
131        private double quadraticTransform(double[] z,
132                                        double xi, double et) {
133            double x1 = 1 - xi;    double x2 = 1 + xi;
134            double e1 = 1 - et;    double e2 = 1 + et;
135
136            return -0.25*(x1*e1*(x2 + et)*z[0]
137                        + x2*e1*(x1 + et)*z[2]
138                        + x2*e2*(x1 - et)*z[4]
139                        + x1*e2*(x2 - et)*z[6])
140                   + 0.5*(x1*x2*e1*z[1]
141                        + e1*e2*x2*z[3]
142                        + x1*x2*e2*z[5]
143                        + e1*e2*x1*z[7]);
144        }
```

Method `midsideNodeShift` takes array `res` containing the smallest elements
on macroelement sides and changes the locations of the midside nodes in arrays
`xip` and `etp` (the ξ and η coordinates of the element, shown on the left of Figure 21.4). For each macroelement side, ratio c/l is calculated according to Equation
21.3 in lines 119–120. Coordinates ξ and η of the midside nodes are estimated in
lines 121–123. Method `quadraticTransform` implements quadratic transformation (21.1). The method returns the interpolated value of scalar z specified at
eight macroelement nodes. The interpolated value corresponds to local coordinates
`xi` (ξ) and `et` (η).

The listing below presents method `generateMesh`, which creates a mesh of
quadratic elements inside a macroelement.

```
146        private void generateMesh() {
147            int ind[] = new int[8];
148            m.nDim = 2;
149
150            // Element connectivities
151            m.nEl = nh*nv;
152            m.elems = new Element[m.nEl];
153            int n = 0;
154
155            for (int iv = 0; iv < nv; iv++) {
156                for (int ih = 0; ih < nh; ih++) {
```

```
157                  m.elems[n] = Element.newElement("quad8");
158                  int in0 = iv*(3*nh+2) + 2*ih;
159                  ind[0] = in0 + 1;
160                  ind[1] = in0 + 2;
161                  ind[2] = in0 + 3;
162                  int in1 = iv*(3*nh+2) + 2*nh + ih + 2;
163                  ind[3] = in1 + 1;
164                  ind[7] = in1;
165                  int in2 = (iv+1)*(3*nh+2) + 2*ih;
166                  ind[4] = in2 + 3;
167                  ind[5] = in2 + 2;
168                  ind[6] = in2 + 1;
169                  m.elems[n].setElemConnectivities(ind);
170                  m.elems[n].setElemMaterial(mat);
171                  n++;
172              }
173          }
174
175      // Shift of midside nodes for element in xi, eta
176      midsideNodeShift();
177
178      // Node coordinate array
179      m.nNod = (3*nh+2)*nv+2*nh+1;
180      m.newCoordArray();
181      n = 0;
182      double dxi = 1.0/nh;
183      double det = 1.0/nv;
184      for (int iv = 0; iv < 2*nv+1; iv++) {
185          for (int ih = 0; ih < 2*nh+1; ih++) {
186              if (iv%2 == 1 && ih%2 == 1) continue;
187              double xi = -1.0 + dxi*ih;
188              double et = -1.0 + det*iv;
189              //  First quadratic transform: xi,et -> s,t
190              double s,t;
191              if (ih%2 == 0)
192                  s = quadraticTransform(xip, xi, et);
193              else
194                  s = 0.5*(quadraticTransform(xip,xi-dxi,et)
195                      + quadraticTransform(xip,xi+dxi,et));
196              if (iv%2 == 0)
197                  t = quadraticTransform(etp, xi, et);
198              else
199                  t = 0.5*(quadraticTransform(etp,xi,et-det)
200                      + quadraticTransform(etp,xi,et+det));
201
202              //  Second quadratic transform: s,t -> x,y
203              m.setNodeCoord(n,0,quadraticTransform(xp,s,t));
204              m.setNodeCoord(n,1,quadraticTransform(yp,s,t));
205              n++;
206          }
207      }
208  }
209
210 }
```

Method `generateMesh` produces element connectivities and nodal coordinates for a mesh of eight-node quadratic elements.

An array of elements is allocated in line 152. Element connectivities are created inside a double loop in lines 155–173. An element of the type `quad8` with eight nodes is constructed in line 157. The connectivity numbers are calculated in lines 158–168. The statements in lines 169–170 set element connectivities and the element material.

Line 179 sets the number of nodes in the mesh. The node coordinate array is allocated in line 180 according to the problem dimension `nDim` and number of nodes `nNod`. Nodal coordinates are set in a double loop along "horizontal" rows in lines 184–207. The double loop `iv − ih` counts rows and columns for all nodes including rows and columns for midside nodes. The statement in line 186 omits nodes at element centers since they are absent in eight-node elements. The first quadratic transformation from ξ, η to s, t is performed in lines 191–200. For midside nodes (`else` conditions), coordinates s and t are interpolated between neighboring corner nodes in order to place midside nodes at side centers. The second quadratic transformation from s, t to x, y and setting of x and y coordinates are done in lines 203–204.

21.3 Example of Mesh Generation

Using mesh generator `genquad8`, a mesh of quadratic elements can be created inside a quarter ring with inner radius $r = 1$ and outer radius $r = 2$, as shown in Figure 21.5a. The mesh should have refinement near the lower left corner with relative sizes 0.1 in the radial direction and 0.07 in the angular direction. The mesh should have 4 elements in the radial direction and 5 elements in the angular direction. Node numbering should be in the radial direction.

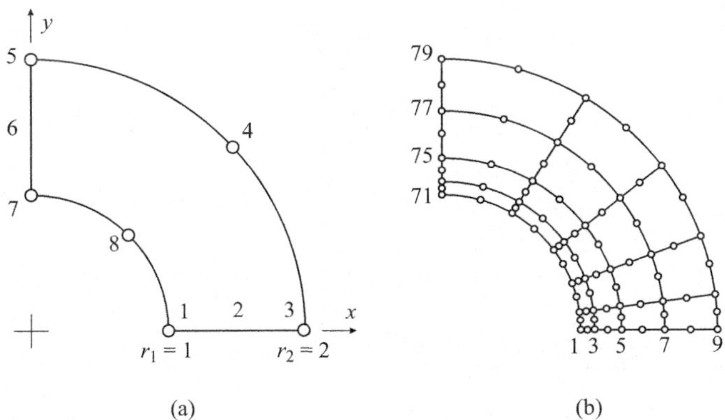

Fig. 21.5 Mesh generation for a quarter of a ring using `genquad8`: specification of nodes for a macroelement (a) and created mesh of quadratic elements (b)

Solution

The quarter ring is approximated by a macroelement with eight nodes. Macroelement node locations and numbers are presented in Figure 21.5a. We start macroelement nodes with the lower left corner of the domain since node numbering is performed in the direction of macroelement nodes 1–2–3. The following data creates the desired mesh.

```
genquad8  b1
  nh = 4   nv = 5
  xyp = 1 0   0 0   2 0   1.414 1.414
        0 2   0 0   0 1   0.7071 0.7071
  res = .1 0.07 0.9 0.93
end
```

The first line specifies the name of a mesh-generator module (`genquad8`) and the name of the mesh block (`b1`). Line 2 sets the numbers of elements in the "horizontal" (radial) and "vertical" (angular) directions. Locations of macroelement corner and midside nodes in anticlockwise order are defined in lines 3 and 4. The coordinates of points 2 and 6 are set to double zeros since they are situated on straight sides and can be determined by linear interpolation. Line 5 assigns relative element sizes for mesh refinement. In order to achieve relative element size 0.1 near the inner radius line, the value of `res` is specified as 0.1 for side 1–2–3 and as 0.9 for side 5–6–7. A quadratic transformation with shifted midside node produces element sizes related by arithmetic progression. For, example, four elements in the radial direction have sizes 0.1, 0.2, 0.3, and 0.4. Similarly, `res` values on sides 3–4–5 and 7–8–1 are 0.07 and 0.93. The final statement `end` indicates the data end for the mesh generator.

The resulting mesh is depicted in Figure 21.5b. Some node numbers are shown indicating the order of node numbering. Nodes are numbered by rows in the direction of side 1–2–3. Element numbering follows the same manner.

Problems

21.1. Derive a relation that provides a lower left node number of an element with number e for node and element numbering shown in Figure 21.1b. Use n_x – number of elements in the x-direction.

21.2. Analyze Equation 21.2, which performs one-dimensional interpolation of x depending on local coordinate ξ. Show that the relation can produce negative x values when the midside node is located too close to the corner node: $c \leq 0.25l$ (see Figure 21.3).

21.3. Modify the data of example in Section 21.3 in such a way that node numbering is in the angular direction.

Chapter 22
Generation of Three-dimensional Meshes by Sweeping

Abstract This chapter presents generation of three-dimensional meshes of hexagonal elements by sweeping a two-dimensional mesh in space. The source two-dimensional mesh consists of eight-node quadrilateral elements. The sweeping path can be straight or circular. The resulting three-dimensional mesh is composed of twenty-node hexagonal elements.

22.1 Sweeping Technique

Because of the difficulty in generating hexahedral meshes for three-dimensional geometry in the general case, methods for special geometry cases are widely used. One of the most common methods is *sweeping* [12]. In the sweeping method, two-dimensional surface source and target meshes with identical topological properties are created. Interpolation between source and target meshes along a sweeping volume produces inner mesh layers. Three-dimensional hexahedral elements are generated between two neighboring mesh layers.

Here, a simplified variant of the sweeping method is selected for implementation. In our implementation, the source mesh and target mesh are the same. The source mesh is located on the xy-plane at $z = 0$. Two possible sweeping paths are shown in Figure 22.1. The first (Figure 22.1a) is moving the two-dimensional source mesh along the positive direction of the z-axis and copying it at specified positions. The second possibility, shown in Figure 22.1b, implies rotation of the source mesh around the y-axis. The source mesh is copied at predetermined positions creating faces for three-dimensional elements.

The two-dimensional mesh consists of eight-node quadrilateral elements. The resulting three-dimensional mesh is composed of twenty-node hexahedral elements. Since both element types contain midside nodes, the mesh-generation process is not straightforward. To help with mesh generation, it is reasonable to create two auxiliary arrays for the two-dimensional mesh. The first array, `nodeType2`, classifies nodes of the two-dimensional mesh into corner nodes and midside nodes. Such an

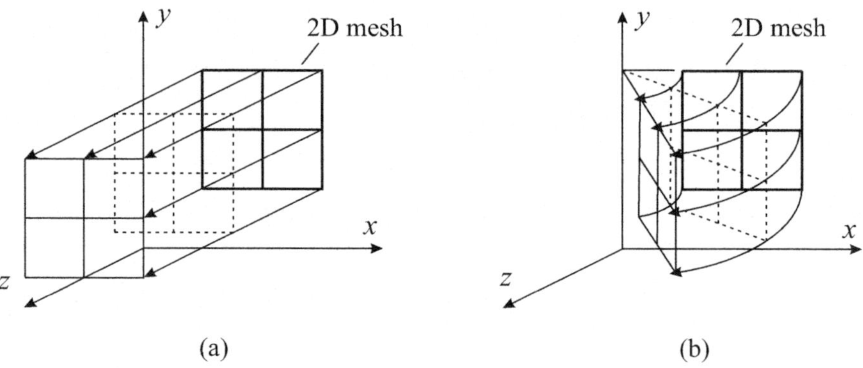

Fig. 22.1 Generation of a three-dimensional mesh by sweeping along the z-axis (a) and around the y-axis (b)

array can be formed using element connectivities. Corner nodes occupy even positions in element connectivities (count starts with zero) and midside nodes are at odd positions. The second array, `nodeNum2`, contains numbers of nodes on the first layer of the three-dimensional mesh. This array is formed with the use of array `nodeType2` and the number of element layers. It is adopted that nodes are numbered starting from the first layer in the direction of mesh sweeping. Both auxiliary arrays are employed for generating element connectivities and nodal coordinates of the resulting three-dimensional mesh.

22.2 Implementation

22.2.1 Input Data

The following input data should be specified for generating a three-dimensional mesh by sweeping. The first three items should be specified first in data in the given order.

> `sweep` – mesh-generator name;
>
> `modelName2` – name of two-dimensional finite element model consisting of 8-node quadrilateral elements;
>
> `modelName3` – name of resulting three-dimensional model composed of 20-node hexahedral elements;
>
> `nlayers` – number of element layers in the three-dimensional mesh;
>
> `zlayers[nlayers+1]` – z-distances or angles (in degrees) in increasing order for copying the two-dimensional mesh;

rotate $=$ Y $-$ rotate the two-dimensional mesh around y-axis, $=$ N $-$ translate the two-dimensional mesh along z (default `rotate = N`).

A constructor of Java^{TM} class sweep and methods for data input and print follow.

```
 1 package gener;
 2
 3 import model.*;
 4 import fea.*;
 5 import util.*;
 6 import elem.*;
 7
 8 // Generate 3D mesh of hexahedral 20-node elements by sweeping
 9 //      2D mesh of quadrilateral 8-node elements.
10 // Input: modelName2 - name of 2D model;
11 //   modelName3 - name of the resulting 3D model;
12 //   nlayers - number of element layers in 3D mesh;
13 //   zlayers - z-distances or angles (deg) for copying 2D mesh;
14 //   [rotate=Y/N] - rotate mesh around y-axis,
15 //      otherwize translate along z.
16 public class sweep {
17
18     // 2D source model
19     private FeModel m2;
20     // 3D resulting model
21     private FeModel m3;
22
23     enum vars {
24         nlayers, zlayers, rotate, end
25     }
26
27     private vars name;
28     private int nlayers;
29     boolean rotate = false;
30     private double zlayers[];
31     // Node types for 2D mesh,
32     int[] nodeType2;
33     // nodeNum2[i] is a number of i-th 2D node in 3D mesh
34     int[] nodeNum2;
35
36     public sweep() {
37
38         String modelName2 = Jmgen.RD.next();
39         String modelName3 = Jmgen.RD.next();
40         Jmgen.PR.printf(
41                 "Sweep: %s %s\n", modelName2, modelName3);
42         readData();
43         printData();
44
45         if (Jmgen.blocks.containsKey(modelName2))
46             m2 = (FeModel) Jmgen.blocks.get(modelName2);
47         else
48             UTIL.errorMsg("No such mesh block: " + modelName2);
49
50         m3 = new FeModel(Jmgen.RD, Jmgen.PR);
```

```
51          m3.nDim = 3;
52          m3.nEl = m2.nEl*nlayers;
53          m3.nNod = nodeTypesNumbers2D();
54          elementConnectivities3D();
55          nodeCoordinates3D();
56          Jmgen.blocks.put(modelName3, m3);
57          Jmgen.PR.printf("Mesh " + modelName3 +
58                  ": nEl = %d  nNod = %d\n", m3.nEl, m3.nNod);
59      }
60
61    private void readData() {
62          String varName, varname;
63
64        while (Jmgen.RD.hasNext()) {
65
66            varName = Jmgen.RD.next();
67            varname = varName.toLowerCase();
68            if (varName.equals("#")) {
69                Jmgen.RD.nextLine();
70                continue;
71            }
72            try {
73                name = vars.valueOf(varname);
74            } catch (Exception e) {
75                UTIL.errorMsg("Variable name is not found: "
76                        + varName);
77            }
78            switch (name) {
79            case nlayers:
80                nlayers = Jmgen.RD.readInt();
81                break;
82            case rotate:
83                rotate =
84                      (Jmgen.RD.next().equalsIgnoreCase("Y"));
85                break;
86            case zlayers:
87                zlayers = new double[2*nlayers + 1];
88                for (int i = 0; i < 2*nlayers + 1; i += 2)
89                    zlayers[i] = Jmgen.RD.readDouble();
90                // Interpolation for 3D midside nodes
91                for (int i = 1; i < 2*nlayers; i += 2)
92                    zlayers[i] =
93                            0.5*(zlayers[i-1] + zlayers[i+1]);
94                break;
95            case end:
96                return;
97            }
98        }
99    }
100
101    private void printData() {
102        Jmgen.PR.printf(" nlayers =%5d\n", nlayers);
103        Jmgen.PR.printf(
104                " rotate  = %s\n", (rotate ? "Y" : "N"));
```

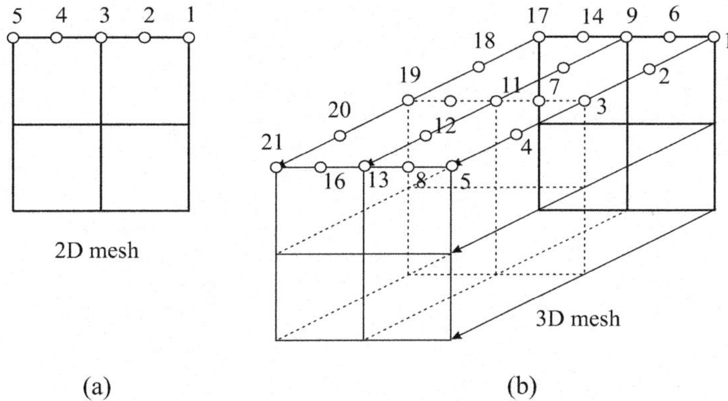

Fig. 22.2 Numbering of nodes in a source two-dimensional mesh (a) and in a resulting three-dimensional mesh (b)

```
105                    Jmgen.PR.printf(" zlay:   ");
106                    for (int i = 0; i <= 2*nlayers + 1; i += 2)
107                        Jmgen.PR.printf("%7.3f", zlayers[i]);
108                    Jmgen.PR.printf("\n");
109            }
```

Lines 38–43 read and print input data. The two-dimensional finite element model m2 is obtained from hash table `blocks` in line 44. If the model with the specified name `modelName2` is absent then the code prints an error message and stops (line 46).

Line 50 constructs the finite element model m3 for generating a three-dimensional mesh. Mesh generation consists of creating auxiliary arrays for the two-dimensional mesh and in generating connectivities and nodal coordinates in lines 54–55. The resulting three-dimensional mesh is placed in the hash table `blocks` under the name `modelName3` (line 56). Methods `readData` (lines 61–99) and `printData` read and print input data.

22.2.2 Node Numbering

Node numbering in a two-dimensional source mesh and in a three-dimensional resulting mesh is shown in Figure 22.2. Node numbering in the three-dimensional mesh is done along a sweeping direction starting from the first (source) layer. Since the number of nodes along the sweeping path is different for corner and midside nodes in the source mesh, it is useful to form two auxiliary arrays for node numbering.

Method `nodeTypesNumbers2D` listed below creates two nodal arrays for the two-dimensional mesh. Array `nodeType2` describes types of nodes (corner or

midside) and array `nodeNum2` contains numbers of three-dimensional nodes in
the first layer.

```
111        // Create arrays nodeType2 and nodeNum2
112        // return number of nodes in 3D mesh
113        private int nodeTypesNumbers2D() {
114
115            // nodeType2 - node types for 2D mesh,
116            //  =0 - midside, =1 - corner/degenerate
117            nodeType2 = new int[m2.nNod];
118            for (int iel = 0; iel < m2.nEl; iel++) {
119                for (int i=0; i<m2.elems[iel].ind.length; i++) {
120                    int in = m2.elems[iel].ind[i] - 1;
121                    if (in != -1) {
122                        nodeType2[in] = (i + 1)%2;
123                    }
124                }
125            }
126
127            // nodeNum2[i] is a number of i-th 2D node in 3D mesh
128            nodeNum2 = new int[m2.nNod];
129            int node = 1;
130            for (int i = 0; i < m2.nNod; i++) {
131                nodeNum2[i] = node;
132                int dn = (nodeType2[i] == 0) ?
133                        nlayers + 1 : 2*nlayers + 1;
134                // If node is located on rotation axis Y
135                if (rotate && m2.getNodeCoord(i, 0) == 0.0) dn = 1;
136                node = node + dn;
137            }
138            return node - 1;
139        }
```

An array of node types `nodeType2` is created in a double loop of lines 117–
125. Corner nodes in the two-dimensional mesh are marked by ones and midside
nodes by zeros. The fact that corner nodes are located at even places in element
connectivities is used.

The *i*th entry of array `nodeNum2` is a node number of the *i*th two-dimensional
node at the first layer of three-dimensional mesh. The array is formed in lines 130–
137. Next, a three-dimensional node at the source layer is computed by incrementing
previous node number by (`nlayers+1`) for midside two-dimensional nodes and
by (`2*nlayers+1`) for corner nodes, where `nlayers` is the number of layers
in the three-dimensional mesh.

The method returns the number of nodes in the three-dimensional mesh.

22.2.3 Element Connectivities and Nodal Coordinates

Element connectivities for the resulting three-dimensional mesh are generated by
method `elementConnectivities3D`, shown below.

```java
141        private void elementConnectivities3D() {
142
143            m3.elems = new Element[m3.nEl];
144
145            int[] ind2 = new int[8], t3 = new int[8],
146                    n3 = new int[8], ind3 = new int[20];
147            int iel3d = 0;
148
149            for (int iel2d = 0; iel2d < m2.nEl; iel2d++) {
150                int nind2 = m2.elems[iel2d].ind.length;
151                String mat = m2.elems[iel2d].matName;
152                for (int i = 0; i < nind2; i++) {
153                    int i2 = m2.elems[iel2d].ind[i] - 1;
154                    t3[i] = nodeType2[i2];
155                    n3[i] = nodeNum2[i2];
156                    ind2[i] = i2;
157                }
158
159                for (int i3 = 0; i3 < nlayers; i3++) {
160                    for (int i = 0; i < nind2; i++) {
161                        int dn = (t3[i] == 0) ? 1 : 2;
162                        int node = n3[i] + i3*dn;
163                        int dn2 = dn;
164                        int dn1 = 1;
165                        // Node at rotation axis
166                        if (rotate &&
167                                m2.getNodeCoord(ind2[i], 0) == 0) {
168                            node = n3[i];
169                            dn2 = 0;
170                            dn1 = 0;
171                        }
172                        ind3[i] = node;
173                        ind3[i + 12] = node + dn2;
174                        if ((i + 1)%2 == 1)
175                            ind3[(i+2)/2-1+nind2] = node + dn1;
176                    }
177                    m3.elems[iel3d] = Element.newElement("hex20");
178                    m3.elems[iel3d].setElemConnectivities(ind3);
179                    m3.elems[iel3d].setElemMaterial(mat);
180                    iel3d++;
181                }
182            }
183        }
```

Line 143 allocates an array of `Element` objects. Finite elements of the three-dimensional mesh are created as prisms of three-dimensional elements (possibly curved) having two-dimensional elements as their base. This is done inside a double loop: an outer loop is over two-dimensional elements and an inner loop is over layers of three-dimensional elements. The loop over two-dimensional elements starts at line 149. Lines 150–151 set `nind2` – number of nodes in the current two-dimensional element and `mat` – material name. Arrays `t3`, `n3`, and `ind2` (lines 154–156) contain node types of two-dimensional elements, three-dimensional node numbers at the first three-dimensional element layer and two-dimensional element connectivities.

Element connectivities of a three-dimensional mesh are formed in a loop of lines 159–181. The increments in node numbers between faces of the three-dimensional element are characterized by dn2. Increment dn1 is used for midside nodes. When a two-dimensional node is located on the rotation axis y a three-dimensional degenerate element is made by setting both increments dn1 and dn2 to zero. The entries of element connectivity array ind3 are set in lines 172–175.

Line 177 creates a new Element object of the type hex20 (20-node hexagonal element). Connectivities of the current three-dimensional element are set in line 178. The name of an element material is specified in line 179.

The nodal coordinates of the three-dimensional mesh are generated and set in method nodeCoordinates3D.

```
185        private void nodeCoordinates3D() {
186
187            m3.newCoordArray();
188
189            for (int i2 = 0; i2 < m2.nNod; i2++) {
190                int step = (nodeType2[i2] == 0) ? 2 : 1;
191                int n = 2*nlayers + 1;
192                // Node at rotation axis
193                if (rotate && m2.getNodeCoord(i2, 0) == 0.0) n = 1;
194                int nodeNum = nodeNum2[i2] - 1;
195                for (int i = 0; i < n; i += step) {
196                    double z = zlayers[i];
197                    double r = m2.getNodeCoord(i2, 0);
198                    double y = m2.getNodeCoord(i2, 1);
199                    if (rotate) {
200                        // Sweeping by rotation around Y
201                        double fi = Math.toRadians(zlayers[i]);
202                        double[] w =
203                            {r*Math.cos(fi), y, r*Math.sin(fi)};
204                        m3.setNodeCoords(nodeNum, w);
205                    }
206                    else {
207                        // Sweeping by translation along Z
208                        double[] w = {r, y, z};
209                        m3.setNodeCoords(nodeNum, w);
210                    }
211                    nodeNum++;
212                }
213            }
214        }
215
216 }
```

The node coordinate array of three-dimensional mesh m3 is initialized in line 187. The node numbering order is shown in Figure 22.2. A row of three-dimensional nodes is generated for each two-dimensional node i2. The number of three-dimensional nodes in a row is (2*nlayers+1) for a corner two-dimensional node. A row that starts at a midside two-dimensional node consists of (nlayers+1) nodes, where `nlayers` is the number of element layers in the three-dimensional mesh. Line 193 sets `n = 1`, so just one node is created in a row if a two-dimensional node lies on the rotational y-axis (`rotate=true`). Node coordinate setting is performed in lines 201–204 for the case of sweeping around the y-axis and in lines 208–209 for the case of sweeping by translation along the z-axis.

22.3 Example of Mesh Generation

Using mesh generator `sweep` create a mesh of brick-type twenty-node elements for a quarter cylinder with inner radius $r = 1$, outer radius $r = 2$ and height $h = 3$, as shown in Figure 22.3a.

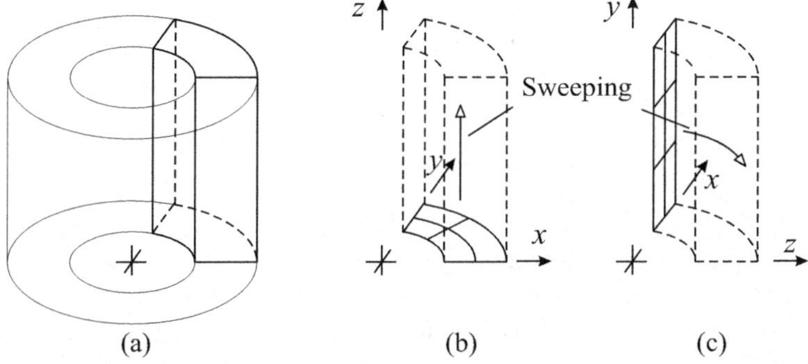

Fig. 22.3 Mesh generation for a quarter of a cylinder (a) by sweeping a two-dimensional mesh along the z-axis (b) and around the y-axis (c)

Solution

Using the sweeping method it is possible to create a three-dimensional mesh for a quarter cylinder using two generation schemes: sweeping a quarter ring along the z-axis as shown in Figure 22.3b and sweeping a rectangle around the y-axis (Figure 22.3c).

For creating the three-dimensional mesh by sweeping along the z-axis, it is possible to prepare and execute the following input data.

```
# 2D mesh for a quarter ring
genquad8 mesh2D
  nh = 2   nv = 3
  xyp = 1 0   0 0
        2 0   1.414 1.414
        0 2   0 0
        0 1   0.7071 0.7071
end
```

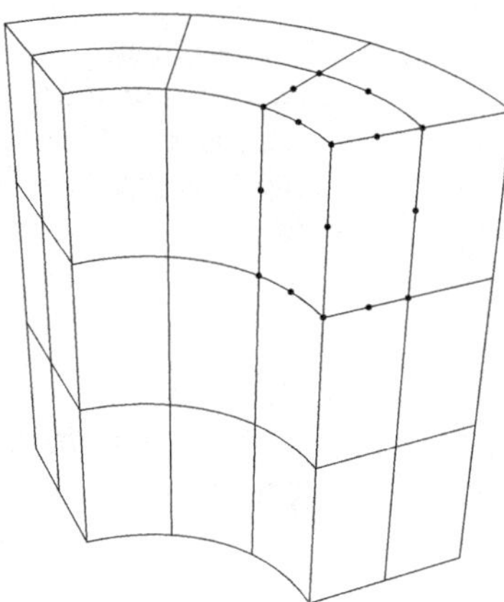

Fig. 22.4 Perspective view of the three-dimensional mesh generated by sweeping

```
# 3D mesh by sweeping along z axis
sweep mesh2D mesh3D
  nlayers = 3
  zlayers = 0 1 2 3
end
```

First, a two-dimensional mesh for a quarter ring is created inside an eight-node macroelement using mesh generator `genquad8`, and placed in a hash table under name `mesh2D`. The mesh has two elements in the radial direction and three elements in the angular direction. Then, this two-dimensional mesh is used for producing three element layers of a three-dimensional mesh by sweeping along the z-axis (statement `sweep mesh2D mesh3D`). The boundaries of layers are defined by `zlayers`. A perspective view of the generated mesh is shown in Figure 22.4.

Another way of creating almost the same three-dimensional mesh is by sweeping a two-dimensional rectangular mesh in the angular direction around the y-axis. Possible input data follows.

```
# 2D mesh for a rectangle
rectangle mesh2D
  nx = 2 ny = 3
  xs = 1 1.5 2
  ys = 0 1 2 3
end
```

```
# 3D mesh by sweeping around y axis
sweep mesh2D mesh3D
  nlayers = 3
  rotate = Y
  zlayers = 0 30 60 90
end
```

A two-dimensional mesh is generated using the `rectangle` module. It ranges from 1 to 2 in the x-direction and from 0 to 3 in the y-direction. The generated mesh block is stored under the name `mesh2D`. It is used by the `sweep` mesh generator to produce mesh block `mesh3D` by rotating the two-dimensional mesh around the y-axis. The two-dimensional mesh is copied at angles 0, 30, 60, and 90 degrees from plane xy.

The generated mesh looks the same as those created by sweeping along the z-direction (Figure 22.4). However, there are some differences between the two generated meshes. The meshes have slightly different nodal coordinates along the circular arcs. The first style of mesh creation approximates these arcs by parabolas, while the second mesh has exact node locations. Another difference is related to the different mesh orientations with respect to the global coordinate axes, as can be seen in Figs. 22.3b and c.

Problems

22.1. Propose possible enhancements for the sweeping technique that can help to handle more complicated cases of three-dimensional mesh generation.

22.2. Suppose that a three-dimensional mesh is generated by sweeping a two-dimensional mesh of n by n eight-node quadrilateral elements. Obtain the number of nodes in a three-dimensional mesh consisting of m layers of twenty-node hexahedral elements.

22.3. Modify the data of example in Section 22.3 to have mesh refinement near the point $x = 0$, $y = 1$, $z = 0$ in Figure 22.3b. An element with a vertex at this point should have edges with size about 0.2.

Chapter 23
Pasting Mesh Blocks

Abstract Pasting mesh blocks allows creation of complicated meshes using relatively simple mesh blocks. Java^{TM} class `connect` implements connection (pasting) two mesh fragments by joining arrays of element connectivities and nodal coordinates.

23.1 Pasting Technique

Pasting mesh blocks is an important step in our procedure of mesh generation. Connection of relatively simple mesh fragments allows creation of a mesh of any complexity. In order to keep the pasting algorithm simple, we require that element faces at a connection surface are compatible. This means that it is possible to find pairs of coincident nodes and connected element faces that have the same interpolation properties. Formally, we can connect a face of the linear element to a face of the quadratic element. However, this leads to discontinuity in displacements and does not guarantee solution convergence. An example of connection of two mesh blocks is shown in Figure 23.1. We paste mesh blocks A and B in order to produce a resulting mesh block C. Let us assume that nodes and elements of the mesh block A will be first in the resulting mesh, i.e., the resulting coordinate array includes nodal coordinates of mesh A, then nodal coordinates of mesh B, and the resulting connectivity array first contains connectivities of mesh A, then connectivities of mesh B. An algorithm for pasting two mesh blocks is as follows:

Pasting: mesh A + mesh B $\rightarrow$ mesh C

Find coincident node pairs in meshes A and B
Copy nodes of mesh A into mesh C
Add nodes of mesh B to mesh C using coincidence information
Copy elements of mesh A into mesh C
Add elements of mesh B to mesh C using coincidence information

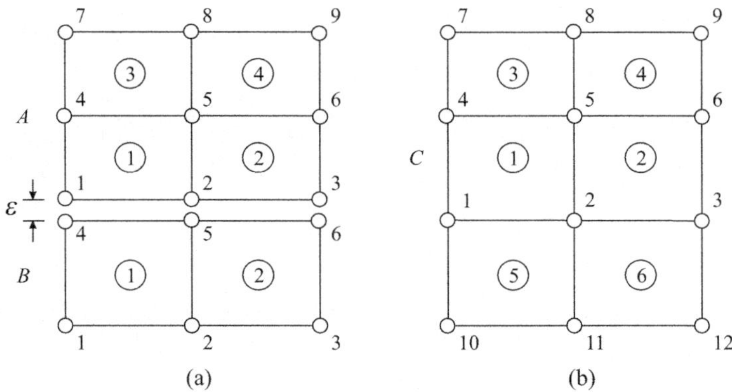

Fig. 23.1 Connection of two separate mesh blocks A and B (a) produces a resulting mesh C (b)

When searching for coincident node pairs we compare coordinates of each node of mesh A with all nodes of mesh B. If the difference in node locations is less than a specified tolerance then two nodes are considered coincident and a node number in mesh B is set equal to a node number in mesh A. Nodal coordinates and element connectivities of mesh A are copied to model C without change. During addition of nodes of mesh B to model C, coincident nodes from mesh block B overwrite nodes in mesh block A. Element connectivities of mesh B are transformed to node numbers in model C during adding process.

23.2 Implementation

23.2.1 Data Input

A call to the connection of two mesh blocks is done with the following statement:

```
connect A B C
```

where A and B are names of two mesh blocks that should be connected, and c is the name of a resulting mesh block.

Input data for the connecting module consists of just a single parameter eps – coordinate tolerance for joining nodes belonging to different mesh blocks. If the default value of the parameter eps = 0.0001 is suitable then it is possible to omit its specification. For compatible block boundaries the default coordinate tolerance is appropriate. When we change this parameter it is necessary to keep it less then a minimum distance between nodes at connecting boundaries. The statement end should be always used in order to mark the end of data for the module.

Constructor connect performing pasting of two mesh blocks and a method for data input are presented below.

```java
1  package gener;
2
3  import fea.*;
4  import model.*;
5  import elem.*;
6  import util.*;
7
8  // Paste two meshes.
9  // Input: modelNameA - name of first mesh to be pasted;
10 //   modelNameB - name of second mesh to be pasted;
11 //   modelNameC - name of resulting mesh;
12 //   [eps] - coordinate tolerance for joining nodes.
13 public class connect {
14
15     private FeModel mA, mB;
16     private double eps = 0.0001;
17     private int nConnected;
18     private int newNodesB[];
19
20     public connect() {
21
22         String modelNameA = Jmgen.RD.next();
23         String modelNameB = Jmgen.RD.next();
24         String modelNameC = Jmgen.RD.next();
25
26         Jmgen.PR.printf("Connect: %s + %s -> %s\n",
27                 modelNameA, modelNameB, modelNameC);
28
29         readData();
30
31         if (Jmgen.blocks.containsKey(modelNameA))
32             mA = (FeModel) Jmgen.blocks.get(modelNameA);
33         else UTIL.errorMsg(
34                 "No such mesh block: " + modelNameA);
35         if (Jmgen.blocks.containsKey(modelNameB))
36             mB = (FeModel) Jmgen.blocks.get(modelNameB);
37         else UTIL.errorMsg(
38                 "No such mesh block: " + modelNameB);
39         if (mA.nDim != mB.nDim) UTIL.errorMsg(
40                 "Models with different nDim");
41
42         findCoincidentNodes();
43         FeModel mC = pasteModels();
44
45         Jmgen.blocks.put(modelNameC, mC);
46         Jmgen.PR.printf(" %d node pairs connected\n",
47                 nConnected);
48         Jmgen.PR.printf("Mesh " + modelNameC +
49                 ": nEl = %d  nNod = %d\n", mC.nEl, mC.nNod);
50     }
51
52     private void readData() {
53         while (Jmgen.RD.hasNext()) {
54             String name = Jmgen.RD.next().toLowerCase();
```

```
55              if (name.equals("#")) {
56                  Jmgen.RD.nextLine();
57                  continue;
58              }
59              if (name.equals("eps"))
60                  eps = Jmgen.RD.readDouble();
61              else if (name.equals("end")) break;
62              else UTIL.errorMsg("Unexpected data: " + name);
63          }
64          Jmgen.PR.printf(
65              "Coordinate error tolerance eps = %10.3e\n", eps);
66      }
```

The constructor first inputs and prints names of two connecting mesh blocks, modelNameA and modelNameB, and the name of the resulting mesh block modelNameC (lines 22–27). The coordinate error tolerance parameter eps and end statement are read by method readData (line 29). Lines 31–40 check the existence of input mesh blocks modelNameA and modelNameB and the condition that they have same dimension nDim. Method findCoincidentNodes (line 42) finds nodes occupying the same positions in models mA and mB. Pasting of two mesh blocks is performed by method pasteModels in line 43. The resulting mesh block is placed in hash table blocks under name modelNameC (line 45).

23.2.2 *Finding Coincident Nodes*

Method findCoincidentNodes given below finds coincident nodes in two connecting nodes and creates array newNodesB containing new numbers for nodes of model mB.

```
68          // Find coincident nodes in models mA and mB,
69          // generate new node numbers for model mB
70      private void findCoincidentNodes() {
71
72          newNodesB = new int[mB.nNod];
73          int ndim = mA.nDim;
74          for (int i = 0; i < mB.nNod; i++) newNodesB[i] = -1;
75
76          // Register coincident nodes of mesh B
77          //     in array newNodesB
78          for (int ia = 0; ia < mA.nNod; ia++) {
79              double xyA[] = mA.getNodeCoords(ia);
80              B:
81              for (int ib = 0; ib < mB.nNod; ib++) {
82                  for (int j = 0; j < ndim; j++) {
83                      if (Math.abs(xyA[j]-mB.getNodeCoord(ib,j))
84                              > eps) continue B;
85                  }
86                  newNodesB[ib] = ia;
87              }
88          }
```

```
 89              nConnected = 0;
 90              int n = mA.nNod;
 91
 92              // New node numbers for nodes of model mB
 93              for (int i = 0; i < mB.nNod; i++) {
 94                      if (newNodesB[i] == -1) newNodesB[i] = n++;
 95                      else nConnected++;
 96              }
 97      }
```

Array newNodesB is initialized with the value −1 in line 71. The double loop
in lines 77–87 compares each node of model mA with all nodes of model mB. If
a difference in nodal coordinates for two nodes is less than the specified tolerance
eps then the node number from model mA is placed in array newNodesB at the
position defined by node from modelmB. This means that coincident nodes from
model mB will have numbers of its pair from model mB. Other nodes from model
mB are assigned node numbers next to node numbers of model mA (lines 92–95).
Parameter nConnected contains number of connected node pairs.

23.2.3 Pasting

Pasting of two mesh blocks is performed by method pasteModels.

```
 99      // Paste two meshes.
100      // Nodes and elements of the first mesh
101      // are first in the resulting mesh.
102      // returns   resulting mesh after pasting.
103      private FeModel pasteModels() {
104
105          FeModel mC = new FeModel(Jmgen.RD, Jmgen.PR);
106          mC.nDim = mA.nDim;
107
108          // Nodal coordinates of model mC
109          mC.nNod = mA.nNod + mB.nNod - nConnected;
110          mC.newCoordArray();
111          // Copy nodes of model mA
112          for (int i = 0; i < mA.nNod; i++)
113              mC.setNodeCoords(i, mA.getNodeCoords(i));
114          // Add nodes of model mB
115          for (int i = 0; i < mB.nNod; i++)
116              mC.setNodeCoords(
117                      newNodesB[i], mB.getNodeCoords(i));
118
119          // Element connectivities of model mC
120          mC.nEl = mA.nEl + mB.nEl;
121          mC.elems = new Element[mC.nEl];
122          // Copy elements of model mA
123          for (int el = 0; el < mA.nEl; el++) {
124              mC.elems[el] =
125                      Element.newElement(mA.elems[el].name);
```

```
126            mC.elems[el].setElemConnectivities(
127                    mA.elems[el].ind);
128            mC.elems[el].matName = mA.elems[el].matName;
129        }
130        // Add elements of mB with renumbered connectivities
131        for (int el = 0; el < mB.nEl; el++) {
132            mC.elems[mA.nEl + el] =
133                    Element.newElement(mB.elems[el].name);
134            int indel[] = new int[mB.elems[el].ind.length];
135            for (int i = 0; i < mB.elems[el].ind.length; i++)
136                indel[i] = newNodesB[mB.elems[el].ind[i]-1]+1;
137            mC.elems[mA.nEl+el].setElemConnectivities(indel);
138            mC.elems[mA.nEl+el].matName = mB.elems[el].matName;
139        }
140        return mC;
141    }
142
143 }
```

A new finite element model mC is constructed in line 104. The number of nodes
in model mC is determined as the sum of nodes in models mA and mB minus the
number of coincident node pairs (line 108). Nodes of model mA are copied into
model mC without change in lines 111–112. Nodes from model mB are placed next
according to node numbers contained in array newNodesB (lines 114–116).

A similar procedure is used for creating a connectivity array for model mC. The
number of elements in the resulting mesh block is equal to the sum of numbers of
elements in the connecting mesh blocks (line 119). The elements of model mA are
copied to model mC in lines 122–128. Connectivities of elements from model mB
are renumbered using array newNodesB and placed in the resulting model mC in
lines 130–138.

The method returns model mC, which is created by pasting models mA and mB.

Problems

23.1. Figure 23.1 depicts the resulting mesh block C after pasting mesh blocks A and
B along a common boundary: $A + B \rightarrow C$. Determine the element and node numbers
in mesh block D after pasting $B + A \rightarrow D$.

23.2. Determine the element and node numbers in the resulting meshes C and D after
pasting of meshes A and B shown below: $A + B \rightarrow C$ and $B + A \rightarrow D$.

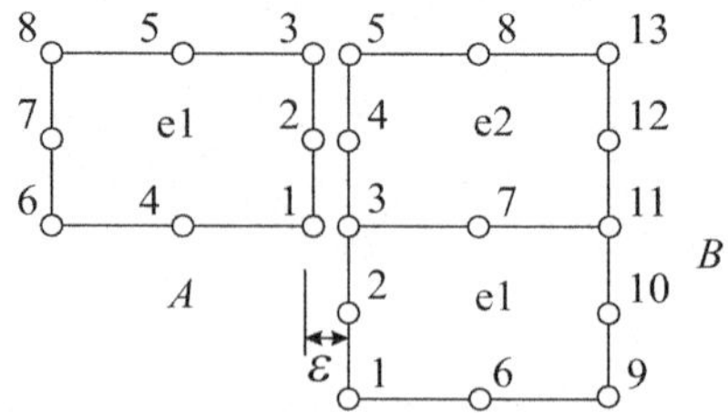

Chapter 24
Mesh Transformations

Abstract Java^{TM} class `transform` implements several coordinate transformations for a mesh block. Transformations include translation, scaling, rotation, and mirroring.

24.1 Transformation Relations

Standard coordinate transformations – translation, scaling and rotation – can help in creating finite element meshes. In addition to these transformations, we implement a mirror transformation that reflects a mesh with respect to a plane. All transformations are shown in Figure 24.1.

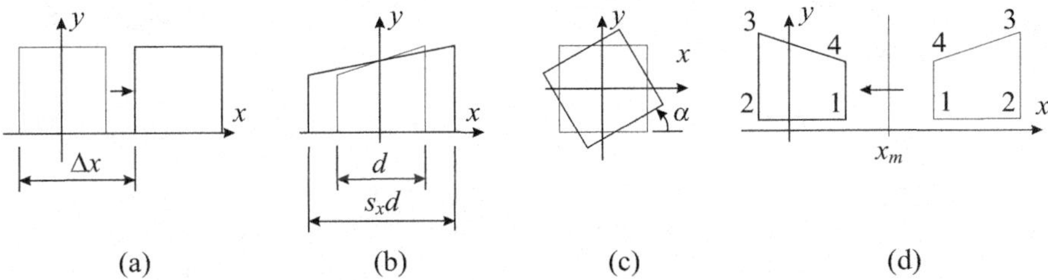

Fig. 24.1 Mesh transformations: translation (a), scaling (b), rotation (c), and mirror transformation (d)

A translation transformation (Figure 24.1a) changes the coordinates of all nodes by the same value:

$$x_i^n = x_i^n + \Delta x_i,$$ (24.1)

where x_i^n is the ith coordinate component of the nth node and Δx_i is a displacement value.

A scaling transformation (Figure 24.1b) multiplies a specified coordinate component of all nodes by the same value:

$$x_i^n = x_i^n s_i,$$
(24.2)

where s_i is a scaling coefficient.

Rotation of the finite element mesh around a coordinate axis (Figure 24.1c) is performed as the following matrix-vector multiplication:

$$x_i^n = \alpha_{ij} x_j^n.$$
(24.3)

Here α_{ij} is the matrix of direction cosines. The matrix of direction cosines has the following appearance for rotations around three coordinate axes:

around x

$$[\alpha] = \begin{bmatrix} 1 & 0 & 0 \\ 0 & \cos\alpha & -\sin\alpha \\ 0 & \sin\alpha & \cos\alpha \end{bmatrix},$$
(24.4)

around y

$$[\alpha] = \begin{bmatrix} \cos\alpha & 0 & \sin\alpha \\ 0 & 1 & 0 \\ -\sin\alpha & 0 & \cos\alpha \end{bmatrix},$$
(24.5)

around z

$$[\alpha] = \begin{bmatrix} \cos\alpha & -\sin\alpha & 0 \\ \sin\alpha & \cos\alpha & 0 \\ 0 & 0 & 1 \end{bmatrix},$$
(24.6)

where α is a rotation angle. Positive angles correspond to rotation in an anticlockwise direction as seen from the top of the axis.

A mirror transformation, shown in Figure 24.1d, reflects a mesh block with respect to a plane normal to a coordinate axis:

$$x_i^n = -x_i^n + 2x_{i0},$$
(24.7)

where x_i is an axis normal to the reflecting plane and x_{i0} is a location of the reflecting plane. In addition to transformation for nodal coordinates it is necessary to modify the order of connectivity numbers. Suppose that nodes $1, 2, 3, 4$ are element connectivities listed in anticlockwise order before the mirror transformation. From Figure 24.1d it is evident that after the transformation, element node numbers are in clockwise order. If the connectivities are changed to $1, 4, 3, 2$ then the correct order of nodes is restored.

24.2 Implementation

24.2.1 Input Data

The four transformations (translation, scaling, rotation, and mirroring) are implemented in the class `transform`. A call to the transformation module is done with the following statement:

```
transform modelName
```

where `modelName` is the name of the mesh block that should be transformed.

The rest of the input data consists of any number of the following statements in any order:

```
translate axis distance
scale axis factor
rotate axis angle
mirror axis plane
```

Here, `translate` is a translation operation, `axis` = `x/y/z` is a coordinate for translation, `distance` is the distance of translation. Scaling operation `scale` is performed for coordinate `axis` by multiplying nodal coordinates with `factor`. Rotation `rotate` is done around axis `axis` on an angle `angle` specified in degrees. Positive angle is counted in a counterclockwise direction looking from the top of the rotation axis. Operation `mirror` reflects a mesh with respect to a plane normal to `axis`. The plane is located at the coordinate `plane`.

The end of data for the `transform` module is marked by statement end. The order of transformation operations is determined by the order of data.

Input of data and call to methods performing transformations is contained in the constructor presented next.

```java
 1 package gener;
 2
 3 import model.*;
 4 import util.*;
 5 import fea.*;
 6
 7 // Make transformations for a specified model.
 8 // Input: modelName - name of the finite element model;
 9 // [translate axis value] - translate axis coordinates;
10 // [scale axis value] - scale axis coordinates;
11 // [rotate axis value] - rotate around axis by angle (degrees);
12 // [mirror axis value] - mirror along axis around axis value
13 public class transform {
14
15     private FeModel m;
16     enum vars {
17         translate, scale, rotate, mirror, end
18     }
19
```

```
20      enum elementMirror {
21          quad8 {int permutation(int i) {
22              int[] p = {1,8,7,6,5,4,3,2};
23              return p[i]-1;}
24          },
25          hex20 {int permutation(int i) {
26              int[] p = {1,8,7,6,5,4,3,2, 9,12,11,10,
27                         13,20,19,18,17,16,15,14};
28              return p[i]-1;}
29          };
30          abstract int permutation(int i);
31      }
32
33      private vars opName;
34      private char axis;
35      private double value;
36
37      public transform() {
38
39          String modelName = Jmgen.RD.next();
40          Jmgen.PR.printf("Transform: %s\n", modelName);
41
42          if (Jmgen.blocks.containsKey(modelName))
43              m = (FeModel) Jmgen.blocks.get(modelName);
44          else UTIL.errorMsg("No such mesh block: " + modelName);
45
46          while (Jmgen.RD.hasNext()) {
47              String name = Jmgen.RD.next().toLowerCase();
48              if (name.equals("#")) {
49                  Jmgen.RD.nextLine(); continue;
50              }
51              if (name.equals("end")) break;
52
53              try {
54                  opName = vars.valueOf(name);
55              } catch (Exception e) {
56                  UTIL.errorMsg("Operation name is not found: "
57                          + name);
58              }
59
60              String Axis = Jmgen.RD.next().toLowerCase();
61              axis = Axis.charAt(0);
62              if (axis!='x' && axis!='y' && axis!='z')
63                  UTIL.errorMsg("Incorrect axis: " + axis);
64              value = Jmgen.RD.readDouble();
65              Jmgen.PR.printf("%10s  %s  %10.4f\n",
66                      opName, Axis, value);
67
68              switch (opName) {
69                  case translate:   doTranslate();
70                      break;
71                  case scale:       doScale();
72                      break;
73                  case rotate:      doRotate();
```

```
74                         break;
75                 case mirror:        doMirror();
76                         break;
77             }
78         }
79         Jmgen.PR.printf("Mesh " + modelName +
80                 ": nEl = %d  nNod = %d\n", m.nEl, m.nNod);
81     }
```

Lines 39–44 read the name of the model, which should undergo transformations
and check that this name exists in hash table `blocks`. Input of data and performing
transformation operations are placed inside the `while` loop in lines 46–78. An
operation name is read in line 47. If, instead of the operation name, the word `end`
is in the input stream then the `transform` module finishes its work. Otherwise,
parameters `axis` and `value` are read and translation, scaling, rotation, or mirroring
is performed for the mesh block, depending on the operation name.

24.2.2 *Performing Transformations*

Methods `doTranslate`, `doScale` and `doRotate` perform translation, scaling
and rotation transformations, respectively.

```
83     private static int getIntAxis(char axis) {
84         int iAxis = 0;
85         if (axis=='y') iAxis = 1;
86         else if (axis=='z') iAxis = 2;
87         return iAxis;
88     }
89
90     private void doTranslate() {
91
92         if (m.nDim == 2 && axis == 'z') return;
93         int iax = getIntAxis(axis);
94
95         for (int i=0; i<m.nNod; i++)
96             m.setNodeCoord(i, iax,
97                     m.getNodeCoord(i, iax) + value);
98     }
99
100    private void doScale() {
101
102        if (m.nDim == 2 && axis == 'z') return;
103        int iax = getIntAxis(axis);
104
105        for (int i=0; i<m.nNod; i++)
106            m.setNodeCoord(i, iax,
107                    m.getNodeCoord(i,iax) * value);
108    }
109
110    private void doRotate() {
```

```
111
112        if (m.nDim == 2 && (axis=='x' || axis=='y')) return;
113
114        double sina = Math.sin(Math.toRadians(value));
115        double cosa = Math.cos(Math.toRadians(value));
116        double a[][] = new double[3][3];
117        double x[] = new double[3];
118
119        if (axis=='x') {
120            a[0][0]= 1;      a[0][1]= 0;      a[0][2]=0;
121            a[1][0]= 0;      a[1][1]= cosa; a[1][2]=-sina;
122            a[2][0]= 0;      a[2][1]= sina; a[2][2]= cosa;
123        }
124        else if (axis=='y') {
125            a[0][0]= cosa; a[0][1]= 0;      a[0][2]= sina;
126            a[1][0]= 0;      a[1][1]= 1;      a[1][2]= 0;
127            a[2][0]=-sina; a[2][1]= 0;      a[2][2]= cosa;
128        }
129        else {  // around z
130            a[0][0]= cosa; a[0][1]=-sina; a[0][2]= 0;
131            a[1][0]= sina; a[1][1]= cosa; a[1][2]= 0;
132            a[2][0]= 0;      a[2][1]= 0;      a[2][2]= 1;
133        }
134
135        for (int inod=0; inod<m.nNod; inod++) {
136            for (int j=0; j<m.nDim; j++)
137                x[j] = m.getNodeCoord(inod,j);
138            for (int i=0; i<m.nDim; i++) {
139                double s = 0;
140                for (int j=0; j<m.nDim; j++) s += a[i][j]*x[j];
141                m.setNodeCoord(inod, i, s);
142            }
143        }
144    }
```

Method `getIntAxis` translates axis symbols x, y and z into numerical values 0,
1 and 2. Implementation of translation (lines 90–98) and scaling (lines 100–108) is
straightforward. Methods `getNodeCoord` and `setNodeCoord` are used to get
the necessary components of nodal coordinates and to set them. Rotation transformation is done by matrix-vector product in lines 135–143.

Method `doMirror` performs a mirror transformation.

```
146        private void doMirror() {
147
148            if (m.nDim == 2 && axis == 'z') return;
149            int iax = getIntAxis(axis);
150
151            // Mirror nodal coordinates
152            for (int i=0; i<m.nNod; i++)
153                m.setNodeCoord(
154                        i, iax, -m.getNodeCoord(i,iax)+2*value);
155
156            // Change order of element connectivities
157            for (int e=0; e<m.nEl; e++) {
```

```
158                  elementMirror em = null;
159                  try {
160                      em = elementMirror.valueOf(m.elems[e].name);
161                  } catch (Exception el) {
162                      UTIL.errorMsg("Mirror: element not supported "
163                              + m.elems[e].name);
164                  }
165
166                  int nind = m.elems[e].ind.length;
167                  int[] ind = new int[nind];
168                  for (int i=0; i<nind; i++)
169                      ind[em.permutation(i)] = m.elems[e].ind[i];
170                  m.elems[e].setElemConnectivities(ind);
171              }
172          }
173
174  }
```

Transformation of nodal coordinates is done in lines 152–154. The element loop in lines 157–171 changes the order of entries in element connectivities. Enumerated `elementMirror` (lines 20–31) is used for connectivity permutations. It provides function `permutation` that returns new positions for element connectivity numbers depending on element types (`quad8` or `hex20`).

24.3 Example of Using Transformations

A square mesh block A of size 2 is centered at the coordinate origin, as shown in Figure 24.2. Transform mesh block A into mesh block A1, which is a square of size 4 rotated by 30 degrees in the anticlockwise direction and centered at point $x = 6$, $y = 2$.

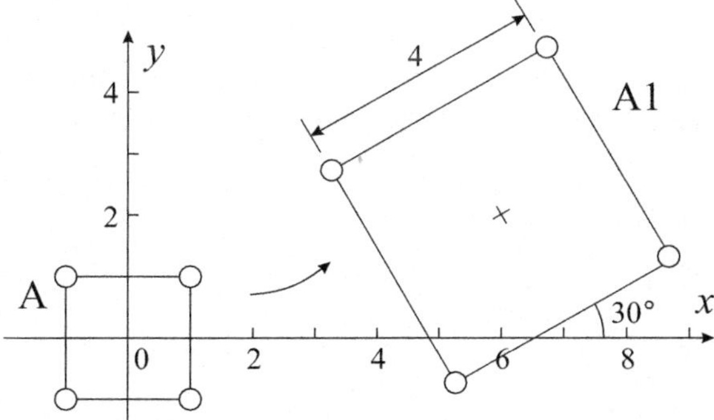

Fig. 24.2 Mesh block A can be transformed into block A1 using scaling, rotation and translation

Solution

Such a transformation can be performed using module `transform` with the following data.

```
transform A
  scale x 2
  scale y 2
  rotate z 30
  translate x 6
  translate y 2
end
```

Mesh block `A` is first scaled in two directions, then rotated around the *z*-axis by 30 degrees, and finally translated along the *x*- and *y*-axes. It should be noted that the operation order is important. While the order of translations along *x* and *y* is not significant, changing the order of rotation and translation leads to a quite different result.

Problems

24.1. Prepare input data for module `transform` that performs transformation of mesh block `A` into block `A1`. The location and sizes of the mesh blocks are shown below.

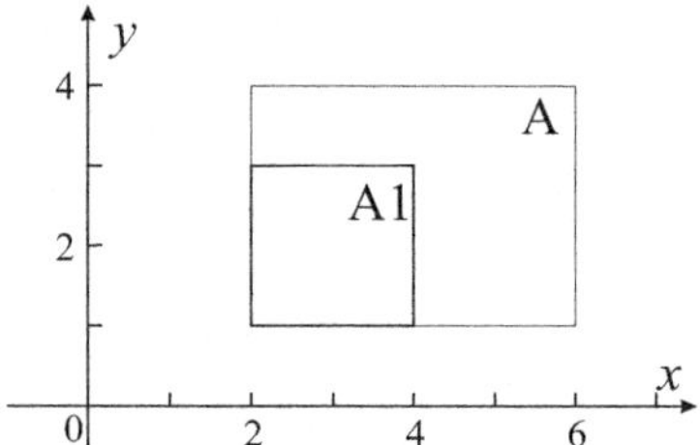

24.2. Initially, mesh block `A` is a square with edge size 1 centered at point $x = 1$, $y = 1$. Determine the locations of block vertices after transformations caused by the following data.

```
transform A
  scale x 2
  scale y 2
  translate x 1
  rotate z 45
  mirror y 2
end
```

24.3. Propose other useful transformations or modifications for mesh blocks. Provide examples of mesh creations that support the usefulness of the proposed transformations.

Chapter 25
Copying, Writing and Reading Mesh Blocks

Abstract This chapter presents classes that perform useful simple operations on mesh blocks. Module `copy` allows creation of a copy of a mesh block under a new name. Writing and reading of mesh blocks to/from text files are performed by modules `writemesh` and `readmesh`.

25.1 Copying

Module `copy` allows creation of a copy of a mesh block in the hash table. To perform copying it is necessary to place the following statement in the input data stream:

```
copy modelNameA modelNameB
```

Mesh block `modelNameA` is copied into new mesh block `modelNameB`. The source code of Java^{TM} class `copy` is given below.

```
 1 package gener;
 2
 3 import model.*;
 4 import elem.*;
 5 import fea.*;
 6 import util.*;
 7
 8 // Copy model.
 9 // Input: modelNameA - name of the model to be copied;
10 //   modelNameB - name of the resulting model.
11 public class copy {
12
13     private FeModel mA;
14
15     public copy() {
16
17         String modelNameA = Jmgen.RD.next();
18         String modelNameB = Jmgen.RD.next();
```

```java
19          Jmgen.PR.printf("Copy: %s -> %s\n",
20                  modelNameA, modelNameB);
21          if (modelNameA.equals(modelNameB)) return;
22          if (Jmgen.blocks.containsKey(modelNameA))
23              mA = (FeModel) Jmgen.blocks.get(modelNameA);
24          else
25              UTIL.errorMsg("No such mesh block: " + modelNameA);
26          FeModel mB = copyMesh();
27          Jmgen.blocks.put(modelNameB, mB);
28          Jmgen.PR.printf(
29                  "nEl = %d  nNod = %d\n", mB.nEl, mB.nNod);
30      }
31
32      private FeModel copyMesh() {
33
34          FeModel mB = new FeModel(Jmgen.RD, Jmgen.PR);
35          mB.nDim = mA.nDim;
36
37          mB.nNod = mA.nNod;
38          mB.newCoordArray();
39          for (int i = 0; i < mB.nNod; i++) {
40              for (int j = 0; j < mB.nDim; j++)
41                  mB.setNodeCoords(i, mA.getNodeCoords(i));
42          }
43
44          mB.nEl = mA.nEl;
45          mB.elems = new Element[mB.nEl];
46          for (int el = 0; el < mB.nEl; el++) {
47              mB.elems[el] = Element.newElement(
48                      mA.elems[el].name);
49              mB.elems[el].setElemConnectivities(
50                      mA.elems[el].ind);
51              mB.elems[el].matName = mA.elems[el].matName;
52          }
53
54          return mB;
55      }
56
57 }
```

Mesh block names for input and output are read in lines 17–18. A copy of finite element model mA is created in line 26 by calling method copyMesh. Line 27 puts output model mB into hashtable blocks under the name modelNameB.

The finite element model during mesh generation contains the following data

nNod – number of nodes;

nEl – number of elements;

nDim – number of dimensions (2 or 3);

xyz[] – array of nodal coordinates;

elems[] – array of Element objects.

Method `copyMesh` explicitly copies the scalar mesh parameters, nodal coordinates (lines 39–42) and element connectivities (lines 46–52) into a new finite element model mB. After copying, both mesh blocks can be used independently. For example, one of two mesh blocks can be transformed and pasted with the other mesh block.

25.2 Writing Mesh to File

The prepared finite element mesh is written in a text file. The file is used as a part of the input data for problem solution (main method `Jfem`) and for visualization (main method `Jvis`).

Writing the resulting mesh block in a file is a result of interpreting the data statement

```
writemesh modelName fileName
```

Here, `modelName` is the name of the resulting finite element mesh block and `fileName` is the file name where text information about the mesh block should be written. The source code of module `writemesh` follows.

```
 1 package gener;
 2
 3 import model.*;
 4 import fea.*;
 5 import util.*;
 6
 7 import java.io.PrintWriter;
 8
 9 // Write mesh to file.
10 // Input:  modelName - name of the finite element model;
11 //   fileName - name of the file.
12 public class writemesh {
13
14     FeModel m;
15
16     public writemesh() {
17
18         String modelName = Jmgen.RD.next();
19         String fileName = Jmgen.RD.next();
20         Jmgen.PR.printf("WriteMesh: %s     %s\n",
21                 modelName, fileName);
22
23         PrintWriter WR =
24                 new FePrintWriter().getPrinter(fileName);
25
26         if (Jmgen.blocks.containsKey(modelName))
27             m = (FeModel) Jmgen.blocks.get(modelName);
28         else UTIL.errorMsg("No such mesh block: " + modelName);
29
30         WR.printf("# Model name: %s\n", modelName);
```

```
31        WR.printf("nNod = %5d\n", m.nNod);
32        WR.printf("nEl  = %5d\n", m.nEl);
33        WR.printf("nDim = %5d\n", m.nDim);
34
35        WR.printf("nodCoord\n");
36        for (int i = 0; i < m.nNod; i++) {
37            for (int j = 0; j < m.nDim; j++)
38                WR.printf("%20.9f", m.getNodeCoord(i, j));
39            WR.printf("\n");
40        }
41
42        WR.printf("\nelCon");
43        for (int iel = 0; iel < m.nEl; iel++) {
44            WR.printf("\n%s %6s", m.elems[iel].name,
45                         m.elems[iel].matName);
46            int nind = m.elems[iel].ind.length;
47            for (int i = 0; i < nind; i++)
48                WR.printf("%6d", m.elems[iel].ind[i]);
49        }
50        WR.printf("\n\nend\n");
51        WR.close();
52        Jmgen.PR.printf("Mesh " + modelName +
53                    ": nEl = %d   nNod = %d\n", m.nEl, m.nNod);
54    }
55
56 }
```

The `PrintWriter` object for writing a mesh block in text format is created
in lines 23–24. Lines 26–28 check the existence of the requested mesh block with
the name `modelName`. The first line of the file contains a comment with the mesh
block name (line 30). The following lines of the file hold the number of nodes,
number of elements, and problem dimension. Then, nodal coordinates are written
under the title `nodCoord` and element connectivities under the title `nelCon`. The
last line of the file is the `end` statement.

The written mesh file can be later read by any of three finite element programs –
`Jmgen`, `Jfem`, and `Jvis`.

25.3 Reading Mesh from File

Reading a mesh block in program `Jmgem` is performed by the module `readmesh`
when the following statement appears in the input data stream:

```
readmesh modelName fileName
```

where `modelName` is the name of the finite element mesh block, and `fileName`
is the file name in which text information about the mesh block resides.

Implementation of the module `readmesh` is rather straightforward. The source
code is presented below.

```java
 1 package gener;
 2
 3 import model.*;
 4 import fea.Jmgen;
 5 import util.FeScanner;
 6
 7 // Read mesh data from text file.
 8 // Input: modelName - name of the finite element model;
 9 // fileName - name of the file.
10 public class readmesh {
11
12     public readmesh() {
13
14         String modelName = Jmgen.RD.next();
15         String fileName = Jmgen.RD.next();
16         Jmgen.PR.printf("ReadMesh:  %s     %s\n\n",
17                 modelName, fileName);
18         FeScanner RD = new FeScanner(fileName);
19         FeModel model = new FeModel(RD, Jmgen.PR);
20         model.readData();
21         Jmgen.blocks.put(modelName, model);
22         Jmgen.PR.printf("nEl = %d      nNod = %d\n",
23                 model.nEl, model.nNod);
24     }
25
26 }
```

Usually, the mesh block (which is read by this module) was previously written by
the module `writemesh`. It is possible to prepare a text file for a mesh block using
any text editor. The file should contain the following information

```
nNod =   <number of nodes>
nEl  =   <number of elements>
nDim =   <number of dimensions, =2/3>

nodCoord
  <coordinates of nodes>

elCon
  <element data>

end
```

For each node i, coordinates are specified as two x_i, y_i or three x_i, y_i, z_i numbers. For
each element, the data contains element type (`quad8`/`hex20`), material name and
element connectivities.

Problems

25.1. Prepare a text file containing data for the mesh block shown below.

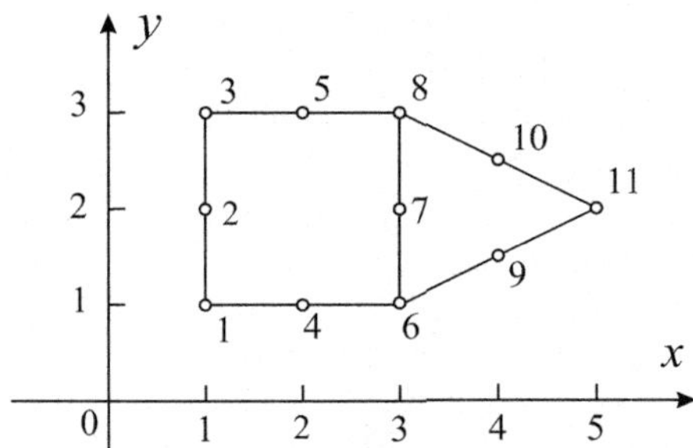

Using program Jmgen, read the mesh block from a file. Then write the mesh block into another text file. Compare the text files (containing information about the same mesh block).

25.2. Create a data file for the program Jmgen that can be used for generating a mesh for the following two-dimensional computational domain.

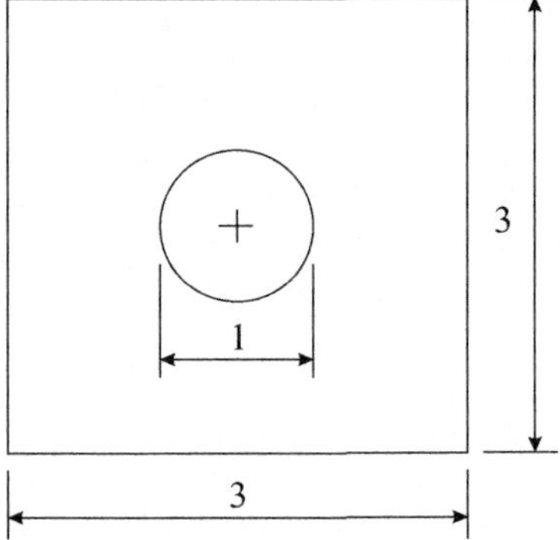

Decompose the domain into blocks. Generate a mesh for a typical block. Use copying, transformations, and pasting to create a mesh for the whole domain.

Part IV
Visualization of Meshes and Results

Chapter 26
Introduction to Java 3D^{TM}

Abstract This chapter opens the concluding part of the book devoted to visualization of finite element models and results. The Java 3D^{TM} API is used for producing three-dimensional real-time graphics. Java 3D is introduced with a brief explanation of a scene graph and specification of the geometry of visualized objects.

26.1 Rendering Three-dimensional Objects

Visualization of three-dimensional finite element models requires rendering three-dimensional objects of an arbitrary shape. Showing the results of finite element solutions as contours implies complex coloring of an object's surface.

Rendering of 3D scenes is an extremely complicated task that is impractical to implement by ourselves. Usually, graphics libraries with an API are used for this purpose. The most famous API for real-time three-dimensional rendering is OpenGL$^{\circledR}$ in the C programming language and its Java^{TM} wrapper JOGL [5]. However, OpenGL and hence JOGL are based on a procedural approach.

Since our finite element code is based on an object-oriented approach it is desirable to follow this approach for development of the visualization part of the code. Such a graphics library exists and has the name Java 3D [30].

Java 3D is an object-oriented API developed by Sun Microsystems for rendering three-dimensional interactive real-time graphics. It is a set of several Java packages designed for easier development of applications and applets with three-dimensional graphics capabilities. A graphical program contains Java 3D objects, which compose a scene graph. Java 3D provides a high-level programming interface by hiding implementation details. The developer specifies the geometry of visual objects, their appearance and behavior, and light sources as Java 3D objects. After compiling, the scene is rendered automatically with "quasi"-photographic quality. The latter means that the effects of light-source shading are shown, but visual object shading and reflections are ignored. Introduction to programming graphics with Java 3D API is given in [28, 31].

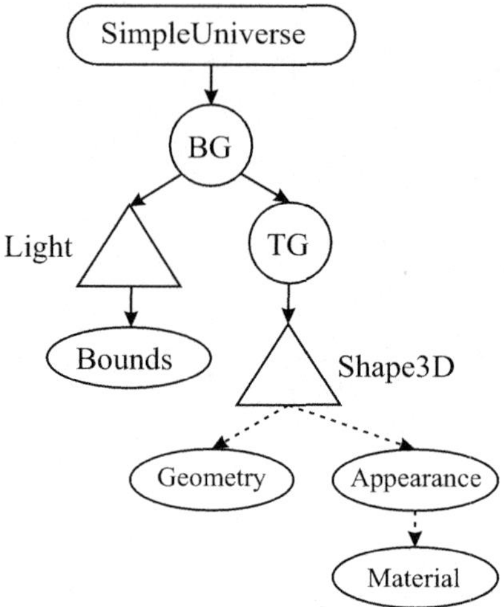

Fig. 26.1 Example of a simple scene graph

26.2 Scene Graph

In Java 3D a scene graph is used to organize all the objects of the scene. The scene
graph is a tree-like data structure. The graph contains nodes connected by links. The
nodes of a scene graph can be of two kinds: group nodes and leafs. Group nodes have
children and can be used for branching, switching, transformations, etc. Leaf nodes
usually represent geometric objects, lights, behaviors, etc. They have no children.
However, leaf nodes often have references to node components. Node components
compose a bundle of attributes for a node. Node components include the geometry
of a shape, appearance, material, texture, and various attributes.

An example of a simple scene graph is shown in Figure 26.1. In many cases
the SimpleUniverse object can be used as the root of the scene graph. Utility
class SimpleUniverse is a Java 3D convenience class for simplification of scene
graphs. It contains all the objects necessary for a standard view of visual objects.

The Java 3D coordinate system assumes that the x-axis has horizontal orientation,
the y-axis has vertical (gravitational) orientation, and the z-axis is directed to the
viewer. Usually, visual objects are located near the coordinate origin. The default
view plane also passes through the origin. In order to be able to see visual objects it
is necessary to move the viewer position along the positive direction of the z-axis.

Group node BG (BranchGroup object) is used for attachment of two children.
One child is a light (class Light and its subclasses) with Bounds object. The other
is used for defining a region where the light acts.

The second child of the branch group BG is a transform group TG, which performs an affine transform for its children. Typically, it can be a translation, rotation,
or scaling of the visual object. Next in the hierarchy of this branch is a Shape3D

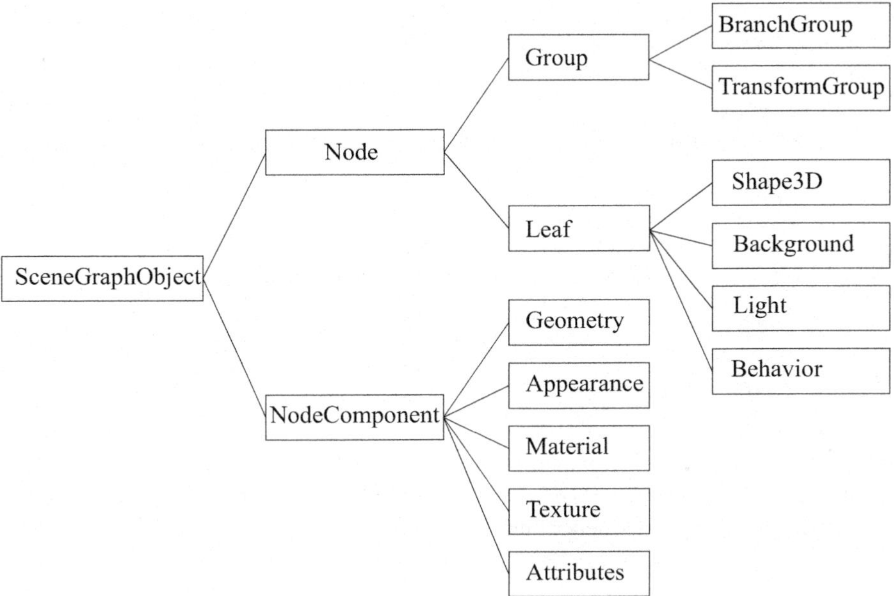

Fig. 26.2 Some Java 3D classes

object that represents the visual object and provides references to its geometry and appearance. The appearance has a reference to a `Material` object.

The hierarchy of some useful Java 3D classes is shown in Figure 26.2. It is worth noting that this class hierarchy is not full. Only classes that will be used further are included. Let us briefly consider these classes in order to ease the understanding of a visualization fragment that will be described later.

26.3 Scene Graph Nodes

`Node` objects are used for creating the scene graph. There are two kinds of scene graph nodes: `Group` nodes and `Leaf` nodes. The group nodes serve for branching and operations on child nodes. The leaf nodes represent certain graphics entities.

26.3.1 Group Nodes

Both classes `BrachGroup` and `TransformGroup` extend class `Group`. Object `BranchGroup` serves as a pointer to the root of a scene graph branch. The `TransformGroup` node specifies a three-dimensional transformation using a `Transform3D` object that can translate, rotate, and scale all its children. The specified transformation must be affine. The effect of transformations in the scene graph is cumulative. The concatenation of the transformations in a direct path produces a composite model transformation that transforms points from a leaf's local coordinates into world coordinates.

To add a child node to a `Group` node it is possible to employ the following method:

```
void addChild(Node child)
```

where `child` is the child to add to this node's list of children.

26.3.2 Leaf Nodes

Class `Leaf` is an abstract subclass of `Node`. Leaf nodes represent geometric objects, lights, and behavior. They cannot have children. Usually, leaf nodes contain references to node component objects.

The `Shape3D` leaf node specifies geometric objects. It can contain one or more `Geometry` component object and a single `Appearance` component object. The geometry objects define the shape node's geometric data. The appearance object specifies the object's appearance attributes, including color, material, texture, etc. Typical statement for constructing shape object is:

```
new Shape3D(geometry, appearance);
```

The `Background` leaf defines a background color or a background image that is used for filling the window at the beginning of each new frame. If there is no `Background` node, then the window is filled by black color.

The `Light` leaf node is an abstract class that has subclasses for different types of light. The light in a scene may come from several light sources. There are four types of lights in Java 3D: ambient, directional, point, and spot.

An ambient light comes from no particular direction or source. It imitates background light resulting from multiple reflections in the real world. A directional light has a specific direction. Its rays are parallel, so it is possible to think of the directional light as being located at infinity. A point light comes from a specific position. A spot light is similar to a point light, but it acts inside a specified cone. Java 3D supports an arbitrary number of lights.

Light color is defined in terms of the red, green, and blue components. The three color components represent the amount of light emitted by the source. Each of the three colors is represented by a floating-point value with a range from 0.0 to 1.0. Black color is defined as (0.0, 0.0, 0.0). A combination (1.0, 1.0, 1.0) creates a white light with maximum brightness. In a scene with multiple lights, the effect of the light on the object is the sum of the lights. If the sum of any of the color values is greater than 1.0, the color value is clipped to 1.0.

Absorption and reflection of light from an object's surface are determined by the object's color and material properties. The Java 3D lighting model specifies the material properties for four color types: emitted color, ambient color, diffuse color, and specular color. The material properties are specified in the `Material` class.

To economize the amount of computations for rendering a scene, it is reasonable to limit the influence of lighting to a region that is determined by the influencing bounds of a `Bounds` object. Lighting should be explicitly enabled with the `setEnable` method.

The `Behavior` leaf node provides a framework for adding user-defined interactions with objects in the scene graph. It is possible to create custom interaction behavior using the basic `Behavior` class. However, it is usually easier to employ predefined Java 3D utility classes, especially, `MouseBehavior` classes. These classes operate on an associated `TransformGroup` node.

`MouseRotate` is a Java 3D behavior object that allows the user to control the rotation of an object by dragging the mouse with the left button down. Object `MouseTranslate` lets the user perform the translation along x and y of an object using the mouse drag motion with the right button pressed. `MouseZoom` is a behavior object that controls the z-axis translation of an object via the mouse drag motion with the middle mouse button or with the left mouse button chorded with the Alt key.

26.4 Node Components

`NodeComponent` is a common superclass for all scene graph node component objects such as: `Geometry`, `Appearance`, `Material`, `Texture`, etc.

26.4.1 Geometry

Three-dimensional objects are typically modeled as a combination of surface patches, line segments, and points. Java 3D provides direct support for arrays of points, lines, triangles, and quadrilaterals. Curved surface patches should be represented as a set of simple polygons.

An abstract class `Geometry` specifies the geometry component information required by a `Shape3D` node. Geometry objects describe both the geometry and topology of the `Shape3D` nodes.

The `Geometry` class has a subclass `GeometryArray`, which in turn has subclasses:

`PointArray` defines a set of points;

`LineArray` defines a set of straight-line segments;

`TriangleArray` defines a set of triangle patches; and

`QuadArray` defines a set of quadrilateral patches.

The `GeometryArray` object contains separate arrays of vertex coordinates, colors, normals, and texture coordinates that describe point, line, or polygon geometry. Vertex data may be supplied to this geometry array in one of two ways: by copying the data into the array, or by passing a reference to the data.

The default mode is "By Copying" when specified data is stored inside Java 3D. In the "By Reference" mode, references are set to user supplied arrays, and the data is not copied inside Java 3D.

A constructor for a geometry object has the following appearance:

```
GeometryArray(int vertexCount, int vertexFormat)
```

where parameters are:

$vertexCount$ – the number of vertex elements in this `GeometryArray`;

$vertexFormat$ – a mask indicating which components are present in each vertex.

The second parameter contains one or more flags that can be combined together using a bitwise OR operator. The flags signal the presence of a particular type of vertex data:

`COORDINATES` – vertex positions (always present);

`NORMALS` – vertex normals;

`COLOR_3` or `COLOR_4` – vertex colors (without or with alpha channel);

`TEXTURE_COORDINATE_2` – vertex two-dimensional texture coordinates;

`BY_REFERENCE` – data is passed by reference.

Below is an example of the construction of a triangle array with specification of vertex coordinates:

```
ta = new TriangleArray(nVertices, TriangleArray.COORDINATES);
ta.setCoordinates(0, coordinates);
```

The first statement constructs a triangle array `ta`, which should have `nVertices` vertices. Each vertex contains just coordinates. The second statement specifies the coordinates of all vertices. Data from `double` or `float` array `coordinates` by default is copied inside Java 3D.

The following example illustrates initialization of a triangle array with specification of vertex coordinates, normals and two-dimensional texture coordinates:

```
ta = new TriangleArray(nVertices,
        TriangleArray.COORDINATES | TriangleArray.NORMALS |
        TriangleArray.TEXTURE_COORDINATE_2 |
        TriangleArray.BY_REFERENCE);
ta.setCoordRefFloat(coordinates);
ta.setNormalRefFloat(normals);
ta.setTexCoordRefFloat(0,texCoords);
```

Here, arrays `coordinates`, `normals` and `texCoords` have float precision. All arrays are passed by reference. Passing arrays by reference economizes on memory since an array copy is not created.

Creation of geometry arrays for points and lines is done with constructors `PointArray` and `LineArray`. Methods for specification of vertex data are the same as for the triangle array.

26.4.2 *Appearance and Attributes*

The `Appearance` object defines a rendering state that can be referenced by a `Shape` node. The appearance can include the following objects:

`ColoringAttributes` – attributes for color selection and shading. The defined color is used when lighting is not enabled and vertex colors are not defined. A shading model can be flat (constant polygon color) or shading can follow the Gouraud method (smooth color change for more realistic appearance).

`LineAttributes` – attributes for line definition, including the pattern, width, and antialiasing.

`PointAttributes` – attributes used to define points, including the size and antialiasing.

`PolygonAttributes` – attributes for defining polygons, including culling, rasterization mode (filled, lines, or points), constant offset, offset factor, and flipping of back-face normals.

`Material` – defines the appearance of an object under illumination, such as the ambient color, diffuse color, specular color, emissive color, and shininess.

`Texture` – defines the texture image and filtering parameters when texture mapping is enabled.

`TextureAttributes` – defines the attributes for texture mapping, such as the texture mode, texture transform, blend color, and perspective correction mode.

Specification of the correct appearance helps to create naturally looking visual objects in the three-dimensional space.

Problems

26.1. Explain the usefulness of the Java 3D scene graph for visualization of three-dimensional objects.

26.2. What is the difference between nodes and leafs in a Java 3D scene graph?

26.3. Suppose that we are going to visualize a mesh of three-dimensional finite elements showing element faces, edges and nodes. What Java 3D geometry objects should be used for such visualization?

Chapter 27
Visualizer

Abstract The visualization algorithm and visualizer class structure are discussed in this chapter. The main class `Jvis` and class `VisData` for storage of visualization parameters are presented.

27.1 Visualization Algorithm

Input data for the visualization consists of a set of nodes defined by spatial co-ordinates, a set of elements specified by nodal connectivities, and a set of result values. Primary results (displacements) are obtained at nodes after solution of the global equation system. Secondary results (stresses), expressed through derivatives of the primary results, have the best precision at reduced integration points inside elements.

As seen in the previous chapter, Java 3D^{TM} can render three-dimensional objects composed of simple polygons (triangle and quadrilaterals), straight-line segments, and points. An appearance can be specified for polygons including material and textures. To include three-dimensional shading effects it is necessary to specify normals at polygon vertices.

We can understand that a difficulty in visualization is caused by differences in geometric properties of element faces of a finite element model and polygons, which are used for rendering. While element faces are curved biquadratic surfaces with curved quadratic edges, polygons are flat triangles. Results are also interpolated quadratically over element faces. In order to have a good appearance of a finite element model in its visualization, it is necessary to subdivide element faces into triangles. The density of triangular subdivision should depend upon surface curvature and on the results gradient.

Taking into account the visualization aspects of Java 3D it is possible to propose the following visualization algorithm.

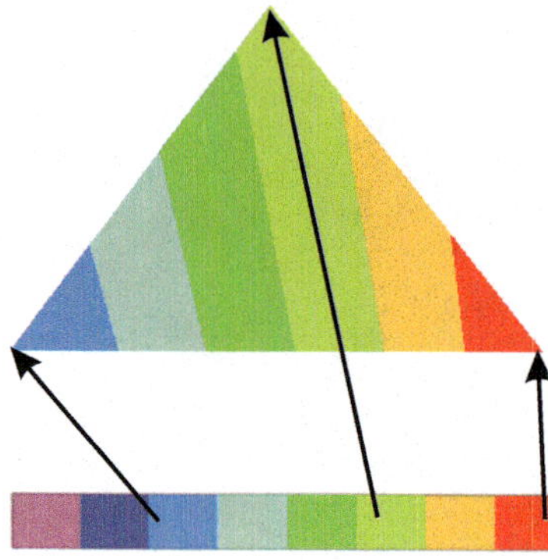

Fig. 27.1 Generating contours using interpolation of a one-dimensional texture

1. Obtain a continuous field of results by extrapolation from integration points inside elements to element nodes with subsequent averaging (omit this step for primary results defined at nodal points).
2. Select surface faces, surface element edges, and surface nodes of the finite element model (omit this step for two-dimensional problems).
3. Subdivide curved faces into flat triangles on the basis of face curvature and gradient of results. Generate normals to finite element faces at triangular vertices.
4. Create contour pictures inside triangles by specifying texture coordinates at vertices.
5. Represent element edges as a set of straight-line segments.
6. Submit arrays of triangles, lines, and points to Java 3D for rendering.

The contour-creation step of the visualization algorithm is performed with the use of texture interpolation available in Java 3D API. Instead of explicit contour drawing we prepare one-dimensional color texture and specify texture coordinates at triangle vertices. Color contours inside triangular polygons are generated using interpolation of one-dimensional texture (gradation strip) as illustrated in Figure 27.1.

27.2 Surface of the Finite Element Model

To visualize a finite element model or results at the surface of the model, it is necessary to select finite element faces that belong to the surface of the finite element model (or part of the finite element model). A topology of a finite element model is described by element connectivities. Connectivities for each finite element are global (model) node numbers listed in order of local (element) node numbers. The twenty-node brick-type element has six element faces. Connectivity numbers for each face can be easily extracted from an element connectivity array.

The surface of the finite element model is created from external element faces. External faces are mentioned in the model connectivity array only once, while inner faces are mentioned exactly twice. Checking that faces from different finite elements are the same can be done by comparison of the face connectivity numbers.

27.3 Subdivision of Quadratic Surfaces

Subdivision of quadratic element faces is necessary for producing smooth representation of faces and edges during three-dimensional rendering. The subdivision depends on two factors: curvature of the surface and range of result function over the surface. We want to create a compatible mesh of triangles, and it is desirable to generate mesh locally, i.e., considering one element face at a time. The latter constraint means that the number of subdivisions along a face edge should be the same in element faces sharing the edge. Thus, the number of edge subdivisions should be based only on nodal values of coordinates and results for this edge.

Curvature-based edge subdivision depends upon the edge curvature radius. On the basis of locations of three nodes defining an edge, it is possible to determine the edge curvature. Multiplication of the curvature by the empirical curvature coefficient provides the number of subdivisions due to curvature.

Results-based edge subdivision is used in addition to the curvature-based subdivision when contour drawing is performed. Such subdivision should limit the number of color intervals inside one triangle. Thus, the number of subdivisions at the face edge should be proportional to the ranges of function between two neighboring nodes.

When both curvature-based and results-based subdivision are used the final number of subdivisions at the edge is selected as the maximum of two numbers.

After determining the number of subdivisions for all edges, subdivision of element faces is performed in two steps. First, vertex locations for subdivision triangles are generated. Next, a triangular mesh is created using Delaunay triangulation.

27.4 Class Structure of the Visualizer

Class structure of the visualizer is presented in Figure 27.2. Class `Jvis` contains the main method, which creates object `J3dScene`. Class `J3dScene` describes a scene graph for visualization of the finite element model with the possibility of results contours at its surface. Class `MouseAndLight` includes methods for setting background and lights, and organizing mouse behavior for interactive transformations of the visualized finite element model. Simple standard cases of mouse behavior are used – rotation, translation and zooming of the finite element model. Class `ContourTexture` generates texture with a color strip used in drawing results contours.

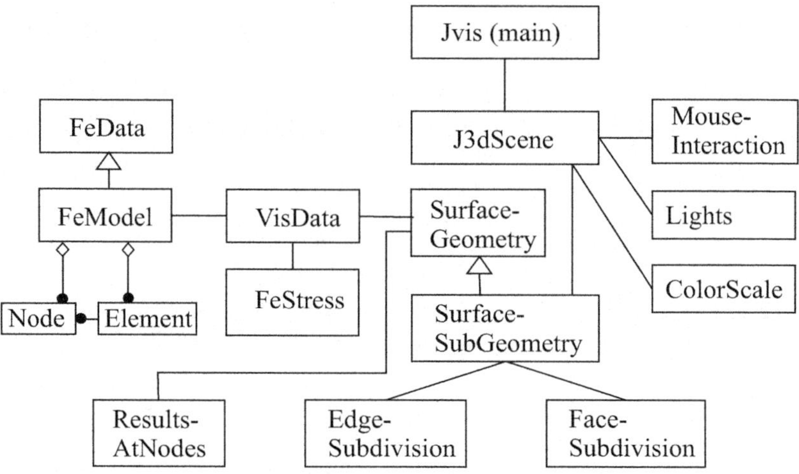

Fig. 27.2 Class structure of the visualizer

Class `J3dScene` uses subdivided faces and edges (`SurfaceSubGeometry`) in geometry attributes of the scene graph. Class `SurfaceSubGeometry` extends class `SurfaceGeometry`, which creates a list of element faces and edges for the surface of the finite element model. Classes `EdgeSubdivision` and `FaceSubdivision` perform subdivision of element edges into line segments and subdivision of element faces into triangles. Classes `VisData`, `FeModel` and `FeStress` are responsible for the input of visualization data, finite element model data and results data. Class `ResultsAtNodes` extrapolates stress values from reduced integration points to nodes.

Finite element visualizer `Jvis` generates only a three-dimensional picture of the finite element model with results contours. Interaction is restricted to rotation, translation and zooming of the model using a mouse. We intentionally do not implement the usual elements of the GUI like menus and toolbars. Implementation of such interface elements is not the subject of this book. In addition, their implementation requires a large amount of the Java^{TM} code that does not fit the book space.

27.5 Visualizer Class

The main class of the visualizer is presented below.

```
1 package fea;
2
3 import visual.*;
4 import util.*;
5
6 import java.applet.Applet;
7 import com.sun.j3d.utils.applet.MainFrame;
```

```
 8
 9 // Main class of the visualizer
10 public class Jvis extends Applet {
11
12     public static FeScanner RD = null;
13
14     public static void main(String[] args) {
15
16         if (args.length == 0) {
17             System.out.println(
18                     "Usage: java fea.Jvis FileIn \n");
19             return;
20         }
21         FE.main = FE.JVIS;
22
23         RD = new FeScanner(args[0]);
24         System.out.println("fea.Jvis: Visualization." +
25                 " Data file: " + args[0]);
26
27         new MainFrame(new Jvis(), 800, 600);
28     }
29
30     public Jvis() {
31
32         VisData.readData(RD);
33
34         new J3dScene(this);
35
36     }
37
38 }
```

Class `Jvis` belongs to package `fea` where all the main classes of the finite element program are located. The class extends `Applet` and uses utility class `MainFrame` for simpler programming of a window displaying graphics.

Lines 16–20 check the existence of the program argument, which should be the name of the data file. If the argument is absent a diagnostic message is printed and the program returns. Line 21 sets the static variable `main` in class `FE` to value `VIS`, thus signaling that the finite element model will be used in the visualizer. The scanner for reading finite element data `RD` is created in line 23. A window of size 800 by 600 pixels with graphical content is initialized in line 27 by passing object `Jvis` to utility class `MainFrame`.

The constructor of class `Jvis` contains data input by method `readData` and construction of object `J3dScene`, which creates the Java 3D scene graph for visualization of a finite element model and results.

27.6 Input Data

27.6.1 Input Data File

In order to run the visualizer it is necessary to prepare an ASCII file with input data. Input data for visualization includes:

`meshFile = <text>` – name of the file containing a finite element mesh;

`resultFile = <text>` – name of the results file (if not specified then results visualization is not done);

`parm = <text>` – results parameter that should be visualized;

`showEdges = Y/N` – draw element edges: Y – yes, N – no;

`showNodes = N/Y` – draw nodes: N – no, Y – yes;

`nDivMin = <number>` – minimum number of element edge subdivisions (default value is 1);

`nDivMax = <number>` – maximum number of element edge subdivisions (default value is 16);

`fMin = <number>` – minimum value of results parameter (if not specified then computed);

`fMax = <number>` – maximum value of results parameter (if not specified then computed);

`nContours = <number>` – number of contours used for results visualization (2..256, default value is 256);

`deformScale = <number>` – if not zero, show deformed shape of the finite element model. Nodal displacements are scaled such that the ratio of a maximum scaled displacement to the largest model size is equal to `deformScale` (default value is 0);

`end` – signal of end of data.

The following results parameters can be visualized:

`Ux, Uy, Uz` – one of components of a displacement vector along coordinate axes x, y or z;

`Sx, Sy, Sz` – normal stresses;

`Sxy, Syz, Szx` – shear stresses;

`S1, S2, S3` – principal stresses;

`Si` – equivalent stress;

`S13` – difference between first and third principal stresses;

`none` – do not visualize results as contours (default value).

The minimum information in the data input file for visualizer is the name of the file with a finite element mesh. In this case the finite element mesh will be visualized with edges but without nodes. If the name of a results file is specified in addition to the mesh file then the visualizer will show the finite element model with color results contours depending on the requested results parameter. The finite element model can be drawn with exaggerated deformation by applying scaled displacements.

27.6.2 Class for Data Input

Visualization data is stored in class `VisData`.

```java
 1 package visual;
 2
 3 import model.*;
 4 import util.*;
 5 import elem.Element;
 6
 7 import javax.vecmath.*;
 8
 9 public class VisData {
10
11     static FeModel fem;
12     // Global vectot of nodal displacements
13     static double displ[];
14
15     // Input data names
16     enum vars {
17         meshfile, resultfile, parm, showedges, shownodes,
18         ndivmin, ndivmax, fmin, fmax, ncontours, deformscale,
19         end
20     }
21
22     // Parameters that can be visualized: displacements,
23     // stresses, principal stresses and equivalent stress
24     enum parms {
25         ux, uy, uz, sx, sy, sz, sxy, syz, szx,
26         s1, s2, s3, si, s13, none
27     }
28
29     static String meshFile = null, resultFile = null;
30     static parms parm = parms.none;
31     static boolean showEdges = true, showNodes = false,
32         showDeformShape = false, drawContours = false;
33     static double deformScale = 0.0;
34     static int nDivMin = 2, nDivMax = 16;
35     static double fMin = 0, fMax = 0;
36     static int nContours = 256;
37
38     static float offset = 500.0f;
39     static float offsetFactor = 1.0f;
40
41     static Color3f bgColor       = new Color3f(1.0f, 1.0f, 1.0f);
42     static Color3f modelColor    = new Color3f(0.5f, 0.5f, 0.9f);
```

```
43        static Color3f surTexColor = new Color3f(0.8f, 0.8f, 0.8f);
44        static Color3f edgeColor   = new Color3f(0.2f, 0.2f, 0.2f);
45        static Color3f nodeColor   = new Color3f(0.2f, 0.2f, 0.2f);
46
47        // Size of the color gradation strip
48        static int textureSize = 256;
49        // Coefficient for curvature: n = 1 + C*ro
50        static double Csub = 15;
51        // Coefficient for contours: n = 1 + F*abs(df)/deltaf
52        static double Fsub = 20;
53
54        public static void readData(FeScanner RD) {
55
56            readDataFile(RD);
57
58            FeScanner fes = new FeScanner(meshFile);
59            fem = new FeModel(fes, null);
60            Element.fem = fem;
61            fem.readData();
62
63            if (resultFile != null) {
64                displ = new double[fem.nNod*fem.nDf];
65                FeStress stress = new FeStress(fem);
66                stress.readResults(resultFile, displ);
67                if (deformScale > 0) showDeformShape = true;
68                drawContours = VisData.parm != VisData.parms.none;
69
70            }
71        }
```

The names of data items, which are described in the previous subsection are places in enum object vars (lines 16–19). Object parms contains the names of results parameters for visualization.

Variables for storing data values are declared in lines 29–36. Most variables have default values. Lines 38–39 specify the offset bias and offset factor, which are used for shifting pixels during rendering polygons. This helps proper drawing of lines and points on the polygon surface.

Statements 41–45 define the following colors:

bgColor – background color of the drawing canvas;

modelColor – element faces color when the finite element model is visualized;

surTexColor – color of element faces, which is modulated by texture used for results contours;

edgeColor – color for drawing element edges;

nodeColor – color for drawing node points.

The size of the texture containing the gradation strip for contouring results is specified in line 48. Lines 50 and 52 define empirical coefficients used for calculating the number of subdivisions for element edges due to their curvature and range of results. Method `readData` reads the visualization data file (line 56), a file containing a finite element mesh (lines 59–61) and the results file with displacements at nodes and stresses at reduced integration points (line 66).

Method `readDataFile` inputs the visualization data placed in the visualization file.

```
73      static void readDataFile(FeScanner RD) {
74
75          vars name = null;
76
77          while (RD.hasNext()) {
78
79              String varName = RD.next();
80              String varNameLower = varName.toLowerCase();
81              if (varName.equals("#")) {
82                  RD.nextLine();    continue;
83              }
84              try {
85                  name = vars.valueOf(varNameLower);
86              } catch (Exception e) {
87                  UTIL.errorMsg(
88                      "Variable name is not found: " + varName);
89              }
90
91              switch (name) {
92
93              case meshfile:
94                  meshFile = RD.next();
95                  break;
96              case resultfile:
97                  resultFile = RD.next();
98                  break;
99              case parm:
100                     try {
101                         varName = RD.next();
102                         parm = parms.valueOf(varName.toLowerCase());
103                     } catch (Exception e) { UTIL.errorMsg(
104                         "No such result parameter: " + varName); }
105                 break;
106             case showedges:
107                 showEdges = RD.next().equalsIgnoreCase("y");
108                 break;
109             case shownodes:
110                 showNodes = RD.next().equalsIgnoreCase("y");
111                 break;
112             case ndivmin:
113                 nDivMin = RD.readInt();
114                 break;
115             case ndivmax:
116                 nDivMax = RD.readInt();
117                 break;
118             case fmin:
119                 fMin = RD.readDouble();
120                 break;
```

```
121                case fmax:
122                     fMax = RD.readDouble();
123                     break;
124                case ncontours:
125                     nContours = RD.readInt();
126                     break;
127                case deformscale:
128                     deformScale = RD.readDouble();
129                     break;
130                case end:
131                     return;
132                }
133            }
134        }
135
136 }
```

Visualization data is read inside the `while` loop in lines 77–133. This loop continues until there are input items in the scanner RD or until data item end is met. The text data item is read to string and transformed to lower case (lines 79–80). If the data item is not the comment sign then we try to find name corresponding to the string `varname` among enumerated `vars` (line 85). An error is generated if `varname` does not correspond to any predetermined value in `vars`. The `switch` construction in lines 91–132 contains case statements for reading the second part of an input statement. Lines 101–102 input the result parameter that should be visualized. The parameter name is sought among the enumerated `parms`. Case end just returns to the calling method.

Problems

27.1. A triangular polygon with vertices 1–3 is used for visualization of results available at vertices.

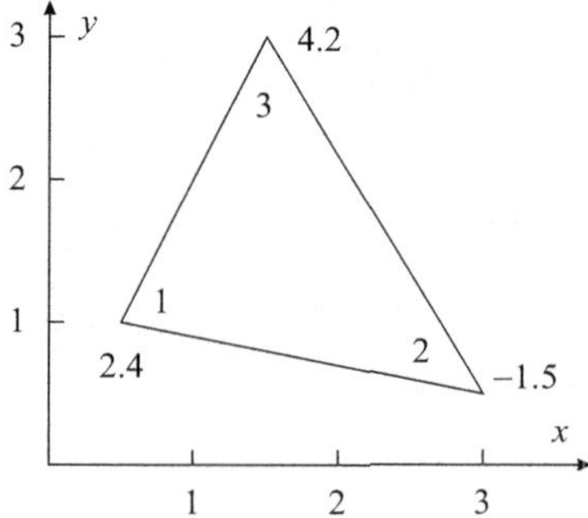

For the result values shown in the figure, determine isolines (contours) where the result values are equal to 1 and 3.

27.2. Develop an algorithm for finding the element faces located on the surface of a finite element model composed of three-dimensional elements. Express the algorithm in the form of pseudocode.

27.3. The curvature radius of an element edge is defined by the coordinates of its three nodes. Propose an algorithm for determining the edge curvature radius or any other approximate curvature parameter using the locations of three nodes.

Chapter 28
Visualization Scene Graph

Abstract Creation of a scene graph determining the hierarchy of visualized objects is discussed. Visual objects of the scene graph are finite element faces, edges and nodes. A Java 3D^{TM} scene graph for visualization of a finite element model and results is generated by class `J3dScene`.

28.1 Schematic of the Scene Graph

Visualization of the finite element model requires rendering its visual components – finite element faces, edges and nodes. A scene graph is used for organizing visual components (objects). The scene graph determining the hierarchy of the objects of the visualizer `Jvis` is shown in Figure 28.1. Visual objects of the scene graph are finite element faces, edges and nodes. They are represented by `Shape3D` objects. The geometry of the faces is described by array of triangles. Appearance references polygon attributes, material and texture. Texture is used for drawing results in the form of contours. Edges are represented as line segments defined in a line array. Appearance includes line attributes and color. Nodes are described by a point array with references to point attributes and color.

All `Shape3D` objects (faces, edges and nodes) are attached to a transform group `TG`, which defines the initial transformation for the finite element model. In our case, we need scaling of the finite element model that is necessary to place the model inside a window used for visualization.

The next (upper) transform group is used for providing interaction with the user. Mouse behavior, which includes rotation, translation and zooming is connected to this transform group. Lights, background and transform group are attached to branch group `BG`, which is connected to utility class `SimpleUniverse`. Lights, background and mouse behavior have references to bounds that define the influence of the corresponding scene graph nodes.

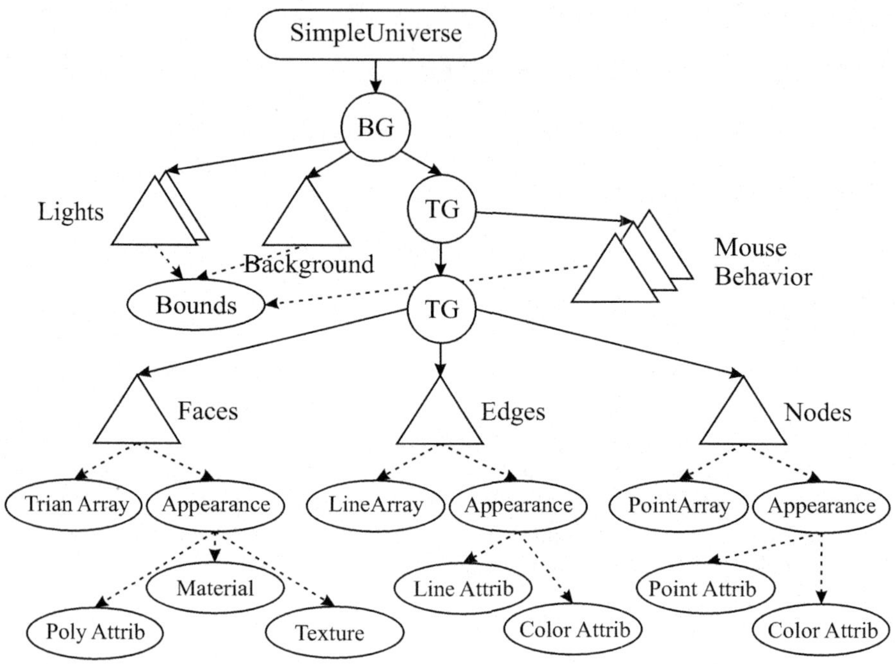

Fig. 28.1 Scene graph for visualization of finite element model and results

28.2 Implementation of the Scene Graph

A Java 3D scene graph for visualization of the finite element model and results is
created by class `J3dScene`. The source code of the class constructor and a method
for adding the shape of a visual object is given below.

```
 1 package visual;
 2
 3 import javax.media.j3d.*;
 4 import javax.vecmath.*;
 5 import java.awt.*;
 6 import java.applet.Applet;
 7 import com.sun.j3d.utils.universe.SimpleUniverse;
 8
 9 // Scene graph for visualization.
10 public class J3dScene {
11
12     private SurfaceSubGeometry subGeometry;
13
14     // Construct Java3D scene for visualization.
15     public J3dScene(Applet c) {
16
17         GraphicsConfiguration config =
18                 SimpleUniverse.getPreferredConfiguration();
19
```

```
20            Canvas3D canvas = new Canvas3D(config);
21            c.setLayout(new BorderLayout());
22            c.add("Center", canvas);
23
24            // Element subfaces, subedges and nodes
25            subGeometry = new SurfaceSubGeometry();
26
27            BranchGroup root = new BranchGroup();
28            Lights.setLights(root);
29            TransformGroup tg =
30                    MouseInteraction.setMouseBehavior();
31
32            // Add finite element model shape
33            tg = addModelShape(tg);
34
35            root.addChild(tg);
36            root.compile();
37
38            System.out.println(" Number of polygons = " +
39                    subGeometry.nVertices/3);
40            if (VisData.showDeformShape) System.out.printf(
41                    " Deformed shape: max displacement ="+
42                    " %4.2f max size\n", VisData.deformScale);
43            if (VisData.drawContours) {
44                System.out.printf(" Contours: %d colors" +
45                        " (Magenta-Blue-Cyan-Green-Yellow-Red)\n",
46                        VisData.nContours);
47                System.out.printf(" %s: Fmin = %10.4e, " +
48                        "Fmax = %10.4e\n", VisData.parm,
49                        subGeometry.fmin, subGeometry.fmax);
50            }
51
52            SimpleUniverse u = new SimpleUniverse(canvas);
53            u.getViewingPlatform().setNominalViewingTransform();
54            u.addBranchGraph(root);
55        }
56
57    // Add model shape to the Java 3D scene graph.
58    // tg - transform group of the scene graph.
59    // returns  transform group of the scene graph
60    TransformGroup addModelShape(TransformGroup tg) {
61
62        Transform3D t3d = new Transform3D();
63        t3d.setScale(subGeometry.getScale());
64        tg.setTransform(t3d);
65
66        // Element faces composed of triangular subfaces
67        tg.addChild(facesShape());
68
69        // Edges composed of line segments
70        if (VisData.showEdges) tg.addChild(edgesShape());
71
72        // Nodes located at the model surface
73        if (VisData.showNodes) tg.addChild(nodesShape());
```

```
74
75              return tg;
76        }
```

Statements 3–7 import Java 3D packages, an abstract windowing toolkit package, an applet package and the package for simple universe object. Constructor `J3dScene` obtains `Applet` object `c` as an argument. Class `Applet` was used in the main class of the visualizer for creating the main frame of the application. `Applet` extends class `Panel`, which is the simplest container class. We use container `c` for placement of a drawing canvas.

The graphics configuration is set as the preferred configuration of a simple universe object (lines 17–18). Lines 20–22 create `Canvas3D` object where visualization will be performed, and place the canvas at the center of panel `c`.

An object of the type `ModelSubGeometry` is constructed in line 25. During construction, the surface geomentry of the finite element model is created and subdivision of element surfaces into triangles and element edges into line segments is performed.

Object `root` of class `BranchGroup` is initialized by statement 27. Lights are set by static method `setLights` in line 28. Lines 29–30 set mouse behaviors that allow interactive transformations of a visualized finite element model. Shape objects of the finite element model are added to transform group `tg` in line 33. This transform group is added to the root transform group (line 35). Lines 38–50 print information about visualization: the number of polygons (triangles) used for rendering; the number of colors for drawing contours and the results parameter and its minimal and maximal values used for visualization. Class `SimpleUniverse` is used for setting an appropriate viewpoint and as a parent of the root branch group (lines 52–54).

Method `addModelShape` shown in lines 60–76 adds element faces, element edges and nodal points to the scene graph. Lines 56–58 set a scaling transform in such a way that the image of the finite element model has a reasonable size inside the drawing canvas.

Lines 67, 70, and 73 add subdivided element faces, element edges and nodal points to the transform group. The method returns the transform group `tg` in line 75.

28.3 Shape Objects

Methods `facesShape`, `edgesShape` and `nodesShape`, which produce shape objects for element faces, element edges and nodes are considered below. All these methods call methods of `subGeometry` to get geometry and then set appearance for visual objects.

Method `facesShape` returns a `Shape3D` object containing the geometry and appearance for element faces.

```
78        // Shape object for element faces
79        private Shape3D facesShape() {
80
81            TriangleArray faces = subGeometry.getModelTriangles();
82
83            Appearance facesApp = new Appearance();
84
85            // Polygon Attributes
86            PolygonAttributes pa = new PolygonAttributes();
87            pa.setCullFace(PolygonAttributes.CULL_BACK);
88            pa.setPolygonOffset(VisData.offset);
89            pa.setPolygonOffsetFactor(VisData.offsetFactor);
90            facesApp.setPolygonAttributes(pa);
91
92            // Material
93            Color3f darkColor = new Color3f(0.0f, 0.0f, 0.0f);
94            Color3f brightColor = new Color3f(0.9f, 0.9f, 0.9f);
95            Color3f surfaceColor = VisData.modelColor;
96            if (VisData.drawContours)
97                surfaceColor = VisData.surTexColor;
98            Material facesMat = new Material(surfaceColor,
99                    darkColor, surfaceColor, brightColor, 16.0f);
100           facesMat.setLightingEnable(true);
101           facesApp.setMaterial(facesMat);
102
103           if (VisData.drawContours) {
104               // Texture for creating contours
105               ColorScale scale = new ColorScale();
106               Texture2D texture = scale.getTexture();
107               facesApp.setTexture(texture);
108               TextureAttributes ta = new TextureAttributes();
109               ta.setTextureMode(TextureAttributes.MODULATE);
110               facesApp.setTextureAttributes(ta);
111           }
112
113           // Create Shape using Geometry and Appearance
114           return new Shape3D(faces, facesApp);
115       }
```

Method `getModelTriangles` provides element faces subdivided into triangles, which are placed in object `faces` (line 81). Polygon attributes for drawing triangles are set in lines 86–90. By default, polygons are rendered by filling the interior between the vertices. Statement 87 specifies that back faces of polygons are not drawn.

Lines 88–89 set an offset, which changes the depth values of all pixels generated by polygon rasterization. We need offset for proper visualization of element edges and nodes. If a polygon surface and a line segment have exactly the same z device coordinates then the line visibility is not guaranteed. The following two values are used to specify the offset. Offset bias is the constant polygon shift that is added to the final device coordinate z value. The offset factor is the factor to be multiplied by the slope of the polygon and then added to the final, the device coordinate z value of the polygon primitives.

Material defined in lines 93–101 is used for specifying the reflecting properties of the surface. Different materials are specified for mesh visualization and for results visualization. The following constructor for the material is employed:

```
Material(Color3f ambientColor, Color3f emissiveColor,
         Color3f diffuseColor, Color3f specularColor,
         float shininess)
```

Here, `ambientColor` is the ambient color reflected off the surface of the material; `emissiveColor` is the color of the light the material emits (like a light source); `diffuseColor` is the color of the material when illuminated (the light bounces off objects in random directions); `specularColor` is the specular color of the material (highlights); `shininess` is the material's shininess.

We use black color as the emissive color. The model color specified in class `Data` is used for both ambient and diffuse colors. If results are drawn by applying texture then an almost white color is used as the surface color (line 94). Specular effects are modeled by a color that is close to white. In order to have light effects line 100 enables lighting for the material. Line 101 sets material properties for the appearance of the faces.

If the results file name is specified in the input data and, consequently, we are going to draw results contours then a texture and its appearance are set in lines 105–110. A one-dimensional color gradation strip is created by method `getTexture` of the class `ColorScale` (line 105). It is set as a texture for the faces appearance in line 106. A texture mode in the texture attributes is set to MODULATE. Choosing this parameter means that polygon material colors will be modulated by the specified texture. Lights effects including shading and bright spots are produced for the polygon material and are visible through the texture since it is applied as modulation of material colors.

Line 114 creates a `Shape3D` object using faces geometry and appearance and returns it to the calling method.

A shape object for the edges of the finite element model surface is created by method `edgesShape`.

```
117     // Shape object for element edges
118     private Shape3D edgesShape() {
119
120         LineArray edges = subGeometry.getModelLines();
121
122         Appearance edgesApp = new Appearance();
123
124         LineAttributes la = new LineAttributes();
125         la.setLineAntialiasingEnable(true);
126         edgesApp.setLineAttributes(la);
127
128         ColoringAttributes ca = new ColoringAttributes();
129         ca.setColor(VisData.edgeColor);
130         edgesApp.setColoringAttributes(ca);
131
132         return new Shape3D(edges, edgesApp);
133     }
```

Geometry array `edges` is provided by method `getModelLines` as straight-line segments sufficient for visually smooth representation of curved element edges. Line attributes are determined in lines 124–126. Setting antialiasing to true value (line 125) produces a smooth appearance of lines on the screen. An edge line color is set in line 129 using the color defined in class `VisData`.

Finally, a shape object for surface nodes of the finite element model is made by method `nodesShape` given below.

```
135        // Shape object for nodes
136        private Shape3D nodesShape() {
137
138            PointArray nodes = subGeometry.getModelPoints();
139
140            Appearance nodesApp = new Appearance();
141
142            PointAttributes pa = new PointAttributes();
143            pa.setPointAntialiasingEnable(true);
144            pa.setPointSize(3.0f);
145            nodesApp.setPointAttributes(pa);
146
147            ColoringAttributes ca = new ColoringAttributes();
148            ca.setColor(VisData.nodeColor);
149            nodesApp.setColoringAttributes(ca);
150
151            return new Shape3D(nodes, nodesApp);
152        }
153
154 }
```

Line 138 creates a geometry object as a point array `nodes`. Point attributes contain antialiasing for drawing points (line 143) and point size in pixels (line 146). Line 148 sets the line color to that specified in class `VisData`. Object `Shape3D` for the nodes of the finite element model is returned in line 151.

Problems

28.1. Analyze the scene graph shown in Figure 28.1. Currently, lights are attached to branch group `BG`. What changes occur in visualization of the finite element model if lights are attached to the upper transform group `TG` through an additional branch group?

28.2. The scene graph of Figure 28.1 contains shape objects for element edges and nodes. Is it possible to refer to the same appearance for both edges and nodes? Are references to the same color attributes possible for edges and nodes (through their appearances)?

28.3. An antialiasing technique is used for smoothing lines and points in methods `edgesShape` and `nodesShape`. Explain how antialiasing is performed on a computer screen consisting of pixels.

Chapter 29
Surface Geometry

Abstract Since only the surface of a three-dimensional finite element model is visible it is useful to create geometry items describing a surface. An algorithm for determining surface faces is based on the fact that faces located on the surface are mentioned in the model connectivity array only once, while internal faces are referenced twice. Lists of element surface edges and surface nodes are formed using a list of element surface faces. Surface geometry is produced by methods of class `SurfaceGeometry`.

29.1 Creating Geometry of the Model Surface

When visualizing a three-dimensional finite element model it is useful to take into account that only the surface of the model is visible. While, in principle, it is possible to render all element faces, edges and nodes, such visualization requires too many resources. So, we are going to create the geometry of the model surface that should lead to efficient real-time visualization. As we already saw, a Java $3D^{TM}$ scene graph includes three shape object – faces, edges and nodes. Figgure 29.1 illustrates creation of element faces, element edges and nodes located on the surface of a finite element mesh. Element faces lying on a model surface are mentioned in element connectivities just once. This allows one to select surface faces. Surface edges and nodes are found using the surface faces.

Creation of the surface geometry is performed by class `SurfaceGeometry`. The source code of the class constructor is given below.

```
1 package visual;
2
3 import model.*;
4
5 import java.util.*;
6
7 // Geometry: surface faces, edges and nodes
8 class SurfaceGeometry {
```

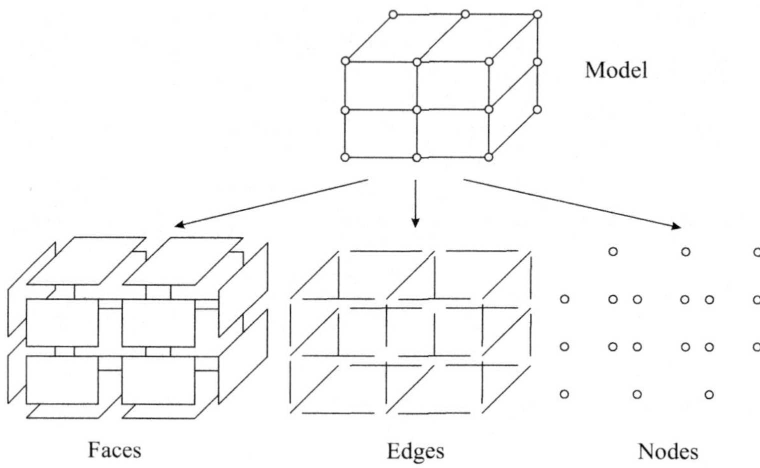

Fig. 29.1 Creation of element faces, element edges and nodes from a finite element mesh

```
 9
10        FeModel fem;
11
12        // Numbers of surface element faces, edges and nodes
13        int nFaces, nEdges, nsNodes;
14        // Surface element faces and edges, surface nodes
15        LinkedList listFaces;
16        LinkedList listEdges;
17        int sNodes[];
18
19        double fun[], fmin, fmax, deltaf;
20        private static double xyzmin[] = new double[3];
21        private static double xyzmax[] = new double[3];
22        static double sizeMax;
23
24        SurfaceGeometry() {
25
26            fem = VisData.fem;
27            listFaces = new LinkedList();
28            listEdges = new LinkedList();
29            sNodes = new int[fem.nNod];
30
31            // Create element faces located at the surface
32            createFaces();
33            nFaces = listFaces.size();
34
35            // Create element edges located at the surface
36            createEdges();
37            nEdges = listEdges.size();
38
39            // Create nodes located at the surface
40            createNodes();
41
```

```
42            if (VisData.drawContours) {
43                fun = new double[fem.nNod];
44                ResultAtNodes ran = new ResultAtNodes(this, fem);
45                ran.setParmAtNodes(VisData.parm, VisData.displ);
46            }
47
48            modifyNodeCoordinates();
49        }
```

Line 10 declares finite element model object fem, which contains data on the finite element mesh and will be used to produce the surface geometry. The numbers of surface faces and surface edges is not known in advance. Because of this we employed linked lists listFaces and listEdges (lines 15–16) for element faces and element edges. Array sNodes is used to register surface nodes (line 17). Variables nFaces, nEdges and nsNodes will contain numbers of surface element faces, edges and nodes.

Constructor SurfaceGeometry initializes objects for storing information on the surface geometry. Calls to methods createFaces, createEdges and createNodes in lines 32, 36 and 40 create surface faces, edges and nodes correspondingly. When contour drawing is requested then line 43 allocates memory for array fun where the requested result parameter will be stored. Values of this parameter are set in array fun by method setParmAtNodes in line 45. Statement 48 modifies the nodal coordinates of the finite element model by taking into account exaggerated nodal displacements (if a deformed shape is requested) and centering the model.

29.2 Surface Faces

An algorithm for determining surface faces is based on the following observation: surface faces are mentioned in the model connectivity array only once, while internal faces are mentioned exactly twice. The algorithm of creating surface faces can be described by the following pseudocode.

Surface faces
Initialize list of faces
for each element in model
    for each element face in element
        for each face in list
            if element face **equal** face in list
            then remove face from list
        end for
        if equal face not found
        then add element face to list
    end for
end for

The above algorithm of creating surface element faces is implemented in methods `createFaces` and `equalFaces`.

```
51      // Create linked list listFaces containing element faces
52      // located on the model surface. 2D case: element = face
53      void createFaces() {
54
55          if (fem.nDim == 3) {   // 3D mesh
56              for (int iel = 0; iel < fem.nEl; iel++) {
57                  int elemFaces[][]
58                          = fem.elems[iel].getElemFaces();
59                  for (int[] elemFace : elemFaces) {
60                      int nNodes = elemFace.length;
61                      int[] faceNodes = new int[nNodes];
62                      for (int i = 0; i < nNodes; i++) {
63                          faceNodes[i]
64                              = fem.elems[iel].ind[elemFace[i]];
65                      }
66                      // Zero area degenerated 8-node face
67                      if (nNodes == 8 &&
68                              (faceNodes[3] == faceNodes[7] ||
69                               faceNodes[1] == faceNodes[5]))
70                          continue;
71                      ListIterator f = listFaces.listIterator(0);
72                      boolean faceFound = false;
73                      while (f.hasNext()) {
74                          int[] faceNodesA = (int[]) f.next();
75                          if (equalFaces(faceNodes,faceNodesA)) {
76                              f.remove();
77                              faceFound = true;
78                              break;
79                          }
80                      }
81                      if (!faceFound) f.add(faceNodes);
82                  }
83              }
84          }
85          else {  // 2D - faces = elements
86              ListIterator f = listFaces.listIterator(0);
87              for (int iel = 0; iel < fem.nEl; iel++) {
88                  f.add(fem.elems[iel].ind);
89              }
90          }
91      }
92
93      // Compare two element faces.
94      // Surface has 8 or 4 nodes, corners are compared.
95      // f1 - first face connectivities.
96      // f2 - second face connectivities.
97      // returns  true if faces are same.
98      boolean equalFaces(int[] f1, int[] f2) {
99
100         // Quadratic elements or linear elements
101         int step = (f1.length > 4) ? 2 : 1;
```

```
102
103            for (int j = 0; j < f1.length; j += step) {
104                int n1 = f1[j];
105                boolean nodeFound = false;
106                for (int i = 0; i < f2.length; i += step) {
107                    if (f2[i] == n1) {
108                        nodeFound = true;
109                        break;
110                    }
111                }
112                if (!nodeFound) return false;
113            }
114            return true;
115        }
```

Method `createFaces` contains two cases of face creation. The first one considers three-dimensional models and the second two-dimensional models. Line 55 checks the number of dimensions variable nDim. If it is equal to 3 then lines 56–83 perform surface face creation for three-dimensional problems. Array elemFaces containing information on local numbers of element faces is provided by the element method getElemFaces in lines 57–58. In a loop over element faces (line 59), we consider each element face and compare it with faces in list listFaces.

The indices of face nodes are extracted from element connectivities in a loop of lines 62–65. Lines 67–69 check the element face indices for the case of a quadratic face with eight nodes. If opposite midside nodes have the same node numbers the face is degenerated into a line with zero area and is omitted from further consideration.

List iterator f is initialized in line 71. It is used to perform operations with entries of list listFaces. In a while loop started in line 73, an array of face nodes faceNodesA is obtained from the list (line 74) and compared with nodes of the current element face faceNodes in line 75. If method equalFaces returns a true value, which means equal node numbers for two element faces then the face is removed from the list in line 76 as the internal face of the finite element model. However, if no face equal to the current element face is found then the element face is placed in the list as a candidate for a surface face (line 81).

In two-dimensional problems, element faces are elements; all of them lie on the model surface. Because of this, a list of surface faces for two-dimensional models is created by simple copying of element connectivities to list listFaces.

Method equalFaces compares the connectivities of two quadrilateral faces f1 and f2 in a double loop of lines 103–113. If the current node of the first face does not match any node of the second face then the faces are different and the method returns false. If the double loop finishes then the faces are considered coincident and the method returns a true value.

29.3 Surface Edges and Nodes

An algorithm for determining surface edges is similar to the algorithm for surface
faces. We take the list of surface faces, extract edges (which automatically lie on the
surface) and place them in a list of edges avoiding repeatedly mentioned edges. The
following pseudocode illustrates the algorithm for creating surface edges.

> **Surface edges**
> Initialize list of edges
> **for each** face in list of faces
>   **for each** face edge in face
>   **for each** edge in list of edges
>   **if** face edge **equal** edge in list of edges
>   **then** break loop
>   **end for**
>   **if** equal edge not found
>   **then** add face edge to list of edges
>   **end for**
> **end for**

The list of surface edges is created by method `createEdges`.

```
117      // Create linked list listEdges containing element edges
118      // located on the model surface
119      void createEdges() {
120
121          for (int iFace = 0; iFace < nFaces; iFace++) {
122
123              int faceNodes[] = (int[]) listFaces.get(iFace);
124              int nFaceNodes = faceNodes.length;
125              int step = (nFaceNodes > 4) ? 2 : 1;
126
127              for (int inod=0; inod < nFaceNodes; inod += step) {
128                  int[] edgeNodes = new int[step + 1];
129                  for (int i = inod, k = 0; i <= inod+step;
130                       i++, k++)
131                      edgeNodes[k] = faceNodes[i%nFaceNodes];
132
133                  ListIterator ea = listEdges.listIterator(0);
134                  boolean edgeFound = false;
135                  while (ea.hasNext()) {
136                      int[] edgeNodesA = (int[]) ea.next();
137                      if (equalEdges(edgeNodes, edgeNodesA)) {
138                          edgeFound = true;
139                          break;
140                      }
141                  }
142                  if (!edgeFound) ea.add(edgeNodes);
143              }
144          }
145      }
```

```
146
147        // Compare two element edges.
148        // e1 - first edge connectivities.
149        // e2 - second edge connectivities.
150        // returns  true if edges have same node numbers at ends
151        boolean equalEdges(int[] e1, int[] e2) {
152
153            int len = e1.length - 1;
154            return (e1[0] == e2[0] && e1[len] == e2[len]) ||
155                   (e1[0] == e2[len] && e1[len] == e2[0]);
156        }
157
158        // Fill out array of surface nodes sNodes (0/1).
159        void createNodes() {
160
161            for (int i = 0; i < sNodes.length; i++) sNodes[i] = 0;
162
163            ListIterator e = listEdges.listIterator();
164
165            for (int iEdge = 0; iEdge < nEdges; iEdge++) {
166                int edgeNodes[] = (int[]) e.next();
167                int nEdgeNodes = edgeNodes.length;
168                for (int i = 0; i < nEdgeNodes; i++)
169                    sNodes[edgeNodes[i] - 1] = 1;
170            }
171            nsNodes = 0;
172            for (int sNode : sNodes)
173                if (sNode > 0) nsNodes++;
174        }
```

The loop over surface element faces starts in line 121. Face node numbers and the number of face nodes are obtained from list `listFaces` in lines 123–124. We are going to compare edges using numbers of their end nodes. Since we take into account linear and quadratic elements, an edge can contain two or three nodes. Variable `step` (line 125) equals 2 for edges with three nodes, otherwise 1. The loop over element edges starts with statement 127. Node numbers for current element edge `edgeNodes` are selected in lines 129–131. Comparison of the current edge with edges stored in list `listEdges` is performed in the loop of lines 135–141. If an equal edge is found in the list (method `equalEdges`) then variable `edgeFound` becomes true and the loop is broken. After the loop end, we check if an equal edge was found. If not, the current face edge is added to the list `listEdges`. Method `equalEdges` compares two edges e1 and e2. If the end node numbers of edges are equal to each other then the edges are considered equal and the method returns a true value.

Method `createNodes` registers surface nodes in array `sNodes`, which has a length equal to the total number of nodes in the finite element model. The presence of a surface node with number i is marked by placement of one at position $i - 1$ in the array. All other entries of array `sNodes` are zeros. Statement 161 initializes the array `sNodes` with zeros. The loop over surface element edges in list `listEdges` starts in line 163. All edge nodes are registered in array sNodes by ones. Lines 171–173 determine the number of surface nodes nsNodes.

29.4 Modification of Nodal Coordinates

Method modifyNodeCoordinates alters an array of nodal coordinates of a finite element model. If drawing of the deformed finite element model is specified, then scaled nodal displacements are added to the nodal coordinates. Another modification centers the finite element model at the coordinate origin. Method setBoundingBox determines the maximum and minimum values of nodal coordinates. Method getScale estimates a scale that allows a reasonable screen size of the model image.

```
176     // Add scaled displacements to nodal coordinates and
177     //  center finite element mesh
178     void modifyNodeCoordinates() {
179
180         // Deformed shape: add scaled displacements
181         // to nodal coordinates
182         if (VisData.showDeformShape) {
183             setBoundingBox();
184             double displMax = 0;
185             for (int i = 0; i < fem.nNod; i++) {
186                 double d = 0;
187                 for (int j = 0; j < fem.nDim; j++) {
188                     double s =  VisData.displ[i*fem.nDim+j];
189                     d += s*s;
190                 }
191                 displMax = Math.max(d, displMax);
192             }
193             displMax = Math.sqrt(displMax);
194             // Scale for visualization of deformed shape
195             double scaleD =
196                     sizeMax*VisData.deformScale/displMax;
197             for (int i = 0; i < fem.nNod; i++) {
198                 for (int j = 0; j < fem.nDim; j++)
199                     fem.setNodeCoord(i, j,
200                         fem.getNodeCoord(i, j) +
201                         scaleD*VisData.displ[i*fem.nDim+j]);
202             }
203         }
204
205         setBoundingBox();
206         // Translate JFEM model to have the bounding
207         //  box center at (0,0,0).
208         double xyzC[] = new double[3];
209         for (int j = 0; j < 3; j++)
210             xyzC[j] = 0.5*(xyzmin[j] + xyzmax[j]);
211         for (int i = 0; i < fem.nNod; i++)
212             for (int j = 0; j < fem.nDim; j++)
```

```
213                     fem.setNodeCoord(i, j,
214                         fem.getNodeCoord(i, j) - xyzC[j]);
215         }
216
217         // Set min-max values of xyz coordinates of JFEM model
218         // xyzmin[] and xyzmax[].
219         void setBoundingBox() {
220
221             for (int j = 0; j < fem.nDim; j++) {
222                 xyzmin[j] = fem.getNodeCoord(0, j);
223                 xyzmax[j] = fem.getNodeCoord(0, j);
224             }
225             for (int i = 1; i < fem.nNod; i++) {
226                 if (sNodes[i] >= 0) {
227                     for (int j = 0; j < fem.nDim; j++) {
228                         double c = fem.getNodeCoord(i, j);
229                         xyzmin[j] = Math.min(xyzmin[j], c);
230                         xyzmax[j] = Math.max(xyzmax[j], c);
231                     }
232                 }
233             }
234             if (fem.nDim == 2) {
235                 xyzmin[2] = -0.01;
236                 xyzmax[2] =  0.01;
237             }
238             sizeMax = 0;
239             for (int i = 0; i < 3; i++) {
240                 double s = xyzmax[i] - xyzmin[i];
241                 sizeMax = Math.max(s, sizeMax);
242             }
243         }
244
245         // Compute scale for the finite element model.
246         // returns  scale value.
247         double getScale() {
248
249             if (sizeMax > 0) return 0.8/sizeMax;
250             else return 1.0;
251         }
252
253 }
```

Method `modifyNodeCoordinates` begins with alteration of the nodal co-ordinates in order to show the deformed finite element model (lines 183–202). The loop over nodes in lines 185–192 and statement 193 determines the maximum total displacement. Displacement scale `scaleD` is computed in lines 195–196 such that the maximum displacement is depicted on the screen as the `deformScale` fraction of the maximum model size. For example, if `deformScale` is specified as 0.2 then the maximum displacement will be shown as 20% of the model size. In lines 208–214, the center of the model bounding box is translated to the coordinate origin.

In method `setBoundingBox`, the maxima and minima for surface nodal coordinates x, y, and z are determined in a loop of lines 225–233 and placed into arrays `xyzmin` and `xyzmax`. For two-dimensional models, small artificial thickness is created in the z-direction (lines 235–236). The maximum model size `sizeMax` is calculated in lines 238–242.

The standard size of the Java 3D drawing canvas is comparable to unit length. Method `getScale` provides a scale value to be used in scaling transformation of the finite element model. A scale factor is determined such that the maximum screen size of the model is equal to 0.8 (line 249).

Problems

29.1. A linked list data structure is used for storage of surface element faces and edges. Describe how data is stored in the linked list. What are the advantages of a linked list for storage of surface faces in comparison to an array?

29.2. Develop an algorithm for determining surface element edges when element connectivities for the whole finite element model are used as input data. Express the algorithm in the form of pseudocode. Explain the difficulties with such an algorithm.

29.3. A finite element model consists of eight-node hexahedral elements with the local node numbering shown below.

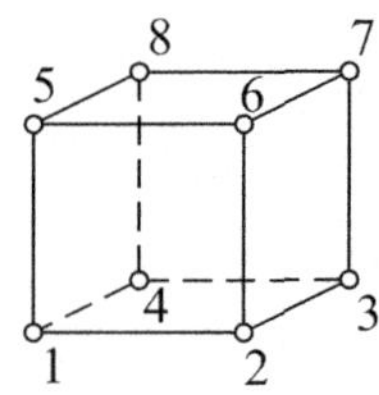

Find the inner element faces for the model of four elements with the following element connectivities:

```
1)   1   3   9   7    2    4  10    8
2)   3   5  11   9    4    6  12   10
3)   7   9  15  13    8   10  16   14
4)   9  11  17  15   10   12  18   16
```

Chapter 30
Edge and Face Subdivision

Abstract Algorithms for edge and face subdivision are considered. An element edge is divided into line segments. The number of edge subdivisions depends upon its curvature and upon the results range. Element face subdivision into triangles is determined by subdivisions of its edges. The positions of triangle vertices are generated and face triangulation is performed using the Delaunay method. Class `FaceSubdivision` implements edge and face subdivision.

30.1 Subdivision for Quality Visualization

The surface of the finite element model consists of element faces, edges, and nodes. In general, element faces are not planar and element edges are not straight. Visualization in Java 3D^{TM} is based on rendering flat polygons and straight-line segments.

To obtain visualization of good quality, element surface faces and edges should be subdivided into a sufficient number of polygons and line segments. First, edge subdivisions are performed that depend on curvature of the edge and on the range of a result parameter along the edge. Then, using subdivisions for four edges, points are generated inside a quadrilateral face, and the face is subdivided into triangles. Class `FaceSubdivision` deals with subdivision of an element face.

Methods of class `SurfaceGeometry` help to produce geometric objects of the model surface: faces, edges, and nodes. For visualization purposes, it is necessary to subdivide the faces and edges into triangles and straight-line segments. Class `FaceSubdivision` presented here deals with one element face. It provides methods for calculating a number of subdivisions for an edge and for subdivision of an element face into triangles. The number of triangles at each element edge is determined by the number of subdivisions for the edge.

We consider edges of quadratic elements. The edge is defined by location of three nodes. When results visualization is performed, result values are related to nodes. The number of edge subdivisions depends on curvature of the edge and on the range

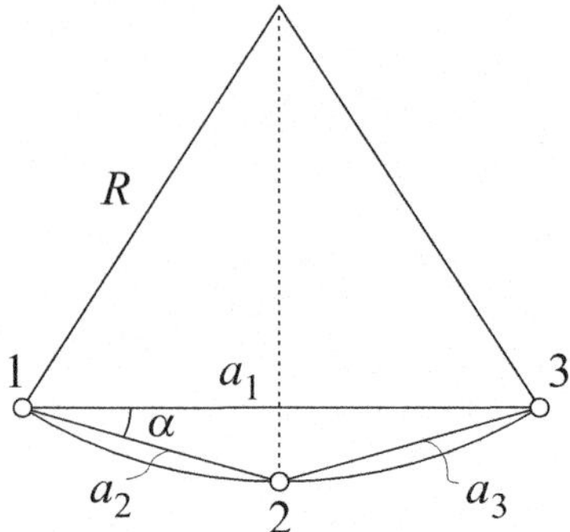

Fig. 30.1 Curvature radius of an element side defined by three nodes

of result parameters for the edge. Subdivision numbers for four edges determine face subdivision into triangles.

30.2 Edge Subdivision

Curvature-based edge subdivision depends on the edge curvature radius. Let us consider three nodes that determine an element edge (see Figure 30.1). Connecting three nodes 1, 2 and 3 on the curved side by straight lines produces a triangle with sides a_1, a_2 and a_3. In the ordinary finite and boundary elements midside node 2 has equal distances to corner nodes 1 and 3. Dimensionless curvature of the element edge can be characterized by the value of $\sin\alpha$.

Sine of the angle α can be found using dual representation of the area of triangle 123:

$$s = \sqrt{p(p-a_1)(p-a_2)(p-a_3)} \ , \quad p = \frac{1}{2}(a_1 + a_2 + a_3), \qquad (30.1)$$

$$s = \frac{1}{2}a_1 a_3 \sin\alpha. \qquad (30.2)$$

Area s is calculated using Equation 30.1. The value of $\sin\alpha$ is then estimated as

$$\sin\alpha = \frac{2s}{a_1 a_3}. \qquad (30.3)$$

It is possible to adopt that the number of geometry side subdivisions n_g is proportional to the curvature parameter ρ:

$$n_g = k_g \rho \ , \quad \rho = \sin\alpha = \frac{s}{a_1}\left(\frac{1}{a_2} + \frac{1}{a_3}\right), \qquad (30.4)$$

where k_g is the empirical coefficient. In the above relation the average value of the curvature parameter estimated with distances a_2 and a_3 is employed.

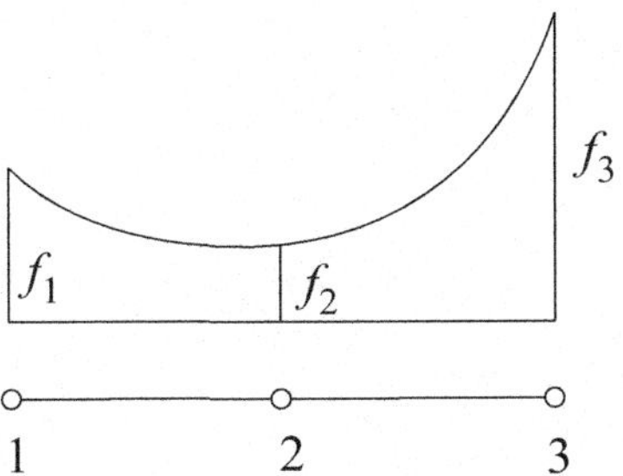

Fig. 30.2 Quadratic results function defined by its values at three nodes

Results-based edge subdivision depends on the range of results along this edge. We are going to use pattern interpolation for creating contours of scalar results on element faces. An element face is subdivided into triangles and pattern interpolation inside triangles is performed linearly as depicted in Figure 27.1. In order to obtain good quality of color contours the size parameter controlling triangulation should be selected such that each triangle contains a limited number of color intervals.

Figure 30.2 shows the quadratic results function defined by its values at three nodes of the element edge. It is reasonable to have the number of subdivisions along the finite element edge proportional to ranges of the function between two neighboring nodes. If result values at the nodes of the element edge are denoted as f_1, f_2 and f_3 then the number of side subdivisions n_c due to color change is determined by relations:

$$n_c = k_c \frac{\max\{abs(f_2 - f_1),\ abs(f_3 - f_2)\}}{df},$$

$$df = \frac{f_{\max} - f_{\min}}{c},$$

(30.5)

where c is the number of color intervals with linear change of color function (=5), k_c is the empirical coefficient. The final number of subdivisions on the particular face edge is selected as the maximum of the number of edge subdivision due to geometry curvature n_g and due to results range n_c.

Edge subdivision is done by method `numberOfEdgefDivisions` of Java[TM] class `FaceSubdivision`. The first part of the class is presented below.

```
1  package visual;
2
3  // Subdivision of an edge and a face
4  public class FaceSubdivision {
5
6      // Number of points used for subdivision
7      public int nFacePoints;
8      // Local coordinates of points
9      public double[] xi, et;
10     // Number of triangles and their indexes
11     public int nTrigs, trigs[][];
```

```
12
13       private double[] ze;
14
15       public FaceSubdivision() {
16
17           int npMax = (VisData.nDivMax +1)*(VisData.nDivMax +1);
18           xi = new double[npMax];
19           et = new double[npMax];
20           ze = new double[npMax];
21
22           trigs = new int[2*VisData.nDivMax*VisData.nDivMax][3];
23       }
24
25       // Compute number of edge subdivisions.
26       // xyz - array [][3] of face nodal coordinates.
27       // fun - function values at nodes.
28       // deltaf - function range for the whole model.
29       // funDiv - if true perform results-based subdivision.
30       // i1, i2, i3 - indexes of three nodes on the edge.
31       // returns   number of edge subdivisions.
32       static int numberOfEdgefDivisions(double[][] xyz,
33           double[] fun, double deltaf, boolean funDiv,
34           int i1, int i2, int i3) {
35
36           int nDiv = VisData.nDivMin;
37
38           // Curvature-based subdivision
39           double a1 = distance(xyz, i1, i3);
40           double a2 = distance(xyz, i1, i2);
41           double a3 = distance(xyz, i2, i3);
42
43           double p = 0.5*(a1 + a2 + a3);
44           double s = Math.sqrt(p*(p-a1)*(p-a2)*(p-a3));
45           // Curvature parameter
46           double ro = 2*s/a1*Math.abs(1/a2 + 1/a3);
47           nDiv = Math.max (nDiv, (int)(1.5 + VisData.Csub*ro));
48
49           if (!funDiv) return Math.min(nDiv, VisData.nDivMax);
50
51           // Results-based subdivision
52           int n = (int) (1.5 + VisData.Fsub*Math.max(
53                   Math.abs(fun[i1]-fun[i2]),
54                   Math.abs(fun[i2]-fun[i3]))/deltaf);
55
56           return Math.min(VisData.nDivMax, Math.max(nDiv, n));
57       }
58
59       // Distance between two points.
60       // xyz - array [][3] containing point locations.
61       // p1, p2 - point indexes.
62       // returns   distance
63       static double distance(double[][] xyz, int p1, int p2) {
64           return Math.sqrt(
65               (xyz[p1][0]-xyz[p2][0])*(xyz[p1][0]-xyz[p2][0])
```

```
66                   + (xyz[p1][1]-xyz[p2][1])*(xyz[p1][1]-xyz[p2][1])
67                   + (xyz[p1][2]-xyz[p2][2])*(xyz[p1][2]-xyz[p2][2]));
68       }
```

The heading of class `FaceSubdivision` contains declarations of data describing face subdivision:

`nFacePoints` – number of points used for face subdivision;
`xi, et` – local coordinates of points for face subdivision;
`nTrigs` – number of triangles;
`trigs` – indexes of triangles.

The class constructor initializes arrays of point coordinates `xi` and `et`, working array `ze` used for triangulation and an array of triangle indices `trigs`. Array sizes are specified on the basis of the maximum number of edge subdivisions `nDivMax` defined in visualization data.

Method `numberOfEdgefDivisions` estimates the number of edge subdivisions using formulas for curvature-based subdivision and for results-based subdivision. The parameters of the method are:

`xyz` – coordinates of eight nodes, which define an element face;
`fun` – result values at face nodes;
`deltaf` – range of result parameter for the whole finite element model;
`funDiv` – if this parameter is true then the number of subdivisions is determined using both edge curvature and the results range, otherwise just edge curvature is used;
`i1, i2` and `i3` are indices of this edge for face coordinates and result arrays.

The method returns the number of subdivisions for this edge.

First, the number of edge subdivisions `nDiv` is set to the minimum number of subdivisions `nDivMin` specified in the data. The curvature parameter ρ is determined according to (30.1)–(30.4) in lines 39–46. Line 47 estimates the number of subdivisions using the curvature empirical factor `Csub`.

The number of subdivisions with the use of the result range empirical factor `Fsub` is calculated in lines 52–54 according to (30.5). Statement 56 returns the maximum of numbers of subdivisions determined by the curvature and result range but is not larger than the specified in data maximum value `nDivMax`.

30.3 Face Subdivision

Face subdivision should provide polygons (triangles) for three-dimensional rendering of good quality. The subdivision is determined by the edge subdivision, which is based on the edge curvature and results range. Different methods of generating a triangular mesh inside a quadrilateral with specified boundary subdivision can be used. We propose here to generate face points and then create triangles using the Delaunay approach.

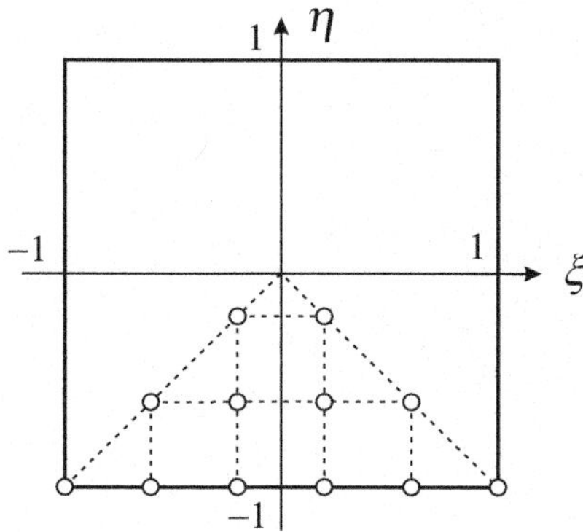

Fig. 30.3 Generating points for face subdivision

An algorithm for generating face points is as follows. A quadrilateral element is subdivided in the local coordinate system $-1 \leq \xi, \eta \leq 1$. The square is partitioned into four triangular areas (sectors) by two diagonals and face points are generated consequently in these areas. Point locations are arranged in rows parallel to the edge as shown in Figure 30.3. During point generation, a new point is added if its distance from any existing point is larger then a predetermined minimum distance. The minimum distance is computed on the basis of minimum triangle size at the quadrilateral boundary. One point is added at the square center. Also, one point is added at the sector center for sectors with the number of edge subdivisions less than three.

Face subdivision is implemented by method subdivideFace. The method gets the numbers of subdivisions for edges of a square face. The results of subdivision are arrays of face points xi and et (in local coordinates) and array trigs containing the indices of triangular polygons.

```
70      // Subdivide 2 x 2 square into triangles.
71      // ndiv - number of subdivisions on edges.
72      public void subdivideFace(int[] ndiv) {
73
74          // Small number to generate slightly imparallel lines
75          final double EPS = 1.e-6;
76          double xiQ, etQ;
77          nFacePoints = 0;
78
79          // Squared min distance for node placement
80          int ndivMax = 1;
81          for (int side = 0; side < 4; side++)
82              ndivMax = Math.max(ndivMax, ndiv[side]);
83          double dMin = 2.0/ndivMax;
84          double minNodeDist2 = Math.min(1.0,0.8*dMin*dMin);
85          // Add point at the face center
86          addPoint(0, EPS, EPS, 0.0);
87
88          // Generate points for triangular sectors
```

```java
89                for (int sector = 0; sector < 4; sector++) {
90                    int n = ndiv[sector];
91                    double d = 2.0/n;
92                    // Points inside quarter of the element
93                    for (int row = 0; row <= n/2; row++) {
94                        for (int i = row; i <= n-row; i++) {
95                            xiQ = -1.0 + d*i;
96                            etQ = -1.0 + d*row;
97                            if (row > 0) {
98                                xiQ += EPS;
99                                etQ += EPS*xiQ;
100                           }
101                           addPoint(sector, xiQ, etQ, minNodeDist2);
102                       }
103                   }
104                   if (n<3) addPoint(sector, EPS, -0.5, minNodeDist2);
105               }
106
107        // Delaunay triangulation
108        nTrigs = triangulateDelaunay();
109    }
110
111    // Add point to seeded points for Delaunay triangulation.
112    // sector - sector number 0..3.
113    // xiQ - xi-sector-coordinate of the point.
114    // etQ - eta-sector-coordinate of the point.
115    // minDistance2 - squared min distance between points.
116    private void addPoint(int sector, double xiQ, double etQ,
117                          double minDistance2) {
118
119        double sin[] = {0, 1,  0, -1};
120        double cos[] = {1, 0, -1,  0};
121
122        double x = xiQ*cos[sector] - etQ*sin[sector];
123        double y = xiQ*sin[sector] + etQ*cos[sector];
124
125        int i;
126        for (i = 0; i < nFacePoints; i++) {
127            double dx = x - xi[i];
128            double dy = y - et[i];
129            if (dx*dx + dy*dy < minDistance2) break;
130        }
131
132        if (i == nFacePoints) {
133            xi[nFacePoints] = x;
134            et[nFacePoints] = y;
135            nFacePoints++;
136        }
137    }
```

Lines 80–84 of method `subdivideFace` calculate the squared minimum distance for adding a new point. A new point is added at its position if its distance from any existing point is larger than the minimum distance. The minimum distance between points is of the order of the smallest triangle at the element edge.

Statement 86 adds a point at the face center. Small shift EPS is used in order to avoid perfectly regular arrangement of points, which may lead to the creation of overlapped triangles during the Delaunay triangulation process.

Generation of points inside four triangular sectors is performed in the loop of lines 89–105. A step d for point generation is estimated in line 91 using the number of subdivisions for the current edge. The double loop of lines 93–103 fills a triangular quarter of the face with points arranged in rows with shrinking horizontal size as shown in Figure 30.3. Small irregularities in point locations are introduced with EPS. Points are added in line 101 using method addPoint.

Method addPoint (lines 116–137) gets the following parameters: sector – sector number from 0 to 3; xiQ, etQ – local coordinates of a point candidate for placement; minDistance2 – squared minimum distance between points.

Point coordinates xiQ, etQ are specified in local coordinates of sector 0. Statements in lines 122–123 perform coordinate transformation according to the sector number. Points in sectors 1, 2 and 3 are rotated around the face center by 90, 180 and 270 degrees around the face center.

The loop of lines 125–130 computes distances from this point xiQ, etQ to already existing points stored in arrays xi and et. If this point is far enough from any existing point then the point is stored in arrays xi and et and the point counter nFacePoints is incremented by one. Otherwise, the method returns without adding the current point.

Method triangulateDelaunay performs face triangulation using a Delaunay approach. Delaunay triangulation is selected because it produces the most equiangular triangles of all triangulations, i.e., it limits the number of slender triangles. It has the property that a triangle circumcircle does not include any other vertices. Here, we use an $O(n^4)$ algorithm since we have a relatively small number of points n and the algorithm allows very compact programming. To improve the performance of the Delaunay triangulation it is possible to employ a more efficient approach, such as, for example, the incremental algorithm given in [26].

```
139        // Create triangular mesh using Delaunay approach.
140        // returns   number of triangles
141        private int triangulateDelaunay() {
142
143            double EPS = 1.e-6;
144            int n = nFacePoints;
145            int nt = 0;
146
147            for (int i = 0; i < n; i++)
148                ze[i] = xi[i]*xi[i] + et[i]*et[i];
149
150            for (int i = 0; i < n; i++) {
151                double pix = xi[i];
152                double pie = et[i];
153                double piz = ze[i];
154
155                for (int j = i + 1; j < n; j++) {
156                    double pjx = xi[j];
157                    double pje = et[j];
```

```
158                    double pjz = ze[j];
159
160                for (int k = i + 1; k < n; k++) {
161                    double pkx = xi[k];
162                    double pke = et[k];
163                    double pkz = ze[k];
164                    double zn = (pjx - pix)*(pke - pie)
165                               - (pkx - pix)*(pje - pie);
166
167                    if (j == k || zn > 0) continue;
168
169                    double xn = (pje - pie)*(pkz - piz)
170                               - (pke - pie)*(pjz - piz);
171                    double en = (pkx - pix)*(pjz - piz)
172                               - (pjx - pix)*(pkz - piz);
173
174                    int m;
175                    for (m = 0; m < n; m++) {
176                        double pmx = xi[m];
177                        double pmy = et[m];
178                        double pmz = ze[m];
179                        if (m != i && m != j && m != k
180                                && (pmx-pix)*xn + (pmy-pie)*en
181                                + (pmz-piz)*zn > 0)
182                            break;
183                    }
184
185                    if (m == n) {
186                        double area = pix*(pke-pje)
187                            + pkx*(pje-pie) + pjx*(pie-pke);
188                        if (Math.abs(area) > EPS) {
189                            // Add triangle polygon
190                            // (unticlockwise node order)
191                            trigs[nt][0] = i;
192                            trigs[nt][1] = k;
193                            trigs[nt][2] = j;
194                            nt++;
195                        }
196                    }
197                }
198            }
199        }
200        return nt;
201    }
202
203 }
```

The method performs Delaunay triangulation in four enclosed loops that start in lines 150, 155, 160 and 175. Triangles with anticlockwise order of three vertices are added into array `trigs`. The method returns the number of generated triangles.

An example of face subdivision for a twenty-node element with one curved edge is shown in Figure 30.4.

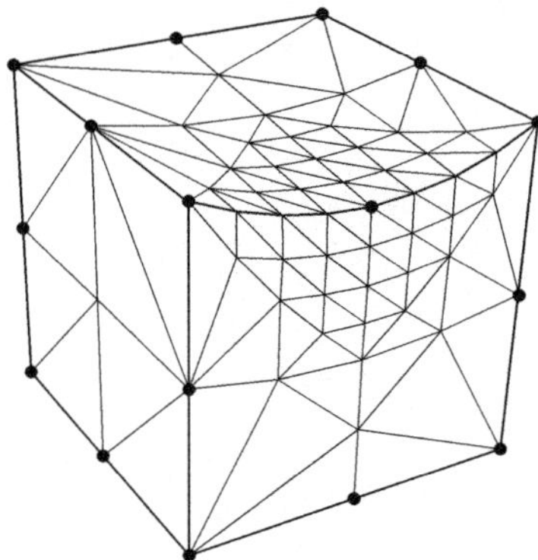

Fig. 30.4 Example of face subdivision for a twenty-node element with a curved edge

Problems

30.1. The curvature parameter ρ for an edge of the quadratic element is determined according to Equation 30.3 and Figure 30.1. Propose another way for estimating the edge curvature parameter.

30.2. Function f is specified at three nodes of the edge of a quadratic element shown below: $f_1 = 0$, $f_2 = 1$ and $f_3 = 4$. The coordinates of the nodal points are: $x_1 = 0$, $x_2 = 0.5$, $x_3 = 1$.

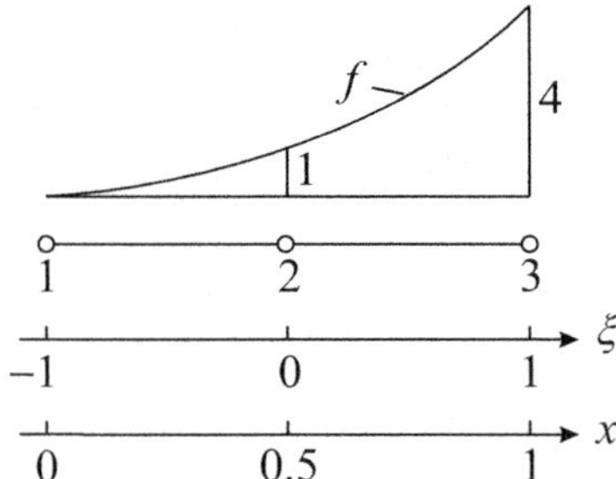

Determine the function gradient df/dx at node 3 where $x = 1$.

30.3. Study the algorithm of generating face points for subsequent Delaunay triangulation described in Section 30.3 and its implementation in Java `subdivideFace`. Propose another algorithm for generating face points when the numbers of points on the edges are specified.

Chapter 31
Surface Subdivision

Abstract In order to have good-quality three-dimensional visualization the faces should be represented with a sufficient number of triangular polygons and edges with a sufficient number of straight-line segments. This chapter describes algorithms and program implementation (class `SurfaceSubGeometry`) for subdividing the surface of a finite element model into graphical components suitable for rendering with Java 3D^{TM}.

31.1 Subdivision of the Model Surface

Subdivision of a surface of a finite element model is required for rendering three-dimensional visual objects with sufficient quality. The surface of a finite element model is represented as sets of element faces, element edges, and nodes. In the process of subdivision, element faces are subdivided into triangles, element edges are separated into line segments, and element nodes are copied without change into a point array, as shown in Figure 31.1, which shows an example of face subdivision into a regular mesh of triangles. In the `Jvis` program, the number of subdivisions at different edges can be different and the triangular mesh can be irregular.

The most complicated subdivision task is subdivision of element faces into triangular polygons. In addition to the necessity of irregular meshes, the normal at each triangle vertex must be determined and put into a Java 3D triangle array. In order to have a smooth image, surface vertex normals should be normals to the model surface, not to the triangle itself. In the case of drawing results as contours of equal values, it is also necessary to determine the texture coordinates of a color scale.

Surface subdivision is implemented in class `SurfaceSubGeometry` presented in this chapter. For extraction of a surface of the finite element model, the constructor of class `SurfaceGeometry` described in Chapter 29 is used. Subdivision of a particular element face or edge is performed by calling methods of class `FaceSubdivision` presented in Chapter 30.

The constructor of Java^{TM} class `SurfaceSubGeometry` is presented below.

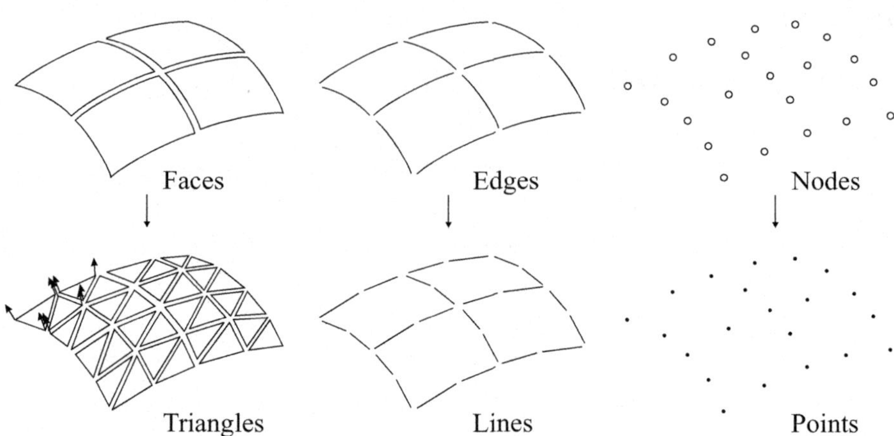

Fig. 31.1 Surface subdivision: element faces are subdivided into triangles, element edges into line segments. Nodes are copied into a point array

```java
1  package visual;
2
3  import elem.*;
4
5  import javax.media.j3d.*;
6  import java.util.ListIterator;
7
8  // Element subfaces, subedges and nodes.
9  class SurfaceSubGeometry extends SurfaceGeometry {
10
11      int nVertices;
12      private FaceSubdivision fs;
13      // Edge subdivisions for faces
14      private int edgeDiv[][];
15      // Coordinates, normals and texture coords for
16      // triangle array of the whole model surface
17      private float xyzSurface[], norSurface[], texSurface[];
18
19      // Arrays for one element face
20      private double xyzFace[][] = new double[8][3];
21      private double funFace[] = new double[8];
22      private double an[] = new double[8];
23      private double deriv[][] = new double[8][2];
24      // Arrays for subdivided surface
25      private double[] xyzFacePoints;
26      private double[] norFacePoints;
27      private double[] texFacePoints;
28
29      // Constructor for subdivision of faces, edges and nodes.
30      SurfaceSubGeometry() {
31
32          // Call to constructor of SurfaceGeometry
33          super();
```

```
34
35              // Constructor edge/face subdivider
36              fs = new FaceSubdivision ();
37
38              edgeDiv = new int[nFaces][4];
39
40              int np = (VisData.nDivMax +1)*(VisData.nDivMax +1);
41              xyzFacePoints = new double[3*np];
42              norFacePoints = new double[3*np];
43              texFacePoints = new double[np];
44
45              // Determine edge subdivisions for element faces
46              int nTrigs = setEdgeDivisions ();
47
48              xyzSurface = new float[9*nTrigs];
49              norSurface = new float[9*nTrigs];
50              texSurface = new float[6*nTrigs];
51
52              // Perform subdivision (triangulation) for faces
53              setFaceTriangles ();
54          }
```

Line 33 calls a constructor of superclass `SurfaceGeometry` that creates a
surface of the finite element model. Then, line 38 creates the `FaceSubdivision`
object `fs` used for subdivision of a single element face. Lines 41–43 allocate memory for arrays of coordinates, normals, and texture coordinates related to an element
face. The maximum allowable number of edge subdivisions `nDivMax` is used to determine the upper limit of array length. Line 46 calls method `setEdgeDivisions`
that computes the number of subdivisions for edges of all surface element faces and
estimates the number of triangles `nTrigs` after face subdivision. The latter is used
for allocating arrays of coordinates, normals, and texture coordinates for the entire
surface of the finite element model.

Method `setEdgeDivisions` follows.

```
56      // Determine numbers of edge subdivisions for all
57      // element faces edgeDiv[nFaces][4].
58      // returns  upper estimate of number of triangles
59      // for the surface of the finite element model.
60      int setEdgeDivisions () {
61
62          int nTriangles = 0;
63          ListIterator f = listFaces.listIterator(0);
64
65          for (int face=0; face<nFaces; face++) {
66              setFaceCoordFun ((int[]) f.next ());
67              for (int i = 0; i < 4; i++) {
68                  int nd = fs.numberOfEdgeDivisions (xyzFace,
69                          funFace, deltaf, VisData.drawContours,
70                          2*i,2*i+1,(2*i+2)%8);
71                  edgeDiv[face][i] = nd;
72                  nTriangles += (int) (0.6*nd*nd + 2);
73              }
74          }
```

```
75              return nTriangles;
76          }
77
78          // Set coordinates and function values for face nodes
79          void setFaceCoordFun(int[] faceNodes) {
80
81              for (int i = 0; i < faceNodes.length; i++) {
82                  int ind = faceNodes[i] - 1;
83                  for (int j = 0; j < 3; j++) {
84                      if (fem.nDim ==2 && j==2) xyzFace[i][j] = 0;
85                      else xyzFace[i][j] = fem.getNodeCoord(ind,j);
86                  }
87                  if (VisData.drawContours) funFace[i] = fun[ind];
88              }
89          }
```

The method populates array `edgeDiv` with the number of subdivisions for all surface element faces. It is supposed that element faces are quadrilateral; therefore the second dimension of array `edgeDiv` is 4. Line 63 initializes list iterator `f` for linked list `listFaces` where connectivities of the surface faces are stored. In a loop over element faces, method `setFaceCoordFun` sets the nodal coordinates and nodal values of the result function for an element face from the linked list `listFaces`. The statement in lines 68–70 determines the number of subdivisions for a current face edge based on edge curvature and function range. The number of triangles for all surface edges is accumulated in variable `nTriangles` (line 72), which is returned in line 75.

Method `setFaceCoordFun` (lines 79–89) puts nodal coordinates into array `xyzFace` and function values into array `funFace`, according to the connectivities `faceNodes` of the current face.

31.2 Subdivision of Faces into Triangles

Method `setModelTriangles` generates a triangulation for all element faces comprising the finite element model surface and populates an array of triangle coordinates, normals, and texture coordinates.

```
91          // Perform triangulation for all faces.
92          void setModelTriangles() {
93
94              int[] faceNodes;
95              nVertices = 0;
96              ListIterator f = listFaces.listIterator(0);
97
98              for (int face = 0; face < nFaces; face++) {
99
100                 faceNodes = (int[])f.next();
101                 setFaceCoordFun(faceNodes);
102
103                 // Subdivide element face into triangles
```

```
104                     // using local coordinates
105                     fs.subdivideFace(edgeDiv[face]);
106
107                     setFaceVertices(faceNodes);
108
109                     // Add triangle coordinates, normals and texture
110                     // coordinates to surface arrays
111                     for (int t=0; t<fs.nTrigs; t++) {
112                         for (int k =0; k <3; k++) {
113                             int ind = fs.trigs[t][k];
114                             for (int i=0; i<3; i++) {
115                                 xyzSurface[3*nVertices+i] =
116                                         (float) xyzFacePoints[3*ind+i];
117                                 norSurface[3*nVertices+i] =
118                                         (float) norFacePoints[3*ind+i];
119                             }
120                             texSurface[2*nVertices] =
121                                     (float) texFacePoints[ind];
122                             nVertices++;
123                         }
124                     }
125                 }
126         }
```

A loop over surface element faces starts in line 98. The node numbers defining
the current element face are taken from linked list listFaces (line 100). Line
101 sets nodal coordinates and nodal function values for the face. In line 105 the
face is subdivided into triangles with the help of method subdivideFace (class
FaceSubdivision). During this subdivision, the face is considered as a square
with edge length 2 and subdivision is done in the local coordinate system.

Method setFaceVertices transforms local coordinates of triangle vertices
into the global coordinate system and additionally determines normals and function values for triangle vertices. The loop in lines 111–124 adds face arrays to
global surface arrays of coordinates, normals, and texture coordinates for triangles.
Triangle data is created using triangle indexes fs.trigs, which refers to face
point numbers. Each triangle is placed in these arrays separately because we do not
use indexing. Type float is used to economize on memory. At the end, variable
nVertices contains the number of triangle vertices, which is thrice the number
of triangles used to represent the whole model surface.

The listing of method setFaceVertices is shown below.

```
128     // Compute global coordinates, normals and texture
129     // coordinates for face triangle vertices.
130     private void setFaceVertices(int[] faceNodes) {
131
132         double e[][] = new double[2][3], en[] = new double[3];
133
134         for (int iv = 0; iv < fs.nFacePoints; iv++) {
135             // Shape functions and their derivatives for
136             // a face of 3D hexahedral quadratic element
137             ShapeQuad3D.shapeDerivFace(fs.xi[iv], fs.et[iv],
138                     faceNodes, an, deriv);
```

```
139                 for (int j = 0; j < 3; j++) {
140                     double s = 0;
141                     for (int i = 0; i < 8; i++)
142                         s += an[i]*xyzFace[i][j];
143                     xyzFacePoints[3*iv + j] = (float) s;
144                 }
145             // Tangents e to the local coordinates
146             for (int i = 0; i < 2; i++) {
147                 for (int j = 0; j < 3; j++) {
148                     double s = 0;
149                     for (int k = 0; k < 8; k++)
150                         s += deriv[k][i]*xyzFace[k][j];
151                     e[i][j] = s;
152                 }
153             }
154             // Normal vector en
155             en[0] = (e[0][1]*e[1][2] - e[1][1]*e[0][2]);
156             en[1] = (e[0][2]*e[1][0] - e[1][2]*e[0][0]);
157             en[2] = (e[0][0]*e[1][1] - e[1][0]*e[0][1]);
158             double s = 1.0/Math.sqrt(en[0]*en[0] +
159                     en[1]*en[1] + en[2]*en[2]);
160             for (int i = 0; i < 3; i++)
161                 norFacePoints[3*iv + i] = (float) (en[i]*s);
162
163             if (VisData.drawContours) {
164                 double f = 0;
165                 for (int i = 0; i < 8; i++)
166                     f += an[i]*funFace[i];
167                 double t = (f - fmin)/(fmax - fmin);
168                 if (t < 0.003) t = 0.003;
169                 if (t > 0.997) t = 0.997;
170                 texFacePoints[iv] = (float) t;
171             }
172         }
173     }
```

The method uses results of face subdivision by method `subdivideFace` located in object `fs: nFacePoints` – the number of points created during face subdivision, `xi`, `et` – the local coordinates of the face points. All computations are performed in a loop over face points. Lines 137–138 determine shape functions an and their derivatives `deriv` for a face of the three-dimensional quadratic element. Global coordinates of the current point are computed in lines 139–144 using shape functions.

Normals to a finite element face are determined according to the algorithm presented in Section 12.5. Two vectors tangent to local coordinates ξ and η are calculated in lines 146–153, using derivatives of the shape functions. A normal to the element face is computed as a vector product of tangent vectors in lines 155–157. The normal vector after its normalization is added to the array of surface normals in line 161.

If drawing of contours is requested, a function value at a current point is determined by interpolation of the nodal values using the shape functions in lines 164–

166. Texture coordinate t is estimated as a function value relative to the range of the function (line 167). Statements 168–169 ensure that the texture coordinate is inside range $0\ldots 1$.

31.3 Arrays for Java 3D

Java 3D shape objects require description of the geometry for three-dimensional visualization. The finite element model is described with three geometry arrays – a triangle array for the surface, a line array for element edges, and a point array for nodes.

The Java 3D triangle array is created by the method `getModelTriangles` presented below.

```
175        // Create TriangleArray containing vertex coordinates,
176        // normals and possibly texture coordinates for contours.
177        // returns   TriangleArray
178        TriangleArray getModelTriangles() {
179
180            int vFormat = TriangleArray.COORDINATES |
181                          TriangleArray.NORMALS |
182                          TriangleArray.BY_REFERENCE;
183            if (VisData.drawContours) vFormat = vFormat |
184                    TriangleArray.TEXTURE_COORDINATE_2;
185
186            TriangleArray triangleArray =
187                    new TriangleArray(nVertices, vFormat);
188
189            triangleArray.setCoordRefFloat(xyzSurface);
190            triangleArray.setNormalRefFloat(norSurface);
191            if (VisData.drawContours)
192                triangleArray.setTexCoordRefFloat(0, texSurface);
193
194            return triangleArray;
195        }
```

The vertex format of the triangle array is specified as an integer that is a combination of predefined static integers combined with bitwise ORs. In lines 180–182 the specified format requires supplying coordinates and normals at triangle vertices. The last integer parameter in the format specifies that arrays will be used by reference, without copying inside Java 3D. Such a vertex format is suitable for visualization of the finite element model. When results visualization is required, texture coordinates for drawing contours are added to the vertex format in lines 183–184.

The constructor of Java 3D class `TriangleArray` (lines 186–187) obtains the number of triangle vertices `nVertices` and the vertex format as parameters. References to arrays of triangle coordinates, normals, and texture coordinates are set in lines 189–192. The method returns the created geometry object in line 194.

The method `getModelLines` prepares a line array for visualization of element edges.

```java
197     // Create array of lines for drawing element edges.
198     // returns  line array for element edges at the surface
199     LineArray getModelLines() {
200
201         float x[] = new float[3];
202         double xys[][] = new double[3][3];
203         double an[] = new double[3];
204         int[] divs = new int[nEdges];
205         int nDivTotal = 0;
206
207         int ii = 0;
208         ListIterator e = listEdges.listIterator(0);
209
210         for (int edge=0; edge<nEdges; edge++) {
211             int edgeNodes[] = (int[])e.next();
212             for (int k = 0; k < 3; k++) {
213                 for (int n = 0; n < fem.nDim; n++)
214                     xys[k][n] = fem.getNodeCoord(
215                             edgeNodes[k]-1,n);
216             }
217             int n = fs.numberOfEdgeDivisions(xys, null,
218                     deltaf, false, 0, 1, 2);
219             divs[edge] = n;
220             nDivTotal += n;
221         }
222         LineArray lineArray =
223             new LineArray(nDivTotal*2, LineArray.COORDINATES);
224
225         e = listEdges.listIterator(0);
226
227         for (int edge=0; edge<nEdges; edge++) {
228             int edgeNodes[] = (int[])e.next();
229             for (int k = 0; k < 3; k++)
230                 for (int n = 0; n < fem.nDim; n++) xys[k][n] =
231                             fem.getNodeCoord(edgeNodes[k]-1,n);
232
233             int ndiv = divs[edge];
234             double dxi = 2.0/ndiv;
235             for (int i=0; i<3; i++) x[i] = (float) xys[0][i];
236
237             for (int k = 1; k <= ndiv; k++) {
238                 lineArray.setCoordinate(ii++, x);
239                 double xi = -1 + k*dxi;
240                 // Quadratic shape functions
241                 an[0] = -0.5*xi*(1 - xi);
242                 an[1] =  1 - xi*xi;
243                 an[2] =  0.5*xi*(1 + xi);
244
245                 for (int i = 0; i < 3; i++) {
246                     double s = 0;
247                     for (int j = 0; j < 3; j++)
248                         s += xys[j][i]*an[j];
249                     x[i] = (float) s;
250                 }
```

```
251                        lineArray.setCoordinate(ii++, x);
252                     }
253                  }
254              return lineArray;
255          }
```

Element edges are defined by three nodal points and therefore are generally curved. For smooth visualization, edges are divided into straight-line segments based on edge curvature.

Element edges are stored in linked list `listEdges`. List iterator e is initialized in line 208. The numbers of subdivisions are estimated in a loop over all element edges (lines 210–221). Nodes for the current edge are obtained from the linked list in line 211. The number of edge subdivisions is estimated in lines 217–218 using method `numberOfEdgeDivisions`. The total number of line segments is accumulated in variable `nDivTotal` (line 220).

A Java 3D line array is initialized in lines 222–223. An inner loop over all edges starts in line 227. Coordinates of three nodes located on the current edge are placed in array `xys` (line 230–231). An element edge is subdivided into line segments in the loop of lines 237–253 using one-dimensional quadratic shape functions. The left and right ends of a line segment are added to the line array in lines 238 and 251. The geometry array containing line segments is returned in line 254.

An array of points for visualization of nodes belonging to the surface of the finite element model is built by method `getModelPoints`.

```
257          // Create array of nodal points.
258          // returns  point array containing nodes at the surface
259          PointArray getModelPoints() {
260
261              float x[] = new float[3];
262              PointArray pointArray =
263                  new PointArray(nsNodes*3, PointArray.COORDINATES);
264              int ii = 0;
265              for (int node = 0; node < sNodes.length; node++) {
266                  if (sNodes[node] > 0) {
267                      for (int i = 0; i < fem.nDim; i++)
268                          x[i] = (float) fem.getNodeCoord(node, i);
269                      pointArray.setCoordinate(ii++, x);
270                  }
271              }
272              return pointArray;
273          }
274
275  }
```

Information on surface nodes is contained in array `sNodes`. Surface nodes are given by the positions of nonzero entries in this array. A point array is filled with the coordinates of surface nodes in a loop for across entries of array `sNodes`. This array is returned to the caller in line 272.

Problems

31.1. In the algorithm for face subdivision presented in this chapter, element faces sharing an edge have the same number of subdivisions on that edge. Explain why this consistency is necessary for finite element model visualization.

31.2. Two unit vectors have the following components:

$$e_1 = \{1 \quad 1 \quad 1\},$$

$$e_2 = \{1 \quad 1 \quad 0\}.$$

Determine vector e_n normal to vectors e_1 and e_2. You may use the algorithm employed for normal calculation in method `setFaceVertices`.

31.3. A Java 3D triangle array `TriangleArray` is used for rendering element faces. In this array, each set of three vertices with specified coordinates forms a triangle. Study Java 3D documentation related to geometry arrays and propose use of a different geometry array that employs less memory.

Chapter 32
Results Field, Color Scale, Interaction and Lights

Abstract The visualization technique for results represented as contours is based upon interpolation of color texture. The results field at nodes is set by methods of class `ResultAtNodes`. Physical quantities are converted to color gradation by computing the texture coordinates. Class `ColorScale` forms a texture with appropriate color gradation. Class `MouseInteraction` contains methods for simple mouse behaviors and for setting light in the scene.

32.1 Results Field

For visualization of results fields as contours on the surface of the finite element model, a requested drawing parameter should be set at nodes. Results of the finite element analysis consist of displacements and stresses. Displacements are related to nodes. Stresses are obtained at reduced integration points inside finite elements. First, they are extrapolated to nodes. Then, the stress contributions from neighboring elements are averaged at nodes.

Setting the requested results parameter at nodes of the finite element model is realized in class `ResultAtNodes`.

```
1  package visual;
2
3  import elem.*;
4  import model.FeModel;
5
6  // Result values at nodes of the finite element model.
7  public class ResultAtNodes {
8
9      private SurfaceGeometry sg;
10      private FeModel fem;
11      private int[] multNod;
12      private double[][] stressNod;
13
14      private double sm, psi, si, f;
```

```java
15      final double THIRD = 1.0/3.0, SQ3 = Math.sqrt(3.0);
16
17      // Constructor for results at nodes.
18      // sg - geometry of model surface.
19      ResultAtNodes(SurfaceGeometry sg, FeModel fem) {
20
21          this.sg = sg;
22          this.fem = fem;
23          multNod = new int[fem.nNod];
24          stressNod = new double[fem.nNod][2*fem.nDim];
25
26          feStressAtNodes();
27      }
28
29      // FE stresses at nodes: global array stressNod.
30      private void feStressAtNodes() {
31
32          double[][] elStressInt = new double[8][6];
33          double[][] elStressNod = new double[20][6];
34
35          for (int i = 0; i < fem.nNod; i++) {
36              multNod[i] = 0;
37              for (int j = 0; j < 2*fem.nDim; j++)
38                  stressNod[i][j] = 0;
39          }
40
41          for (int iel = 0; iel < fem.nEl; iel++) {
42              Element el = fem.elems[iel];
43              for (int ip = 0; ip < el.str.length; ip++)
44                  for (int j = 0; j < 2*fem.nDim; j++)
45                      elStressInt[ip][j] = el.str[ip].sStress[j];
46
47              el.extrapolateToNodes(elStressInt, elStressNod);
48
49              // Assemble stresses
50              for (int i=0; i<fem.elems[iel].ind.length; i++) {
51                  int jind = fem.elems[iel].ind[i] - 1;
52                  if (jind >= 0) {
53                      for (int k = 0; k < 2*fem.nDim; k++)
54                          stressNod[jind][k] += elStressNod[i][k];
55                      multNod[jind] += 1;
56                  }
57              }
58          }
59          // Divide by node multiplicity factor
60          for (int i = 0; i < fem.nNod; i++) {
61              for (int j = 0; j < 2*fem.nDim; j++)
62                  stressNod[i][j] /= multNod[i];
63          }
64      }
```

The class constructor obtains the surface geometry object sg and finite element
model object fem as parameters. It calls method feStressAtNodes in line 26.
This method sets stresses at nodes of the finite element model. Stress extrapolation

to nodes is performed in a loop of lines 41–58 over all finite elements. Stresses at reduced integration points are assigned to entries of array `elStressInt`. Element method `extrapolateToNodes` extrapolates stresses from integration points to nodes in line 47. Element stresses at nodes `elStressNod` are assembled in global nodal array `stressNod` in line 54. When stresses from an element are accumulated at nodes, multiplicity array entries `multNod` are incremented. Stress averaging at nodes is done in lines 60–63 using division by the respective multiplicity factor.

A results parameter requested for visualization is set at all nodes of the finite element model by method `setParmAtNodes`. Possible parameters are:

ux, uy, uz – displacements along global coordinates axes x, y, and z;
sx, sy, sz – normal stresses σ_x, σ_y, and σ_z;
sxy, syz, szx – shear stresses τ_{xy}, τ_{yz}, and τ_{zx};
s1, s2, s3 – principal stresses σ_1, σ_2, and σ_3;
si – equivalent stress σ_i;
s13 – maximum shear stress τ_{max}.

To get principal stresses we calculate deviatoric stresses s_x, s_y, and s_z –

$$s_x = \sigma_x - \sigma_m,$$
$$s_y = \sigma_y - \sigma_m,$$
$$s_z = \sigma_z - \sigma_m, \tag{32.1}$$
$$\sigma_m = \frac{1}{3}(\sigma_x + \sigma_y + \sigma_z),$$

and their second J_2 and third J_3 invariants –

$$J_2 = \frac{1}{2}(s_x^2 + s_y^2 + s_z^2) + \tau_{xy}^2 + \tau_{yz}^2 + \tau_{zx}^2,$$
$$J_3 = s_x s_y s_z + 2\tau_{xy}\tau_{yz}\tau_{zx} - s_x\tau_{yz}^2 - s_y\tau_{zx}^2 - s_z\tau_{xy}^2. \tag{32.2}$$

Equivalent stress is expressed through the second deviatoric invariant,

$$\sigma_i = \sqrt{3J_2}. \tag{32.3}$$

Principal stresses are determined by relations [4]

$$\sigma_1 = \sigma_m + \frac{2}{3}\sigma_i \cos\psi,$$
$$\sigma_{2,3} = \sigma_m - \frac{2}{3}\sigma_i \cos\left(\frac{\pi}{3} \pm \psi\right), \tag{32.4}$$
$$\cos 3\psi = \frac{J_3}{2}\left(\frac{3}{J_2}\right)^{3/2}.$$

The maximum shear stress is calculated using two principal stresses

$$\tau_{max} = \frac{\sigma_1 - \sigma_3}{2}. \tag{32.5}$$

Implementation of setting a results parameter in methods `setParmAtNodes` and `setEquivalentStress` is presented below.

```
66      // Set array sg.fun[] containing requested value at nodes.
67      // parm - requested result value.
68      // displ - displacement vector.
69      void setParmAtNodes(VisData.parms parm, double[] displ) {
70
71          sg.fmin = 1.e77;
72          sg.fmax = -1.e77;
73
74          for (int node = 0; node < fem.nNod; node++) {
75              if (sg.sNodes[node] >= 0) {
76                  if (parm == VisData.parms.s1 ||
77                      parm == VisData.parms.s2 ||
78                      parm == VisData.parms.s3 ||
79                      parm == VisData.parms.si ||
80                      parm == VisData.parms.s13)
81                          setEquivalentStress(node);
82              switch (parm) {
83              case ux:
84                  f = displ[node*fem.nDim];               break;
85              case uy:
86                  f = displ[node*fem.nDim + 1];           break;
87              case uz:
88                  if (fem.nDim == 3)
89                      f = displ[node*fem.nDim + 2];
90                  else f = 0;
91                                                          break;
92              case sx:
93                  f = stressNod[node][0];                 break;
94              case sy:
95                  f = stressNod[node][1];                 break;
96              case sz:
97                  f = stressNod[node][2];                 break;
98              case sxy:
99                  f = stressNod[node][3];                 break;
100             case syz:
101                 f = stressNod[node][4];                 break;
102             case szx:
103                 f = stressNod[node][5];                 break;
104             case s1:
105                 f = sm + 2*THIRD*si*Math.cos(psi); break;
106             case s2:
107                 f = sm - 2*THIRD*si*
108                     Math.cos(THIRD*Math.PI+psi);    break;
109             case s3:
110                 f = sm - 2*THIRD*si*
111                     Math.cos(THIRD*Math.PI-psi);    break;
112             case si:
113                 f = si;                                 break;
114             case s13:
115                 f = THIRD*si*(Math.cos(psi) +
116                     Math.cos(THIRD*Math.PI-psi));
```

```
117                        }
118                        sg.fun[node] = f;
119                        sg.fmin = Math.min(sg.fmin, f);
120                        sg.fmax = Math.max(sg.fmax, f);
121                    }
122                }
123                if (!(VisData.fMin == 0.0 && VisData.fMax == 0.0)) {
124                    sg.fmax = VisData.fMax;
125                    sg.fmin = VisData.fMin;
126                }
127                if (sg.fmax - sg.fmin < 1.e-6) sg.fmax += 1.e-6;
128                sg.deltaf = sg.fmax - sg.fmin;
129            }
130
131            // Compute stress invariants and equivalent stress.
132            private void setEquivalentStress(int node) {
133                // Stresses
134                double sx  = stressNod[node][0];
135                double sy  = stressNod[node][1];
136                double sz  = stressNod[node][2];
137                double sxy = stressNod[node][3];
138                double syz, szx;
139                if (fem.nDim == 3) {
140                    syz = stressNod[node][4];
141                    szx = stressNod[node][5];
142                }
143                else { syz = 0;   szx = 0; }
144                // Mean stress
145                sm = THIRD*(sx + sy + sz);
146                // Deiatoric stresses
147                double dx = sx - sm;
148                double dy = sy - sm;
149                double dz = sz - sm;
150                // Second and third deviatoric invariants
151                double J2 =  0.5*(dx*dx + dy*dy + dz*dz)
152                            + sxy*sxy + syz*syz + szx*szx;
153                double J3 = dx*dy*dz + 2*sxy*syz*szx
154                            - dx*syz*syz - dy*szx*szx - dz*sxy*sxy;
155                // Angle
156                psi = THIRD*Math.acos(1.5*SQ3*J3/Math.sqrt(J2*J2*J2));
157                // Equivalent stress
158                si = Math.sqrt(3*J2);
159            }
160
161 }
```

Method `setParmAtNodes` obtains the requested results parameter `parm` and nodal displacement array `displ`. The results field at nodes `sg.fun` containing values of the requested results parameter is formed in a loop over nodes of the finite element model that starts in line 74. If the requested parameter is one of principal stresses, equivalent stress or maximum shear stress, then method `setEquivalentStress` is called. This method sets the magnitude of the equivalent stress and other parameters necessary for calculations.

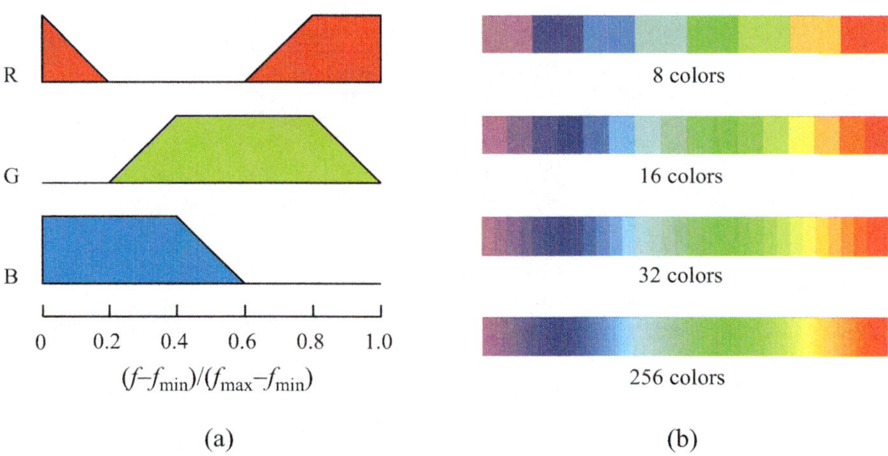

Fig. 32.1 Functions for red, green and blue channels used for color-scale creation (a) and resulting color scales with different number of solid colors (b)

The case statement in lines 82–117 sets the value of parameter f requested for drawing. This value is assigned to an entry of array `sg.fun` in line 118. It is also used for determining the minimum and maximum values of the parameter (lines 119–120). The parameter range is calculated in line 128.

Method `setEquivalentStress` in lines 132–159 calculates the mean stress, deviatoric stress invariants, angle for determining principal stresses, and equivalent stress according to Equations 32.1–32.4.

32.2 Color Scale

Our technique to visualize results as contours is based on interpolation of a one-dimensional color texture. The physical quantity of stresses or displacements is converted to color gradation by computing a texture coordinate in the range 0 to 1.

One-dimensional color texture is created in class `ColorScale` on the basis of functions for red (R), green (G), and blue (B) channels shown in Figure 32.1a. We selected six primary colors for the color scale: magenta (RGB = 101), blue (001), cyan (011), green (010), yellow (110) and red (100). Inside each interval between these basic colors, two color functions remain constant and one function varies linearly. The minimum of the results parameter corresponds to magenta. The maximum value is depicted by red.

The resulting texture depicting the color scale can contain different numbers of solid colors as shown in Figure 32.1b. Specification of the number of solid colors equal to the texture size in pixels leads to a color scale with smooth color change.

Class `ColorScale` producing a one-dimensional texture containing a gradation strip with a specified number of solid colors is given below.

```
1  package visual;
2
3  import javax.media.j3d.*;
4  import java.awt.*;
5  import java.awt.image.MemoryImageSource;
6  import com.sun.j3d.utils.image.TextureLoader;
7
8  // Create 2D texture VisData.textureSize by 1 pixels.
9  // Texture contains VisData.nContours color intervals.
10 class ColorScale extends Component {
11
12     // Returns textture with color gradation
13     Texture2D getTexture() {
14
15         int pix[] = new int[VisData.textureSize];
16         int n2 = 0;
17         double delta =
18             (double) VisData.textureSize/VisData.nContours;
19         for (int i = 0; i < VisData.nContours; i++) {
20             int n1 = n2;
21             n2 = (int) ((i + 1)*delta + 0.5);
22             int color =
23                 getScaleColor(1.0/VisData.nContours*(i+0.5));
24             for (int j = n1; j < n2; j++) pix[j] = color;
25         }
26
27         Image img = createImage(
28                 new MemoryImageSource(VisData.textureSize, 1,
29                         pix, 0, VisData.textureSize));
30         TextureLoader loader = new TextureLoader(img, null);
31         Texture2D texture = new Texture2D(Texture.BASE_LEVEL,
32                 Texture.RGBA, VisData.textureSize, 1);
33         texture = (Texture2D) loader.getTexture();
34
35         return texture;
36     }
```

Method `getTexture` forms a texture with color gradation. Line 15 allocates
an array of pixels for one-dimensional texture size `textureSize` specified in data
(class `VisData`). The loop in lines 19–25 iterates over a number of result contours
`nContours`, which is equal to the number of solid colors in the color scale. Solid
color is provided by the method `getScaleColor` in lines 22–23. Line 24 places
solid color in pixel array `pix`. This array is used for creating image `img` in lines 27–
29. In turn, image `img` is used for creating texture in lines 30–33. Line 35 returns
the resulting color scale texture.

The method `getScaleColor` presented below computes RGBA color according to texture coordinate v.

```
38     // Compute color for color scale texture.
39     // Texture (v=RGB): 0.0=101; 0.2=001; 0.4=011;
40     //                  0.6=010; 0.8=110; 1.0=100.
41     // v - texture coordinate (0..1).
42     // returns  color in RGBA int format.
```

```java
43        private static int getScaleColor(double v) {
44
45            double R, G, B;
46            if (v < 0.2) {            // magenta - blue
47                R = 1 - 5*v;
48                G = 0;
49                B = 1;
50            }
51            else if (v < 0.4) {   // blue - cyan
52                R = 0;
53                G = 5*(v - 0.2);
54                B = 1;
55            }
56            else if (v < 0.6) {   // cyan - green
57                R = 0;
58                G = 1;
59                B = 1 - 5*(v - 0.4);
60            }
61            else if (v < 0.8) {   // green - yellow
62                R = 5*(v - 0.6);
63                G = 1;
64                B = 0;
65            }
66            else {                    // yellow - red
67                R = 1;
68                G = 1 - 5*(v - 0.8);
69                B = 0;
70            }
71            int iR = (int) (R*255 + 0.5);
72            int iG = (int) (G*255 + 0.5);
73            int iB = (int) (B*255 + 0.5);
74            return (255 << 24) | (iR << 16) | (iG << 8) | iB;
75        }
76
77 }
```

The range 0–1 of texture coordinate v is divided into five intervals. Double values R, G and B (0..1) for red, green and blue colors are evaluated in lines 46–70. Statements 71–73 transform double color values into integer values ranged from 0 to 255. Line 74 returns the integer color value composed of R, G, B colors and alpha transparency channel A (equal to 255 that means opaque).

32.3 Mouse Interaction

Visualization programs are usually characterized by rich interaction possibilities for the user including menus, toolbars, keyboard and mouse interactions. Programming such interactions leads to large code, which cannot be published in this book. This is why we restrict ourselves by a simple mouse interaction allowing to rotate, zoom and translate the visual object and requiring very little code for implementation. Our mouse interaction is realized by Java^{TM} class `MouseInteraction` shown below.

```java
 1 package visual;
 2
 3 import javax.media.j3d.*;
 4 import javax.vecmath.Point3d;
 5 import com.sun.j3d.utils.behaviors.mouse.*;
 6
 7 public class MouseInteraction {
 8
 9     // Set mouth behavior (rotate, zoom, translate).
10     // returns  transform group.
11     public static TransformGroup setMouseBehavior() {
12
13         BoundingSphere bounds =  new BoundingSphere(
14             new Point3d(0.,0.,0.), 16*SurfaceGeometry.sizeMax);
15
16         TransformGroup tg = new TransformGroup();
17         tg.setCapability(TransformGroup.ALLOW_TRANSFORM_WRITE);
18         tg.setCapability(TransformGroup.ALLOW_TRANSFORM_READ);
19
20         // Create the rotate behavior node
21         MouseRotate behavior1 = new MouseRotate();
22         behavior1.setSchedulingBounds(bounds);
23         behavior1.setTransformGroup(tg);
24         tg.addChild(behavior1);
25
26         // Create the zoom behavior node
27         MouseZoom behavior2 = new MouseZoom();
28         behavior2.setSchedulingBounds(bounds);
29         behavior2.setTransformGroup(tg);
30         tg.addChild(behavior2);
31
32         // Create the translate behavior node
33         MouseTranslate behavior3 = new MouseTranslate();
34         behavior3.setSchedulingBounds(bounds);
35         behavior3.setTransformGroup(tg);
36         tg.addChild(behavior3);
37
38         return tg;
39     }
40
41 }
```

The class contains just one static method `setMouseBehavior`. First, bounding sphere `bounds` with radius proportional to the maximum model size is defined in lines 13–14. Transform group `tg` is constructed in line 16. Statements 17 and 18 allow changing the transform because of user interaction by setting write and read capabilities. The groups of lines 21–24, 27–30 and 33–36 create mouse behaviors for rotation, zooming and translation. In all three cases there are predetermined standard behaviors in Java 3D^{TM}. In each case, the scheduling bounds are set, then the current transform group is set for the behavior and finally the behavior is added to the transform group. Statement 38 returns the transform group with mouse behaviors. This transform group will affect all nodes that are located lower in the scene graph.

32.4 Lights and Background

Class `Lights` provides background and lights for the scene. The code presented below sets a simple color background and three lights.

```
1  package visual;
2
3  import javax.media.j3d.*;
4  import javax.vecmath.*;
5
6  public class Lights {
7
8      // Set lights and background.
9      // root - branch group.
10     public static void setLights(BranchGroup root) {
11
12         BoundingSphere bounds = new BoundingSphere(
13             new Point3d(0.,0.,0.), 16*SurfaceGeometry.sizeMax);
14
15         // Set up the background
16         Background bgNode = new Background(VisData.bgColor);
17         bgNode.setApplicationBounds(bounds);
18         root.addChild(bgNode);
19
20         // Set up the ambient light
21         Color3f ambientColor = new Color3f(1.f, 1.f, 1.f);
22         AmbientLight light0 = new AmbientLight(ambientColor);
23         light0.setInfluencingBounds(bounds);
24         root.addChild(light0);
25
26         // Set up the directional lights
27         Color3f color1 = new Color3f(0.6f, 0.6f, 0.6f);
28         Vector3f direction1 = new Vector3f(4.f, -7.f, -12.f);
29         DirectionalLight light1
30             = new DirectionalLight(color1, direction1);
31         light1.setInfluencingBounds(bounds);
32         root.addChild(light1);
33
34         Color3f color2 = new Color3f(0.4f, 0.4f, 0.4f);
35         Vector3f direction2 = new Vector3f(-5.f, -3.f, -1.f);
36         DirectionalLight light2
37             = new DirectionalLight(color2, direction2);
38         light2.setInfluencingBounds(bounds);
39         root.addChild(light2);
40     }
41
42 }
```

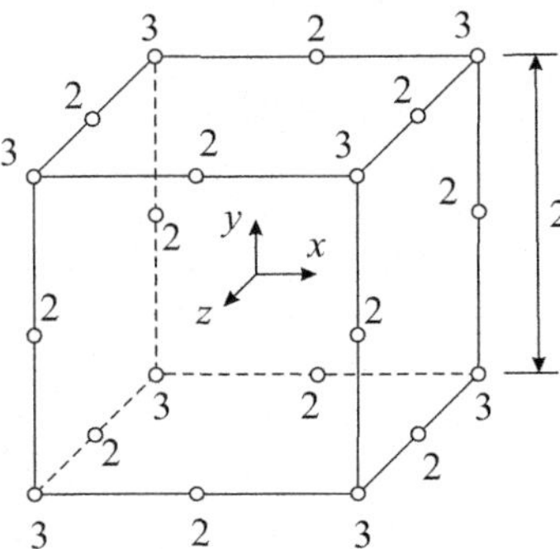

Fig. 32.2 Function $x^2 + y^2 + z^2$ specified at 20 nodal points of hexahedral element

Class `Lights` contains just one static method `setLights`, which gets a reference to `root`, an instance of class `BranchGroup`. Lines 12–13 create bounds as a sphere of radius proportional to the maximum model size centered at the coordinate origin. A scene background is set in lines 16–18. Color `bgColor` specified in class `VisData` is used for the background. The background, as well as all lights, is attached to `root`.

An ambient light of white color is added to the scene graph in lines 21–24. The ambient light is a background light going in all directions. Two directional lights are created in lines 27–32 and 34–39. The directional lights go from two different directions in order to have nonflat light effects.

32.5 Visualization Example

Let us demonstrate the developed visualization techniques on a visualization of function

$$f = x^2 + y^2 + z^2 \tag{32.6}$$

specified at nodes of a single quadratic hexahedral element.

First, we prepare a finite element mesh consisting of one element, shown in Figure 32.2, and place mesh data in file `f.mesh`. The file contains:

```
# File f.mesh
nNod = 20    nEl = 1    nDim = 3


nodCoord
-1 -1  1     0 -1  1      1 -1  1      1 -1  0
 1 -1 -1     0 -1 -1     -1 -1 -1     -1 -1  0
-1  0  1     1  0  1      1  0 -1     -1  0 -1
-1  1  1     0  1  1      1  1  1      1  1  0
 1  1 -1     0  1 -1     -1  1 -1     -1  1  0
```

```
elCon
hex20   1    1  2  3  4  5  6  7  8  9 10
            11 12 13 14 15 16 17 18 19 20
end
```

Here, nNod – number of nodes, nEl – number of elements, nDim – number of
dimensions, nodCoord – nodal coordinates, and elCon – element connectivities.
The element is a cube centered at the coordinate origin, and its edges have length 2.
Therefore, the coordinates of its eight vertices are $(\pm 1, \pm 1, \pm 1)$.

Results are specified at nodes. Computing function f (32.6) at the nodes it appears that it has a value 3 at vertex nodes and a value 2 at midside nodes, as shown
in Figure 32.2. We create file f.res that contains displacements and stresses in a
format compatible with a results file of the finite element processor.

```
# File f.res
Displacements
 Node ux uy uz
     1  3  0  0       2  2  0  0       3  3  0  0       4  2  0  0
     5  3  0  0       6  2  0  0       7  3  0  0       8  2  0  0
     9  2  0  0      10  2  0  0      11  2  0  0      12  2  0  0
    13  3  0  0      14  2  0  0      15  3  0  0      16  2  0  0
    17  3  0  0      18  2  0  0      19  3  0  0      20  2  0  0

Stresses
El 1 sxx syy szz sxy syz szx epi
     0   0   0   0   0   0   0
     0   0   0   0   0   0   0
     0   0   0   0   0   0   0
     0   0   0   0   0   0   0
     0   0   0   0   0   0   0
     0   0   0   0   0   0   0
     0   0   0   0   0   0   0
     0   0   0   0   0   0   0
```

Nodal data shown in Figure 32.2 is set as displacements ux. Next, we create file
f.vis with the following content:

```
# File f.vis
meshFile = f.mesh
resultFile = f.res
parm = ux

showEdges = Y
showNodes = Y
nContours = 8

fmin = 0.85
fmax = 3.0

end
```

We visualize the parameter ux using mesh and results from files s.mesh and
s.res. The color scale consists of eight colors. The range of results fmin-fmax

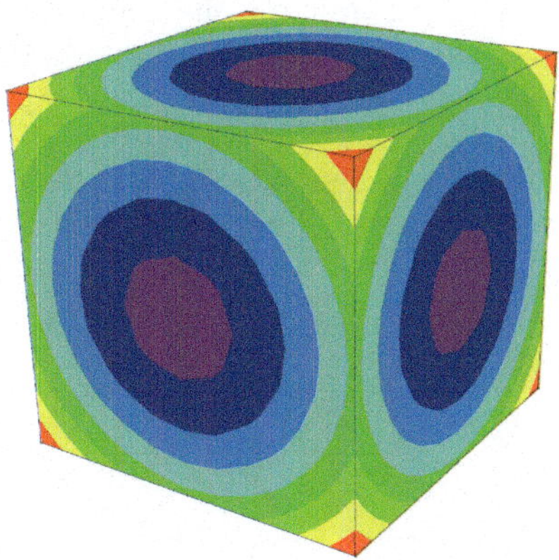

Fig. 32.3 Function $x^2 + y^2 + z^2$ visualized as eight contours on a single hexahedral quadratic element

is specified explicitly in order to have contours located near face centers. Since for each particular value of f Equation 32.6 describes a sphere, visualization of parameter `ux` as contours should look like concentric circular rings of different width.

The visualizer `Jvis` is executed by the command

```
java -cp classes fea.Jvis f.vis
```

After the program starts, an image of the finite element mesh with eight color contours appears on the screen (see Figure 32.3). The Java 3D triangle array used for visualization contains 1200 polygons. It can be seen that the generated contour picture corresponds to the expected result. Some visualized contours are located entirely inside element faces and do not have intersections with face edges. The developed visualization algorithm successfully treats such cases, which would be problematic for algorithms based on fixed face subdivisions or on explicit generation of contour lines.

Problems

32.1. Determine the principal stresses for the following stress vector

$$\{\sigma\} = \{0 \quad 0 \quad 0 \quad 1 \quad 0 \quad 0\},$$

where only shear stress τ_{xy} is nonzero. Use Equations 32.1–32.4.

32.2. Find the value of equivalent stress σ_i for the case of uniaxial tension along the x-axis when the stress vector is

$$\{\sigma\} = \{10 \quad 0 \quad 0 \quad 0 \quad 0 \quad 0\}.$$

32.3. Suppose that for contour drawing we want a gradation strip consisting of just the hue blue with changing intensity. Modify the method `getScaleColor` to produce such a gradation strip.

32.4. Modify the data of the visualization example in Section 32.5 to visualize the function $f = x^2 + y^2$. Use component `uy` for storing nodal function data.

Appendix A
Data for Finite Element Solver

A.1 Data Statements

A.1.1 Data Statement

```
<name> = <data>
```

Item `<name>` is a data name and `<data>` is data content. Data names are not case sensitive. White spaces are a blank and the equals sign. White spaces are not allowed inside data names. Data on the right of an equals sign can contain one or more tokens. Tokens can be numbers or text literals. The number of tokens is predetermined by the data name. Several statements can be placed on one line.

A.1.2 Comment Statement

```
# comment text
```

Everything is ignored at the current line after a comment sign # followed by a blank. If a comment sign plus a blank are at the line beginning then the entire line is a comment.

A.1.3 Including File

```
includeFile <fileName>
```

The statement includes all data contained in the file with name `<fileName>`. Data in the file should be terminated with an end statement.

A.1.4 End Statement

```
end
```

Used as the last statement in the model data and in the load data.

A.2 Model Data

A.2.1 Parameters

Scalar parameters for the problem are specified by the following statements:

`nNod = <number>` – number of nodes;

`nEl = <number>` – number of elements in the finite element model;

`stressState = `**`THREED`**`/PLSTRAIN/PLSTRESS/AXISYM` – type of problem: `THREED` – three-dimensional problem, `PLSTRAIN` – plane strain two-dimensional problem, `PLSTRESS` – plane stress two-dimensional problem, `AXISYM` – axisymmetrical problem;

`physLaw = `**`ELASTIC`**`/ELPLASTIC` – physical law for material behavior: elastic or elastic–plastic;

`solver = `**`LDU`**`/PCG` – equation solver: `LDU` – direct solver based on LDU decomposition, `PCG` – preconditioned conjugate gradient iterative solver;

`thermalLoading = `**`N`**`/Y` – existence of thermal loading: `N` – no, `Y` – yes.

Default parameter values are emboldened. If the default parameter value is suitable for the current problem then it is possible to omit its specification. The order of data specification is arbitrary unless it follows logical relation of data.

A.2.2 Material Properties

For each elastic material the following data should be specified:

```
material = matName E nu alpha
```

Here, `matName` is any name selected for refering to this material, `E` is the elasticity modulus, `nu` is Poisson's ratio, and `alpha` is a thermal-expansion coefficient.

For elastic–plastic material, the data statement contains three additional parameters related to elastic–plastic material behavior:

```
material = matName E nu alpha sY hardCoef hardPower
```

Additional parameters have the following meanings: sY is material yield stress, `hardCoef` is a hardening coefficient, and `hardPower` is a hardening power.

A.2.3 Finite Element Mesh

Two arrays describe the mesh: nodal coordinates and element connectivities (including element type and element material).

Nodal coordinates

```
nodCoord = <array>
```

For three-dimensional problems nodal coordinates are specified as $x_1\ y_1\ z_1\ x_2\ y_2\ z_2$ etc. A two-dimensional coordinate array has the form $x_1\ y_1\ x_2\ y_2\ ...$.

Element data

```
elCon = <array>
```

For each element the following data should be provided:

`ElType = QUAD8/HEX20` – element type (`QUAD8` – two-dimensional element with eight nodes, `HEX20` – twenty-node three-dimensional element),

`ElMat` – material name corresponding to that in material properties,

`ElemNodeNumbers` – node numbers belonging to the element.

A.2.4 Displacement Boundary Conditions

Displacement boundary conditions can be specified in two ways: direct specification of node numbers with constrained displacements, or specification of a box for node definition.

Direct specification of displacement boundary conditions

```
constrDispl = Direct Value nNumbers NodNumbers[]
```

where `Direct = x/y/z` – direction of displacement constraint, `Value` – constrained displacement value, `nNumbers` – number of items in the list of node numbers, `NodNumbers[]` – list of nodes where displacement boundary conditions are specified. The list can include positive integers as node numbers. If an integer pair $n_1 - n_2$ is present in the list, then it is interpreted as a set of nodes from n_1 to n_2 (condition $n_1 < n_2$ should be fulfilled).

Box for specification of displacement boundary conditions

```
boxConstrDispl = Direct Value BoxDiadonal[]
```

where `Direct` = `x/y/z` – direction of displacement constraint, `Value` – constrained displacemet value, `BoxDiadonal[]` – coordinates of two points at ends of a box diagonal. In the three-dimensional case the diagonal is specified as x_{min} y_{min} z_{min} x_{max} y_{max} z_{max}. In the two-dimensional case z-coordinates are absent.

A.3 Load Specification

A load can consist of one or several load steps. Load steps are considered as increments to the previous state.

A.3.1 Load Step Name

```
loadStep = LoadStepName
```

This statement should be the first statement of a load step. The specified load step name is used for identifying this load step. Result files have the load label as their extensions.

A.3.2 Parameters

`scaleLoad = Scale` – load scaling parameter. Load vector of the current load step is obtained by multiplying the load vector from the previous step with the specified parameter `Scale`.

`residTolerance = Tolerance` – tolerance for ratio of a residual norm to the norm of force load. If the relative residual norm becomes less than the specified `Tolerance` then equilibrium iterations are finished. Default value `Tolerance = 0.01`.

`maxiternumber = Number` – maximum allowed number of equilibrium iterations (default value is 100).

Parameters `scaleLoad`, `residTolerance`, and `maxiternumber` are useful for elastic–plastic problems. They are not used in purely elastic problems. Parameters `residTolerance` and `maxiternumber` are valid for the next load step if they are not changed. All other loading parameters, including `scaleLoad`, do not exist at the beginning of each new load step and should be specified if necessary.

A.3.3 Nodal Forces

```
nodForce = Direct Value nNumbers NodNumbers[]
```

This statement is used for specification of nodal forces. Here, `Direct` = `x/y/z` – direction of nodal forces, `Value` – force value, `NodNumbers[]` – list of nodes where nodal forces are applied. The list can include positive integers as node numbers and pairs $n_1 - n_2$ for specifying nodes ranges.

A.3.4 Surface Forces

```
surForce = Direct ElNumber nFaceNodes
           faceNodes[] forceAtFaceNodes[]
```

Specification of a distributed surface load consists of the following items: `Direct` = `x/y/z/n` – direction of surface load (x, y, z – along coordinate axes x, y or z, n – loading in the direction of the external normal to the surface), `ElNumber` – element number, `nFaceNodes` – number of nodes on the element face, `faceNodes[]` – node numbers defining the element face, `forceAtFaceNodes[]` – intensities of distributed load at nodes.

A.3.5 Surface Forces Inside a Box

```
boxSurForce = Direct Value BoxDiagonal[]
```

This statement allows application of a distributed surface load for all element faces that are inside a box. Data include: `Direct` = `x/y/z/n` – direction of surface load (x, y, z – along coordinate axes x, y or z, n – external normal direction), `Value` – intensity of distributed load common to all faces, `BoxDiadonal[]` – coordinates of two points at ends of a box diagonal (x_{min} y_{min} z_{min} x_{max} y_{max} z_{max}).

A.3.6 *Nodal Temperatures*

```
nodTemp = NodeTemperatures[]
```

This statement is used for specifying temperature values at nodes of the finite element model.

Appendix B
Data for Mesh Generation

B.1 Mesh-generation Modules

Mesh generation is performed by main class `Jmgen`. The following modules may be called:

`rectangle` – generate rectangular mesh inside rectangular region;

`genquad8` – generate topologically regular mesh inside curvilinear quadrilateral region;

`sweep` – generate three-dimensional mesh by sweeping two-dimensional mesh in space;

`readmesh` – read mesh from file;

`writemesh` – write mesh to file;

`copy` – copy mesh block;

`transform` – make transformations (translate, scale, rotate) for the mesh block;

`connect` – produce new mesh block by connecting two mesh blocks.

The module is identified by its class name in a data file. The module inputs its data and does operations of mesh generation, transformations, pasting, etc.

The data file for the mesh generation has the following appearance:

```
Head statement A
data A

Head statement B
data B

Head statement C
data C

. . .
```

B.2 Rectangular Mesh Block

Function: Generation of a mesh of quadrilateral quadratic elements inside a rectangular area (Figure 21.1).
Head statement: `rectangle modelName`
`modelName` – name of the model where the generated mesh is placed.
Input data:

`nx, ny` – number of elements along `x` and `y`;

`xs[nx+1], ys[ny+1]` – locations of element boundaries on `x` and `y`;

`mat` – material name (default `mat=1`);

`end` – end of input data.

B.3 Mesh Inside Eight-node Macroelement

Function: Generation of mesh of quadrilateral quadratic elements inside a quadrilateral area with curved edges (Figure 21.2).
Head statement: `genquad8 modelName`
`modelName` – name of the model where the generated mesh is placed.
Input data:

`nh, nv` – number of elements along "horizontal" and "vertical" directions. "Horizontal" direction is along macroelement side 1–2–3;

`x1,y1,x2,y2 ...x8,y8` – locations of eight nodes for macroelement definition in anticlockwise order starting from any corner node. If both coordinates of a midside node are zeros then they are interpolated linearly from neighboring corner nodes;

`res[4]` – relative sizes of the smallest elements on macroelement sides. If the smallest element is located at the side end (anticlockwise order) then it is specified as one minus size (default `res = 0, 0, 0, 0` – equal element sizes);

`mat` – material name (default `mat=1`);

`end` – end of input data.

B.4 Three-dimensional Mesh by Sweeping

Function: Generation of three-dimensional mesh of twenty-node brick-type elements by sweeping of a two-dimensional mesh along the z-axis or around the y-axis (Figure 22.1).

Head statement: `sweep modelName2 modelName3`
`modelName2` – name of two-dimensional finite element model;
`modelName3` – name of resulting three-dimensional model.
Input data:

   `nlayers` – number of element layers in the three-dimensional mesh;

   `zlayers` – z-distances or angles (in degrees) for copying the two-dimensional mesh;

   `rotate = Y` – rotate the two-dimensional mesh around the y-axis, $=$ `N` - translate the two-dimensional mesh along z (default `rotate = N`);

   `end` – end of input data.

B.5 Reading Mesh from File

Function: Read mesh from a file with specified name.
Head statement: `readmesh modelName filelName`
`modelName` – name of the model where the mesh will be placed;
`filelName` – file name containing text information about the mesh.

B.6 Writing Mesh to File

Function: Write mesh to a file with specified name.
Head statement: `writemesh modelName filelName`
`modelName` – name of the model where the mesh is;
`filelName` – file name where the mesh will be written.

B.7 Copying Mesh

Function: Copy mesh from model A to model B.
Head statement: `copy modelNameA modelNameB`
`modelNameA` – name of the model where the mesh is;
`modelNameB` – name of the model where the mesh will be copied.

B.8 Mesh Transformations

Function: Perform translation, scaling, rotation and mirroring of a finite element mesh (Figure 24.1).

Head statement: `transform modelName`
`modelName` – name of the model with the mesh to be transformed.
Input data consists of any number of the following statements in any order:

```
translate axis value

scale axis value

rotate axis value

mirror axis plane
```

Here, `translate`, `scale`, `rotate` and `mirror` are operations of translation, scaling, rotation and mirroring; `axis` is an axis name, which can have values x, y, z; `value` is the translation value, scaling coefficient or rotation angle in degrees. For two-dimensional meshes, `axis` can be x or y for translation and scaling operation and only z for rotation operation. Operation `mirror` reflects a mesh with respect to a plane normal to `axis`. The plane is located at coordinate `plane`. End of data for the `transform` module is marked by the statement `end`.

B.9 Connecting Two Mesh Blocks

Function: Connect mesh blocks A and B and place resulting mesh in model C.
Head statement: `connect modelNameA modelNameB modelNameC`
`modelNameA, modelNameB` – names of models that will be connected;
`modelNameB` – name of the model where the resulting mesh after connection will be placed.
Input data:

eps – coordinate tolerance for joining nodes belonging to different mesh blocks (default value `eps = 0.0001`);
end – end of input data.

Appendix C
Data for Visualizer

C.1 Visualization Data

The input data for the visualization consists of a finite element mesh, a set of finite element results and a set of visualization parameters. When the finite element results are absent just visualization of a finite element mesh is performed.

C.2 Input Data

The input data for visualization includes:

`meshFile = <text>` – name of the file containing a finite element mesh (required);

`resultFile = <text>` – name of the results file (if not specified then results visualization is not done);

`parm = <text>` – results parameter that should be visualized;

`showEdges = **Y**/N` – draw element edges: Y – yes, N – no;

`showNodes = **N**/Y` – draw nodes: N – no, Y – yes;

`nDivMin = <number>` – minimum number of element edge subdivisions (default value is 1);

`nDivMax = <number>` – maximum number of element edge subdivisions (default value is 16);

`fMin = <number>` – minimum value of results parameter (if not specified then computed);

`fMax = <number>` – maximum value of results parameter (if not specified then computed);

`nContours = <number>` – number of contours used for results visualization (2..256, default value is 256);

`deformScale = <number>` – if not zero, show deformed shape of the finite element model. Nodal displacements are scaled such that the ratio of a maximum scaled displacement to the largest model size is equal to `deformScale` (default value is 0);

`end` – signal of end of data.

The following *results parameters* can be visualized:

`Ux, Uy, Uz` – one of the components of a displacement vector along coordinate axes x, y or z;

`Sx, Sy, Sz` – normal stresses;

`Sxy, Syz, Szx` – shear stresses;

`S1, S2, S3` – principal stresses;

`Si` – equivalent stress;

`S13` – difference between first and third principal stresses;

`none` – do not visualize results as contours (default value).

Minimum information in the data input file for visualizer is the name of the file with a finite element mesh.

Appendix D
Example of Problem Solution

D.1 Problem Statement

A rectangular plate with a central circular hole is subjected to tensile loading as shown in Figure D.1a. The plate has the following dimensions: width $W = 4$, height $H = 8$, thickness $t = 2$. The central hole has the radius $R = 1$. The plate is loaded by distributed surface forces $p = 1$ applied at the upper and lower plate edges.

It is necessary to determine the elastic stress state of the plate with a hole by the Java^{TM} finite element system Jfea.

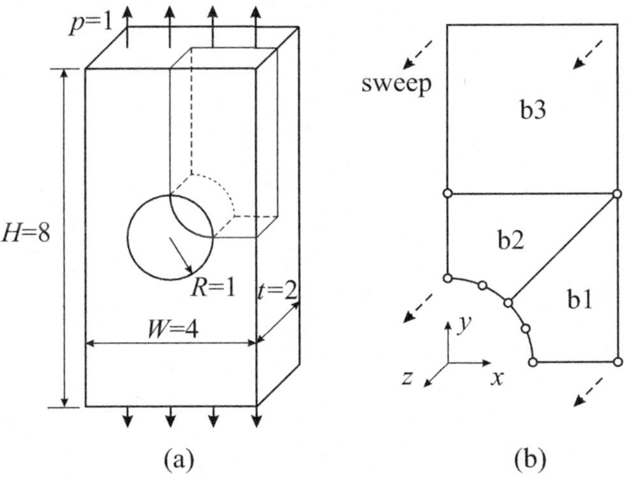

Fig. D.1 Example problem: tensile plate with a hole (a), mesh generation using block decomposition and sweeping (b)

D.2 Mesh Generation

The mesh is generated for one eighth of the specimen (see Figure D.1a). A schematic of the mesh generation is depicted in Figure D.1b. A two-dimensional area is decomposed into blocks b1, b2 and b3 of simple shapes. Two-dimensional meshes inside blocks are created by local mesh generators. Blocks are pasted together in one mesh. Subsequent sweeping produces a three-dimensional mesh. Two-dimensional meshes are composed of eight-node quadrilateral elements. The twenty-node hexahedral element is used in the three-dimensional mesh. Input data for preprocessor Jmgen is created with any text editor and placed in file `hole3d.gen`.

```
# 3D rectangular plate with a central hole
# Mesh generation

GenQuad8  b1
  nh = 4   nv = 4
  xyp = 1 0   0 0   2 0   0 0
        2 2   0 0   0.7071 0.7071   0.9239 0.3827
  res = 0.15 0.15 0.85 0
end


GenQuad8  b2
  nh = 4   nv = 3
  xyp = 0.7071 0.7071   0 0   2 2   0 0
        0 2   0 0   0 1   0.3827 0.9239
  res = 0.15 0 0.85 0
end


Connect  b1 b2 b12
  eps = 0.01
end


Rectangle  b3
  nx = 3   ny = 3
  xs = 0 0.6667 1.3333 2
  ys = 2 2.6667 3.3333 4
end


Connect  b12 b3 b123
  eps = 0.1
end


Sweep  b123 b3
  nlayers = 4
  zlayers = 0 0.25 0.5 0.75 1
end


WriteMesh  b3  hole3d.mesh
```

Mesh blocks b1 and b2 are produced by GenQuad8 generator. The key points shown in Figure D.1b by circles are specified in array xyp to define curved quadrilaterals. The relative sizes of the smallest elements on edges are determined by values in array res. A zero value in array res means that equal-size elements will be generated along that edge. In the absence of array res default zero values are adopted.

Mesh blocks b1 and b2 are connected together by the Connect module. The resulting mesh is stored under the name b12. Parameter eps determines the coordinate tolerance for joining nodes from two mesh blocks.

Upper block b3 is meshed by module Rectangle, which creates a mesh inside a rectangular block using specified locations of corner nodes at block edges. Another pasting of mesh blocks b12 and b3 produces a final two-dimensional mesh b123.

A three-dimensional mesh b3 is created by sweeping the two-dimensional mesh b123 along the z-axis. Module WriteMesh writes the resulting three-dimensional mesh to file hole3d.mesh.

To create a mesh, program Jmgen is executed with the following command:

```
java -cp classes fea.Jmgen hole3d.gen
```

Here, classes is a path to a directory where Java classes of the finite element system Jfea are. After program execution, listing of mesh generation is in file hole3d.gen.lst. The created mesh (file hole3d.mesh) consists of 148 hexahedral twenty-node elements and 908 nodes. It is used by the finite element processor Jfem during problem solution. The mesh can be visualized by the visualization program Jvis.

D.3 Problem Solution

Input data for the finite element processor Jfem is prepared in file hole3d.fem.

```
# 3D rectangular plate with a central hole
# Finite element analysis

  StressState = threeD

  IncludeFile hole3d.mesh

  Solver = LDU

  Material = 1  1000  0.3  1.0

  BoxConstrDispl = x  0.0   -0.01  0.99 -0.01  0.01 4.01 1.01
  BoxConstrDispl = y  0.0    0.99 -0.01 -0.01  2.01 0.01 1.01
  BoxConstrDispl = z  0.0   -0.01 -0.01 -0.01  2.01 4.01 0.01
end

LoadStep = 1
  BoxSurForce = n 1.0   -0.01 3.99 -0.01  2.01 4.01 2.01
end
```

A small amount of data is sufficient since the mesh prepared by the preprocessor is included using instruction `includeFile`. The principle of adopting default values for many parameters also contributes to the reduction of input data. For example, the instruction

```
solver = LDU
```

specifies that the LDU method is employed for the solution of the finite element equation system. This instruction can be omitted since the LDU solver is the default one.

Next, the material properties are specified. For a material with name 1 we determine just the mechanical constants – elasticity modulus, Poisson's ratio and thermal-expansion coefficient that are necessary for the selected type of analysis.

Different options exist for the specification of boundary conditions. In the example, both displacement boundary conditions and force boundary conditions are generated on surfaces, which are identified by a bounding box with given diagonal ends. Instruction `BoxConstrDispl` implies that the following data is given: constraint direction, constraint value, three coordinates of the first diagonal end, and three coordinates of the second diagonal end. First, `BoxConstrDispl` statement constrains displacement u_x at plane $x = 0$, second – u_y at plane $y = 0$, and third – u_z at plane $z = 0$. Specification of a normal distributed force on a surface is performed in an analogous way.

Program `Jfem` is executed with the following command:

```
java -Xmx1000m -cp classes fea.Jfem hole3d.fem hole3d.lst
```

Here, option `-Xmx` sets the maximum heap size that can be allocated by a Java virtual machine. By default, the JVM uses up to 16 MB of RAM. Specification `1000m` requests 1000 MB of the memory. While the current problem requires a small memory amount, our larger problems will not be able to run with the default memory size. Parameter `hole3d.fem` specifies the input data file and parameter `hole3d.lst` a listing file where brief information about the solution is presented. Results consisting of nodal displacements and stresses at reduced integration points are placed in file `hole3d.lst.1` (listing file plus a load step name). The mesh and results files are used for visualization.

To check the accuracy of the finite element model, the same three-dimensional problem under plane strain conditions (additional displacement constraint $u_z = 0$ at plane $z = 1$) has been solved. The stress concentration factor K_{tn} is determined as the maximum σ_y stress at line $x = 0$, $y = 0$ divided by average net stress at cross-section $y = 0$. Our result $K_{tn}^{\text{FEM}} = 2.187$ differs by just 1.25% from the reference value $K_{tn} = 2.16$ [25].

In order to demonstrate the amount of computing time required for finite element analysis let us solve the same problem for a tensile plate with a central circular hole using larger finite element meshes. When creating a finite element mesh for the second problem let us double the number of elements in each direction of the coordinate system. To do this it is possible to double the number of subdivisions nx and ny, as well as the number of layers `nlayers` during sweeping in file `hole3d.gen`. A mesh generated by the program `Jmgen` contains 1184 elements and 5925 nodes.

A mesh for the third problem is created by doubling the number of subdivisions nx and ny in the second mesh while maintaining the same number of layers $nlayers$. The resulting mesh contains 4736 elements and 22193 nodes.

Problem solution (program $Jfem$) was performed on a desktop computer with Intel$^{®}$ Quad-Core Xeon$^{®}$ 3.0 GHz processor using Sun Java^{TM} virtual machine JVM 1.6. The computing time and memory consumption for the three problems are presented in Table D.1.

Table D.1 Solution time and memory for problems of different sizes

Problem	Elems	Nodes	DOFs	LDU solver		PCG solver		
				Mem, MB	Time, s	Mem, MB	Time, s	Iters
1	148	908	2724	6.1	0.4	4.2	0.7	346
2 (client)	1184	5925	17755	130.4	12.2	31.8	8.7	690
2 (server)					15.9		7.2	
3 (client)	4736	22193	66579	912.0	153.0	125.0	52.1	1152
3 (server)					205.5		42.3	

The table shows the number of elements, number of nodes, number of degrees of freedom, memory for storing the global stiffness matrix, computing time and number of iterations (for PCG solver). For the smaller problem 1, the direct LDU solver is faster than the iterative PCG solver. However, for the larger problems, the PCG solver takes less time than the LDU solver.

The juxtaposition compares the efficiency of using client and server variants of the JVM in the larger problems. The server JVM is activated with option $-server$. It appears that the direct LDU solver is faster with the client JVM. The iterative PCG solver shows better speed when the server JVM is used. Such differences in client and server JVM efficiency are related to the fact that direct and iterative solvers have quite different computation characteristics in their critical program sections.

D.4 Visualization

The created finite element model and the results of the problem solution are visualized using postprocessor $Jvis$. Program $Jvis$ is executed with the command:

```
java -cp classes fea.Jvis hole3d.vis
```

Here, $hole3d.vis$ is a text file with data describing the visualization task. For mesh viewing it is necessary to specify just the mesh file name:

```
# Mesh visualization
  meshFile = hole3d.mesh
end
```

An image on the computer screen is shown in Figure D.2a. Using the mouse the user can rotate, zoom and pan the finite element model.

Visualization of a deformed shape of the finite element model presented in Figure D.2b is done with the following data:

```
# Deformed shape without element edges
  meshFile = hole3d.mesh
  resultFile = hole3d.lst.1
  deformScale = 0.2
  showedges = N
end
```

The results file name `hole3d.lst.1` contains nodal displacements and stresses at reduced integration points for load step 1. The value `0.2` of the `deformScale` parameter leads to drawing the deformed shape of the finite element model. Displacements are scaled to have the maximum displacement equal to 20% of the model maximum size.

The `Jvis` data file demonstrated below creates an image of the deformed finite element model with ten contours of stress σ_y.

```
# Contours of stress Sy
  meshFile = hole3d.mesh
  resultFile = hole3d.lst.1

  deformScale = 0.2
  parm = Sy
  nContours = 10
  fmin = 0
  fmax = 3.0
end
```

Statement `parm = Sy` requests visualization of stress σ_y using ten color contours (`nContours`). The parameters `fmin` and `fmax` define the range of the result parameter that corresponds to the entire color gradation strip. Values of σ_y below 0 are depicted by magenta and values above 3.0 by red. If we omit explicit specification of the parameter range then it will be determined automatically. An image with ten contours of σ_y shown in Figure D.3a required 2750 triangular polygons for rendering by the Java 3D^{TM}.

In order to smooth the color picture for σ_y it is necessary to set the number of contours to the size of the one-dimensional texture used as a gradation strip. Changing the statement defining the number of contours to `nContours = 256` produces the image in Figure D.3b.

Visualization of stress field σ_y for the refined finite element model with 1184 hexahedral twenty-node elements using 10 and 256 contours is shown in Figure D.4. Comparing contours of σ_y in Figs D.3 and D.4 it is possible to conclude that finite element models consisting of 148 elements and 1184 elements provide practically identical stress fields.

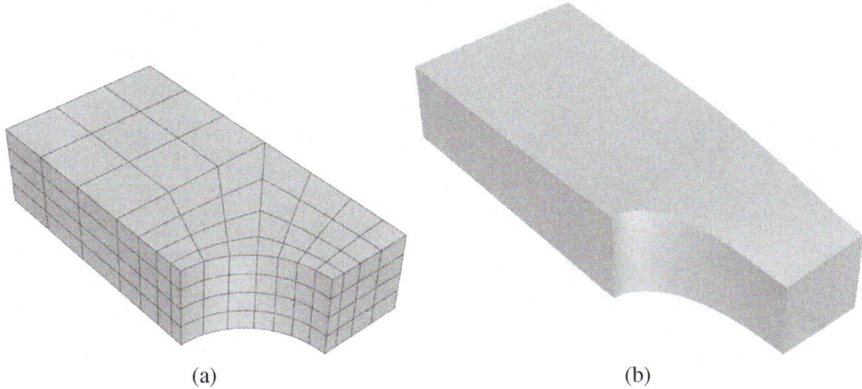

Fig. D.2 (a) Finite element mesh; (b) Deformed finite element model

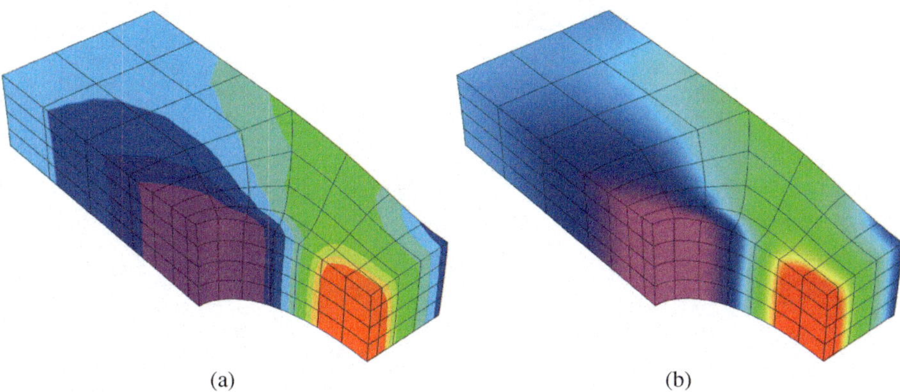

Fig. D.3 Contours of stress σ_y: (a) 10 colors; (b) 256 colors

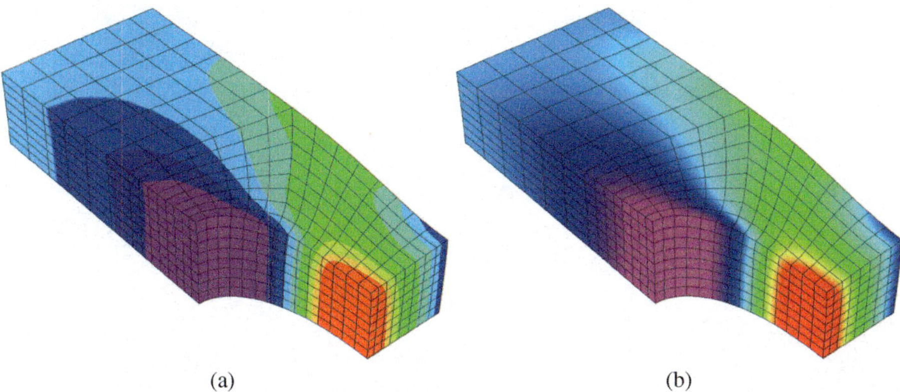

Fig. D.4 Contours of stress σ_y for the refined finite element model: (a) 10 colors; (b) 256 colors

References

1. J.E. Akin, M. Singh, *Object-oriented Fortran 90 P-adaptive finite element method. Advances in Engineering Software*, 2002, 33, 461–468.
2. G.C. Archer, G. Fenves and C. Thewalt, *A new object-oriented finite element analysis program architecture. Computers and Structures*, 1999, 70, 63–75.
3. K.-J. Bathe, *Finite Element Procedures*. Prentice-Hall, Englewood Cliffs, NJ, 1996.
4. J. Chakrabarty, *Theory of Plasticity*. Elsevier Butterworth-Heinemann, Oxford, UK, 2006.
5. J.X. Chen and E.J. Wegman, *Foundation of 3D Graphics Programming Using JOGL and Java3D*. Springer, London, 2006.
6. S. Commend, and T. Zimmermann, *Object-oriented nonlinear finite element programming: a primer. Advances in Engineering Software*, 2001, 32, 611–628.
7. M.A. Crisfield, *Non-linear Finite Element Analysis of Solids and Structures, Vol. 1: Essentials*. Wiley, New York, 1991.
8. Y. Dubois-Pelerin and T. Zimmermann, *Object-oriented finite element programming. III. An efficient implementation in C++. Computer Methods in Applied Mechanics and Engineering*, 1993, 108, 165–183.
9. D. Eyheramendy and D. Guibert, *A Java approach for F.E. computational mechanics. ECCOMAS 2004* (P. Neittaanmaki *et al.* eds.), 2004, 13 p.
10. J. Fish and T. Belytschko, *A First Course in Finite Elements*. Wiley, Chichester, UK, 2007.
11. B.W.R. Forde, R.O. Foschi and S.F. Steimer, *Object-oriented finite element analysis. Computers and Structures*, 1990, 34, 355–374.
12. P.J. Frey and P.-L. George, *Mesh Generation. Application to Finite Elements*. Hermes Science Publishing, Oxford, UK, 2000.
13. G.H. Golub and C.F. Van Loan, *Matrix Computations*. The Johns Hopkins University Press, Baltimore, 1996.
14. M. Hoit and K. Krishnamurthy, *A 14-point reduced integration scheme for solid elements. Computers and Structures*, 1995, 54, 725–730.
15. K.H. Huebner, D.L. Dewhirst, D.E. Smith and T.G. Byrom, *The Finite Element Method for Engineers*. Wiley-Interscience, New York, 2001.
16. T.J.R. Hughes, *The Finite Element Method. Linear Static and Dynamic Finite element Analysis*. Dover, Mineola, NY, 2000.
17. B. Irons and S. Ahmad, *Techniques of Finite Elements*. Ellis Horwood, Chichester, UK, 1980.
18. R.I. Mackie, *Object Oriented Methods and Finite Element Analysis*. Saxe-Coburg, Stirling, Scotland, 2001.
19. J. Mackerle, *Object-oriented programming in FEM and BEM: a bibliography (1990–2003). Advances in Engineering Software*, 2004, 35, 325–336.
20. G.P. Nikishkov, *Generating contours on FEM/BEM higher-order surfaces using Java 3D textures. Advances in Engineering Software*, 2003, 34, 469–476.

21. G.P. Nikishkov, *Object oriented design of a finite element code in Java. Computer Modeling in Engineering and Sciences*, 2006, 11, 81–90.

22. G.P. Nikishkov and S.N. Atluri, *Implementation of a generalized midpoint algorithm for integration of elastoplastic constitutive relations for von Mises' hardening material. Computers and Structures*, 1993, 49, 1037–1044.

23. G.P. Nikishkov, H. Kanda and A. Makinouchi, *Tuning the Schur complement computations for finite element partitions. Advances in Engineering Software*, 2000, 31, 913–920.

24. G.P. Nikishkov, Yu.G. Nikishkov and V.V. Savchenko, *Comparison of C and Java performance in finite element computations. Computers and Structures*, 2003, 81, 2401–2408.

25. W.D. Pilkey and D.F. Pilkey, *Peterson's Stress Concentration Factors*. Wiley, Hoboken, NJ, 2008.

26. P.J. Schneider and D.H. Eberly, *Geometric Tools for Computer Graphics*. Morgan Kaufman, San Francisco, 2003.

27. R. Schneiders, *Quadrilateral and hexahedral element meshes*. In: *Handbook of Grid Generation* (J.F. Thompson *et al.* eds), CRC Press, Boca Raton, FL, 2000, 21.1–26.

28. D. Selman, *Java 3D Programming*. Manning, Greenwich, CT, 2002.

29. I.M. Smith and D.V. Griffiths, *Programming the Finite Element Method*. Wiley, Chichester, UK, 1998.

30. H. Sowizral, K. Rushforth and M. Deering, *The Java 3D API Specification*. Addison-Wesley, Reading, MA, 2000.

31. H. Zhang and Y.D. Liang, *Computer Graphics Using Java 2D and 3D*. Pearson Prentice Hall Upper Saddle River, NJ, 2007.

32. O.C. Zienkiewicz and R.L. Taylor, *Finite Element Method: Volume 1 – The Basis*. Butterworth Heinemann, London, 2000, 712 pp.

33. O.C. Zienkiewicz and R.L. Taylor, *Finite Element Method: Volume 2 – Solid Mechanics*. Butterworth Heinemann, London, 2000, 480 pp.

34. T. Zimmermann, Y. Dubois-Pelerin and P. Bomme, *Object-oriented finite element programming. I. Governing principles. Computer Methods in Applied Mechanics and Engineering*, 1992, 98, 291–303.

Index

Printed in Dunstable, United Kingdom